Cellular Therapies for Retinal Disease

Steven D. Schwartz
Aaron Nagiel
Robert Lanza
Editors

Cellular Therapies for Retinal Disease

A Strategic Approach

Foreword by Mark S. Blumenkranz

Springer

Editors
Steven D. Schwartz
Retina Division Chief
Department of Ophthalmology
Ahmanson Professor of Ophthalmology
Stein Eye Institute
David Geffen School of Medicine at UCLA
Los Angeles, CA, USA

Aaron Nagiel
Retina Division
Department of Ophthalmology
Stein Eye Institute
David Geffen School of Medicine at UCLA
Los Angeles, CA, USA

Robert Lanza
Head of Astellas Global Regenerative Medicine
Chief Scientific Officer at the Astellas Institute for Regenerative Medicine
Marlborough, MA, USA

ISBN 978-3-319-49477-7 ISBN 978-3-319-49479-1 (eBook)
DOI 10.1007/978-3-319-49479-1

Library of Congress Control Number: 2017943126

Cover illustration: Saravanan Karumbayaram, M.Pharm., Ph.D.

Printed on acid-free paper

This Springer imprint is published by Springer Nature
The registered company is Springer International Publishing AG
The registered company address is: Gewerbestrasse 11, 6330 Cham, Switzerland

Foreword

We all look forward to the day when low cost personalized genomics to identify persons at high risk and some combination of small molecules, customized proteins, or gene therapy will severely reduce the incidence or entirely prevent blindness caused by the degeneration of ocular tissues. Unfortunately at the present time, although great strides in all those areas are being made, we are still left with limited options to treat the most severely affected persons with a wide variety of diseases including atrophic age-related macular degeneration, many inherited retinal dystrophies, and acquired optic nerve disease including glaucoma to name a few. But hope is on the way with the possibility of replacement of irreversibly diseased or absent tissues, including the retinal pigment epithelium, photoreceptors, and other neurons, with functionally active cells using the techniques of modern regenerative medicine. The biologic basis for these forms of treatment including the use of stem cells of various origin and degree of differentiation, and practical considerations including the use of scaffolding systems and novel microsurgical tools to enhance engraftment are all addressed in this book, which is one of if not the first on the subject and a welcome addition to the literature.

This volume contains chapters from investigators around the world working in these areas and is edited by translational clinician scientists known for their contributions to ocular regenerative medicine. Robert Lanza is widely acknowledged as one of the fathers of the field of applied stem cell biology. His important collaborations with Drs. Steven Schwartz and Aaron Nagiel at the Jules Stein Eye Institute have led to the first influential publications in the *Lancet* of the use of differentiated stem cell-derived retinal pigment epithelium (RPE), for the treatment of geographic atrophy (GA) of the RPE related to age-related macular degeneration in older adults and Stargardt macular dystrophy in juveniles and young adults. Where there is still much to be learned about the long-term safety and efficacy of these efforts, the early results show some promise. Other chapters include the use of monolayers or scaffolds to implant cells, as well as supra-choroidal delivery systems. While much of the book is devoted to the treatment of retinal disease, there are well-written chapters addressing the possibility of cell transplantation techniques for glaucoma and early research efforts in disease modeling with affected stem cells that are able to be studied more effectively in vitro than by other methods.

To be sure these efforts are still in the earliest stages of development and much more needs to be done before we even begin to talk about the possibility of their becoming standard of care. But every journey begins with a single step and the authors are to be congratulated for a timely volume that lays out the potential promise for this approach to better understanding and treating the most severe forms of ocular degenerative disease.

Mark S. Blumenkranz, M.D., M.M.S.
HJ Smead Professor of Ophthalmology Emeritus,
Stanford University
Stanford, CA, USA

Preface

As a graduate student in 1962, John B. Gurdon created live tadpoles after transplanting nuclei from frog intestinal epithelial cells into enucleated eggs. Thus began 50 years of work in stem cell research culminating in the 2012 Nobel Prize, awarded to Gurdon and Shinya Yamanaka for their contributions to the field. Research in regenerative medicine has progressed inexorably due to the exceptional promise it holds for understanding and potentially treating a variety of otherwise untreatable medical conditions. Some of these therapies are already being studied in patients, and many others are on the horizon.

Alongside this transformative, exponential increase in regenerative medicine research, during the last 50 years we have witnessed an incredible expansion in our knowledge and treatment of retinal disease. Beginning with the advent of fluorescein angiography and ophthalmic lasers in the 1960s, we now find ourselves in an extraordinarily rich scientific and clinical landscape with widespread use of targeted biologic therapeutics, high resolution optical coherence tomography, ultra-widefield fundus imaging, and remarkable advances in vitreoretinal surgery. There is more to come. We are at the dawn of an era of cellular and gene therapy approaches that may provide legitimate hope for many untreatable retinal conditions affecting millions of people worldwide.

These individuals are the inspiration for this book. *Cellular Therapies for Retinal Disease: A Strategic Approach* will systematically address the potential of cell-based therapies for the treatment of several retinal disorders for which there are no known treatments today. Rather than utilizing a disease-centric approach to this topic, this book will focus on strategies for creating, transplanting, and studying the cells. Accordingly, the book is divided into three parts:

Part I Cell Replacement Therapy
Part II Cell-Based Neuroprotection
Part III Disease Modeling Using Induced Pluripotent Stem Cells

In many retinal disorders, dysfunction and loss of particular retinal cell types is a common endpoint in disease pathogenesis. Transplanting replacement cells remains an attractive, "one size fits all" therapeutic ideal, as this approach may be agnostic to the molecular and cellular mechanisms underlying given conditions. For example, the retinal pigment epithelium (RPE) loss in dry age-related macular degeneration (AMD) and Stargardt disease might be addressed with RPE transplantation despite disparate pathophysiologies. Part I describes various aspects of cell replacement therapy for retinal diseases such as AMD, Stargardt disease, and glaucoma. This includes the use of embryonic and induced pluripotent stem cells to produce RPE and retinal ganglion cells, surgical techniques for subretinal transplantation, and scaffold materials to provide support for RPE monolayers. Part II contains two chapters demonstrating the use of cells for neuroprotection and trophic support of existing retinal cells. Rather than replacing retinal cells, this therapeutic strategy relies on stem cells to release trophic signaling factors that may sustain surviving cells and halt ongoing cell loss in slowly degenerative diseases such as dry AMD and retinitis pigmentosa. Part III showcases some of the most promising avenues for retinal research today. Induced pluripotent stem cells derived from skin fibroblasts or peripheral blood leukocytes can be derived from patients with any retinal disease and then differentiated into specific retinal cell types or multicellular retinal organoids to recreate the patient's condition in the laboratory. A powerful research paradigm that could allow us to model human retinal disease and test treatments in a patient-specific, high-throughput fashion, this approach may facilitate important therapeutic interventions.

Each chapter fleshes out the latest advances in cellular therapy with highly esteemed authors contributing from around the world. The goal of this book is to not only update the retinal physician or laboratory scientist on the current state of the field, but also bring together leaders in the field under the umbrella of this book. This is an age of increasingly collaborative and complex science, and a team effort will be necessary to develop novel treatments for retinal disorders. As editors of this text, we are compelled to emphasize the importance of rigorous research, scientific methodology, and responsible communication particularly in this era of uncurated, non-peer-reviewed internet-based hyperbole. As scientists and clinicians, it is our privilege to explore the fundamentals of regenerative biology and try to translate knowledge into clinically meaningful interventions that may reduce human blindness and suffering. Along with this privilege comes the responsibility to protect the public, our patients, and even health care providers from misinformation and those who might take advantage of the excitement surrounding "stem cell therapies."

The editors owe a deep gratitude for the support of the leadership, faculty, and staff at UCLA's Stein Eye Institute, UCLA's Eli & Edythe Broad Center of Regenerative Medicine & Stem Cell Research, and the universities and eye institutes of the many authors who contributed to this book. Furthermore, on behalf of Drs. Schwartz and Nagiel, we wish to singularly thank Bob Lanza for his inspirational work and leadership. Importantly, encouragement from the American Academy of Ophthalmology, the Retina Society, the Macula Society, the American Society of Retina Specialists, and the Pine Ridge Eye Study Club must be mentioned. In addition, this endeavor would not have been possible without the assistance of Michael Koy, Bibhuti Sharma, and Aleta Kalkstein at Springer, who guided us and the authors flawlessly through the production process. Lastly, and most importantly, we dedicate this book to our patients and our families from whom we constantly draw strength.

Los Angeles, CA, USA — Steven D. Schwartz, M.D.
Los Angeles, CA, USA — Aaron Nagiel, M.D., Ph.D.
Marlborough, MA, USA — Robert Lanza, M.D.

Contents

Contributors

Juan Amaral, M.D. National Eye Institute, National Institutes of Health, Bethesda, MD, USA

Jennifer G. Aparicio, Ph.D. Division of Ophthalmology, Department of Surgery, The Vision Center, Children's Hospital Los Angeles, Los Angeles, CA, USA

James S. Baldassarre, M.D. Janssen Cell Therapy, Janssen R&D LLC, Spring House, PA, USA

Kapil Bharti, Ph.D. Unit on Ocular and Stem Cell Translational Research, National Eye Institute, Bethesda, MD, USA

Maria Campos, M.D. National Eye Institute, National Institutes of Health, Bethesda, MD, USA

Dennis Clegg, Ph.D. University of California, Santa Barbara, Molecular, Cellular and Developmental biology, Center for Stem Cell Biology and Engineering, University of California, Santa Barbara, CA, USA

David Cobrinik, M.D., Ph.D. Division of Ophthalmology, Department of Surgery, The Vision Center, Children's Hospital Los Angeles, Los Angeles, CA, USA

Development, Stem Cells, and Regenerative Medicine Program, Keck School of Medicine, University of Southern California, Los Angeles, CA, USA

The Saban Research Institute, Children's Hospital Los Angeles, Los Angeles, CA, USA

Department of Biochemistry & Molecular Medicine, Keck School of Medicine, University of Southern California, Los Angeles, CA, USA

USC Roski Eye Institute, Keck School of Medicine, University of Southern California, Los Angeles, CA, USA

Peter Coffey, Ph.D. Institute of Ophthalmology, University College London, London, UK

Lyndon da Cruz, Ph.D. Vitreoretinal Surgery, NIHR Moorfields Biomedical Research Centre, Moorfields Eye Hospital NHS Foundation Trust, London, UK

Dean Eliott, M.D. Retina Division, Department of Ophthalmology, Massachusetts Eye and Ear Infirmary, Harvard Medical School, Boston, MA, USA

Meena S. George, M.D., Ph.D. USC Roski Eye Institute, Keck School of Medicine, University of Southern California, Los Angeles, CA, USA

Odysseas Georgiadis, M.D. Vitreoretinal Clinical Research, Moorfields Eye Hospital, NHS Foundation Trust, London, UK

Jeffrey L. Goldberg, M.D., Ph.D. Byers Eye Institute at Stanford University, Palo Alto, CA, USA

Ninel Z. Gregori, M.D. Department of Ophthalmology, Bascom Palmer Eye Institute, University of Miami Miller School of Medicine, Miami, FL, USA

Jeffrey S. Heier, M.D. Ophthalmic Consultants of Boston, Boston, MA, USA

David R. Hinton, M.D. USC Roski Eye Institute, Keck School of Medicine, University of Southern California, Los Angeles, CA, USA

Mark S. Humayun, M.D., Ph.D. USC Roski Eye Institute, Keck School of Medicine, University of Southern California, Los Angeles, CA, USA

Balendu Shekhar Jha, Ph.D. Unit on Ocular and Stem Cell Translational Research, National Eye Institute, Bethesda, MD, USA

Anthony Joseph, M.D. Ophthalmic Consultants of Boston, Boston, MA, USA

Vitreoretinal Surgery and Disease, Ophthalmic Consultants of Boston, Boston, MA, USA

Michael Keane, M.S. Janssen Cell Therapy, Janssen R&D LLC, Spring House, PA, USA

Vladimir Khristov, B.S. Section on Epithelial and Retinal Physiology and Disease, National Eye Institute, Bethesda, MD, USA

Yichao Li, M.S. National Eye Institute, National Institutes of Health, Bethesda, MD, USA

Arvydas Maminishkis, Ph.D. Section on Epithelial and Retinal Physiology and Disease, National Eye Institute, Bethesda, MD, USA

Carlos A. Medina, M.D. Department of Ophthalmology, Bascom Palmer Eye Institute, University of Miami Miller School of Medicine, Miami, FL, USA

Debbie Mitra, Ph.D. USC Roski Eye Institute, Keck School of Medicine, University of Southern California, Los Angeles, CA, USA

Elad Moisseiev, M.D. Department of Ophthalmology & Vision Science, University of California Davis Eye Center, Sacramento, CA, USA

Hossein Nazari, M.D. University of Texas Medical Branch, Ophthalmology, Galveston, Texas, USA

Huy V. Nguyen, M.D. Massachusetts Eye and Ear Infirmary, Boston, MA, USA

Susanna S. Park, M.D., Ph.D. Department of Ophthalmology & Vision Science, University of California Davis Eye Center, Sacramento, CA, USA

Haohua Qian, Ph.D. National Eye Institute, National Institutes of Health, Bethesda, MD, USA

Carl D. Regillo, M.D. Midatlantic Retina of Wills Eye Hospital, Philadelphia, PA, USA

Aaron Rising, Ph.D. Section on Epithelial and Retinal Physiology and Disease, National Eye Institute, Bethesda, MD, USA

Mira M. Sachdeva, M.D., Ph.D. Retina Division, Department of Ophthalmology, Massachusetts Eye and Ear Infirmary, Harvard Medical School, Boston, MA, USA

Priya Sharma, M.D. Wills Eye Hospital, Philadelphia, PA, USA

Dominic W.H. Shayler, B.A. Division of Ophthalmology, Department of Surgery, The Vision Center, Children's Hospital Los Angeles, Los Angeles, CA, USA

Development, Stem Cells, and Regenerative Medicine Program, Keck School of Medicine, University of Southern California, Los Angeles, CA, USA

Jayanth Sridhar, M.D. Midatlantic Retina of Wills Eye Hospital, Philadelphia, PA, USA

Stephen H. Tsang, M.D., Ph.D. Jonas Children's Vision Care, and Bernard & Shirlee Brown Glaucoma Laboratory, Department of Ophthalmology, Columbia University Medical Center, New York, NY, USA

Edward S Harkness Eye Institute, New York-Presbyterian Hospital, New York, NY, USA

Department of Pathology & Cell Biology, Institute of Human Nutrition, College of Physicians and Surgeons, Columbia University, New York, NY, USA

Praseeda Venugopalan, Ph.D. Byers Eye Institute at Stanford University, Palo Alto, CA, USA

Xiong Zhang, Ph.D. Byers Eye Institute at Stanford University, Palo Alto, CA, USA

The original version of this book was revised. An erratum to this book can be found at DOI 10.1007/978-3-319-49479-1_11

Part I

Cell Replacement Therapy

Human Embryonic Stem Cell-Derived Retinal Pigment Epithelial Cell Transplantation for Retinal Degeneration

Ninel Z. Gregori, Carlos A. Medina, Mira M. Sachdeva, and Dean Eliott

1 Background

Age-related macular degeneration (AMD) affects over 1.75 million people in the Western countries and represents the third leading cause of blindness worldwide [1, 2]. While targeted molecular therapy inhibiting vascular endothelial growth factor (VEGF) has revolutionized management of exudative (wet) AMD, there is currently no definitive intervention to slow or reverse progression of the nonexudative (dry) form of the disease. Clinically, early dry AMD is characterized by the accumulation of lipid- and protein-containing deposits (drusen) underneath the retinal pigment epithelium (RPE) that can progress to large areas of geographic RPE and photoreceptor atrophy [3]. The pathophysiology of drusen formation is incompletely understood but studies in both human patients and animal models have implicated both activated oxidative stress response pathways and impaired immunological responses, as mutations in the complement factor H (CFH) gene have been linked to AMD in genome-wide association studies [3]. The pathophysiology of geographic atrophy progression also remains an area of active investigation [4]. On fundus autofluorescence imaging, increased autofluorescence signal can be seen at the edges of GA lesions and likely represents sick RPE accumulating lipofuscin at a stage that precedes frank RPE loss [5]. Variability in cell layers has been shown at the edge of GA where in some eyes there is a sharp transition from few cone nuclei to multiple layers of rods and cones at the edge of GA, while in others a large transition zone outside of GA, as large as 1400 microns, with dysfunctional cone cells is seen [4]. Regions where dysfunctional RPE underlies living photoreceptors may provide an opportunity for intervention during a therapeutic window before photoreceptor death occurs.

Stargardt macular dystrophy (SMD) represents the most common form of juvenile-onset macular degeneration, affecting approximately 1 in 8000–10,000 individuals, and likewise has no known treatment [6]. SMD is most commonly inherited in an autosomal recessive fashion due to mutations in the ABCA4 gene encoding an ATP-binding cassette (ABCR) transmembrane transporter for vitamin A metabolites expressed exclusively in the photoreceptor disc membranes [6, 7]. The rare autosomal dominant form of SD is caused by mutations in the ELOVL4 gene, which encodes a protein involved in the elongation of very-long-chain fatty acids [8]. SMD typically presents in the first or second decade with loss of central vision, pigmentary "beaten bronze" changes in the macula associated with pisiform yellow-white flecks in the posterior pole and midperiphery. The presence of dark choroid on fluorescein angiography, due to accumulation of lipofuscin in the RPE cells, is present in the majority of patients; however its absence does not rule out the diagnosis. Macular changes typically progress to pronounced atrophy, and vision eventually declines to the 20/200 or 20/400 level. Onset after the age of 20 years, foveal sparing, milder forms of the disease at presentation, and presence of flecks and atrophic retinal changes limited to the macula are associated with better visual prognosis [6]. AMD and SMD share a common underlying pathophysiology—accumulation of bisretinoids within and underneath RPE cells, leading to RPE dysfunction and ultimate loss, and then subsequent photoreceptor damage [9].

N.Z. Gregori, M.D. (✉) • C.A. Medina, M.D
Department of Ophthalmology, Bascom Palmer Eye Institute, University of Miami Miller School of Medicine, 900 NW 17th Street, Miami, FL 33136, USA
e-mail: ngregori@med.miami.edu

M.M. Sachdeva, M.D., Ph.D. • D. Eliott, M.D.
Retina Division, Department of Ophthalmology, Massachusetts Eye and Ear Infirmary, Harvard Medical School, Boston, MA, USA

S.D. Schwartz et al. (eds.), *Cellular Therapies for Retinal Disease*, DOI 10.1007/978-3-319-49479-1_1

While various novel approaches to treat these diseases are currently under investigation, including molecular targeting of complement regulators in the innate immune system or visual cycle modulation for geographic atrophy in dry AMD and gene therapy or modified vitamin A derivatives for SMD, direct replacement of the damaged or absent RPE with healthy tissue is an attractive paradigm potentially applicable to both patient populations [9–12].

To create healthy RPE cells, the identity of RPE must be defined both structurally and functionally. The human RPE exists as a monolayer of polarized columnar epithelial cells with its basal surface resting on Bruch's membrane and its apical microvilli interdigitating with photoreceptors. Critical RPE functions include maintenance of the blood-retina barrier via intercellular tight junctions, phagocytosis of outer segment discs shed by the photoreceptors, metabolic regeneration of 11-cis-retinal during the visual cycle, and secretion of growth factors [13]. During the past decade, gene expression profiling of RPE cell lines as well as primary and cultured human RPE have defined a set of transcription factors and proteins expressed in RPE [14–16]. Any cells intended to replace abnormal or absent RPE must meet these histologic, physiologic, and molecular criteria.

2 Hypothesis and Strategy: Cellular Replacement Therapy

Because RPE dysfunction and loss may represent an early step in the pathogenesis of diseases such as AMD and SMD, the introduction of healthy RPE cells into the diseased eye presents a logical treatment strategy. Conceived over 20 years ago, this concept of RPE cellular replacement was originally applied experimentally to the treatment of exudative AMD in the pre-anti-VEGF and prephotodynamic therapy era. Failure to improve visual outcomes in patients with exudative AMD after surgical removal of choroidal neovascular membranes (CNVMs) in several studies including the submacular surgery trial suggested the potential utility of replacing diseased RPE with healthy cells [17]. Early pilot studies demonstrated the possibility of using fetal RPE transplants in patients with exudative AMD following surgical removal of CNVMs; however eyes developed macular edema and fluorescein leakage implying graft rejection [18, 19]. In nonexudative AMD patients, transplantation of fetal RPE cells in suspension showed no evidence of rejection and was associated with drusen disappearance, stable visual acuity, and preserved retinal function over the transplanted area by microperimetry [19]. In 1991, a technique for autologous RPE cells and Bruch's membrane graft following surgical removal of a submacular scar was reported in two patients with end-stage exudative AMD [20]. The technique was subsequently refined, and some modest visual stability or improvement up to 7 years postoperatively was documented [20, 21]. Although potentially beneficial in patients with few alternative options, such as in eyes with massive submacular hemorrhage or RPE tears, the ultimate success of full-thickness RPE/choroid autografts is limited by postoperative occurrence of proliferative vitreoretinopathy, macular pucker, or recurrent CNVM [22]. In another strategy, based on in vitro similarities between iris pigment epithelium (IPE) and RPE, autologous transplantation of IPE cells into the subretinal space of patients with exudative AMD was attempted. The IPE grafts remained viable for up to 3 years postoperatively but there was no significant visual improvement [23, 24].

After the anti-VEGF agents transformed exudative AMD into a treatable disease, the focus of RPE replacement research has shifted to geographic atrophy in the nonexudative form of AMD. Recently embryonic stem cells have emerged as a potentially unlimited source of healthy RPE cells.

3 Scientific Basis for Intervention

Theoretically, the eye represents an ideal microenvironment for stem cell therapy given its small size, relative immune privilege, and ability to assess graft survival and function via several noninvasive imaging modalities. As a monolayer of cells with well-characterized physiology and morphology, the RPE represents an attractive target for a stem cell-based approach. Moreover, as RPE dysfunction and loss precede photoreceptor loss, the areas of preserved photoreceptors overlying damaged RPE provide a potential opportunity to intervene with the introduction of healthy RPE. In 2013, the National Eye Institute and NIH Center for Regenerative Medicine hosted a meeting of experts in the field of stem cell replacement therapy for retinal diseases to advance knowledge and promote collaborative efforts, underscoring the anticipated potential of this strategy to have a significant impact on widespread ophthalmologic disease [25]. In order to have clinical application, engineered RPE must mimic normal RPE in gene expression, morphology, and mostly importantly function.

3.1 Human Embryonic Stem Cells

Embryonic stem cells (ESCs) are pluripotent cells harvested from the early mammalian embryo, specifically from the inner cell mass at the blastocyst stage, and have unlimited self-renewal and differentiation potential. The functional definition of ESCs also includes the capability of maintaining a normal karyotype during proliferation [26]. In 1998, work in murine models and primates ultimately led to the isolation and characterization of embryonic stem cells from human blastocysts (hESCs) and the demonstration that these cells could be propagated by in vitro coculture with feeder mouse embryonic fibroblast cells [27]. These hESCs demonstrated high levels of telomerase (correlated with immortality in human cell lines), expressed cell surface markers of undifferentiated cells (including embryonic antigens (SSEA)–3, SSEA-4, TRA-l-60, TRA-1-81, and alkaline phosphatase), and were capable of differentiation into all three germ layers when transplanted into mice [27].

Teratoma formation remains a potential concern when considering transplantation of hESCs to patients. Exposure of cultured hESCs to nonhuman tissues in the feeder scaffolds or growth media has also presented theoretical concerns, such as increased immunogenicity of the cells and exposure to animal pathogens. Various "xeno-free" modifications of the original protocols have been explored [28]. Another significant consideration is the potential need for systemic immunosuppression as with any allogeneic organ transplant, which is a major consideration especially in the elderly population.

3.2 Induced Pluripotent Stem Cells

In 2007, another potential source for cellular replacement therapy emerged when adult human dermal fibroblasts were dedifferentiated into a pluripotent state by retrovirus-mediated transfection of four transcription factors present in hESCs, namely Oct 3/4, Sox2, Klf4, and c-Myc. The first of these human induced pluripotent stem (iPS) cells were derived from human dermal fibroblasts and resembled hESCs in morphology, gene expression, telomerase activity, proliferation, and capacity to differentiate into all three germ layers [29–31]. These iPS cells expressed hES cell-specific surface antigens, including SSEA-3, SSEA-4, tumor-related antigen (TRA)-1–60, TRA-1-81, and TRA-2-49/6E (alkaline phosphatase), and NANOG protein [30-31]. Because iPSCs can be harvested from the same patient into whom the iPS-derived RPE cells will be transplanted (i.e., an autograft), the transplanted tissue may not be subject to immune rejection, thereby potentially avoiding the risks associated with systemic immunosuppression [32]. Furthermore, iPS-derived RPE cells have been shown in vitro to inhibit T-cell proliferation and activation, at least in part by production of the cytokine TGF-beta [33]. Additional advantages of iPS cells over hESCs include several potential source pools (e.g., dermal fibroblasts, adult corneal limbal epithelium) and the avoidance of ethical concerns over the use of human embryonic tissue [25, 26, 34].

Unfortunately, iPS cells require extensive in vitro manipulation compared with hESCs and the expense of harvesting and manipulating autologous cells for transplantation may preclude larger scale production and thus limit the economic feasibility of this approach [25]. Additional drawbacks include a potentially high degree of genomic instability, possibility of viral integration during iPSC generation, and persistence of underlying genetic mutations harbored by the patient's own cells (though this may be modifiable by gene-editing techniques) [35]. Since the experience with iPS cells is more limited than that with hESCs, future research is needed to explore these questions.

3.3 Strategies for Generating RPE from Pluripotent Cells

Cultured hESCs and iPS cells can spontaneously give rise to pigmented cells that phenotypically recapitulate RPE in morphology and function. However, this process occurs with low efficiency and is estimated to be approximately 1% after 1–2 months in cell culture [36]. Further insights into the molecular signals guiding normal RPE and retinal differentiation were applied to optimize the generation of RPE cells from hESCs by exposure to nicotinamide and hiPSCs by serial addition of noggin, basic fibroblast growth factor (bFGF), retinoic acid, and sonic hedgehog [32, 33]. In various studies this "directed differentiation" successfully increased the efficiency of RPE production up to 60–80% [36–38]. Current efforts continue to improve the process of RPE differentiation from pluripotent cells by manipulation of molecular signaling pathways. A recent study reported 97% efficiency of RPE generation from hESCs by activation of the canonical Wnt/beta-catenin signaling cascade [39, 40]. These RPE cells have been rigorously characterized by gene expression profiling, assessment of cell polarity, and rod outer segment phagocytosis capability [41].

RPE cells derived from either human ESCs or iPS cells can be delivered into the subretinal space in one of the two ways: either as a cell suspension or as a RPE monolayer on a scaffold. Although the latter approach presents a greater challenge surgically, early studies demonstrated decreased rates of RPE cell apoptosis when cells were adherent to various scaffolds [42]. Naturally occurring or synthetic polymers can be used as scaffolds for cellular transplantation, and the physical properties of the particular substrate determine key characteristics of the final product, including mechanical flexibility and longevity (e.g., biodegradable vs. nonbiodegradable) [43]. RPE monolayer sheets have been generated in vitro on scaffolds composed of a variety of materials including collagen, gelatin, Descemet's membrane, lens capsule, and amniotic membrane, as well as synthetic polymers such as poly(lactic-co-glycolic acid) (PLGA) and poly L-lactic acid (PLLA) [43]. In addition to serving as a mechanical support to transplanted cells, scaffolds may also provide trophic support by the incorporation of factors promoting survival and differentiation [44]. Despite these advantages, further considerations for the successful placement of scaffolds in the subretinal space include the requirement of a thin sheet (5–90 μm), potential for an inflammatory response to the implanted material, and use of a reliable and safe surgical technique [43]. Stem cell-based clinical trials for diseases of the RPE which are currently either under way or in the advanced planning stages include studies of both transplanted cells in suspension and cells attached to a scaffold [25]. At the present time, no published data exist regarding the efficacy of transplantation of RPE cells on a scaffold or iPSC-derived RPE cells in human subjects; however mid- to long-term trial results in patients with atrophic AMD and SMD undergoing transplantation of hESC-derived RPE cell suspension were published in 2014, and this approach is discussed in detail in the remainder of this chapter [45].

4 Preclinical Data

Once studies demonstrated the ability to generate RPE from embryonic stem cells, scientists quickly moved to evaluate the safety and efficacy of this strategy in animal models of RPE degeneration. These studies utilized strict guidelines from the US Food and Drug Administration in order to be clinically applicable to patients in the future. In the preclinical studies that would become the basis for the clinical trials of hESC- RPE transplantation, undertaken by Advanced Cell Technology (a.k.a Ocata Therapeutics, Inc. and now Astellas Institute of Regenerative Medicine.), subretinal delivery of hESC-derived RPE cells generated in accordance with current Good Manufacturing Practices (cGMPs) and Good Tissue Practices (cGTPs) demonstrated sustained visual function and photoreceptor integrity rescue in a dose-dependent fashion. There was no teratoma formation, hyperproliferation, or evidence of rejection, in both the Royal College of Surgeons (RCS) rat and the Elov14 mouse, which are animal models of retinal degeneration and Stargardt disease, respectively [46, 47]. Furthermore, transplantation of cells into the National Institute of Health (NIH) III immune-deficient mouse model showed no evidence of tumor formation up to 9 months [48]. In the more recent of these studies, the RPE cells were generated from the MA01 or MA09 hESC lines, which were both derived from a single blastomere isolated from an early human embryo, cocultured and passaged with mitomycin-treated, mitotically inactivated mouse embryonic fibroblasts. These hESCs were shown to express molecular markers of pluripotency (Oct-4, SSEA-3, SSEA-4, TRA-1-60, TRA-1-81, Nanog, and alkaline phosphatase), and demonstrated to be able to differentiate into all three germ cell layers while maintaining normal karyotype [41].

These hESC lines produced RPE clusters whose cellular identity was confirmed by morphology and expression of RPE-specific genes (*RPE65*, *Bestrophin*, *CRALBP*, *PEDF*, *MitF*, *Otx-2*, *Tyr*, and *Pax2*) [46]. Quantitative real-time reverse-transcription polymerase chain reaction (qPCR) was used to confirm a 10- to 100-fold enrichment in RPE-specific transcripts and proteins with a corresponding decrease in stem cell markers during the in vitro maturation process [46]. RPE cells thus generated were thoroughly tested for contamination and karyotype according to US Food and Drug Administration and International Conference of Harmonization guidelines [46]. These mature cells were then suspended in balanced salt solution in different doses ranging from 5000 to 100,000 cells, and injected into the subretinal space of RCS rats [46]. Spatial acuity and luminance threshold (sensitivity) were improved compared with sham or control-injected mice as the dose increased from 5000 to 50,000 cells, but no additional benefit was seen increasing the dose to 75,000 or 100,000 cells [46]. Similar findings were observed when injecting hESC-RPE into the subretinal space of *Elovl4* mice [46]. Histologic analyses of retinal/RPE tissue in transplanted RCS rats demonstrated preservation of retinal anatomy and rescue of photoreceptors which was optimal at a dose of 50,000 cells, correlating with the functional data [46]. Notably, the area of photoreceptor rescue extended beyond the distribution of the transplanted hESC-RPE, suggesting a diffusible effect [46]. No recipients of the hESC-RPE cells in these preclinical studies developed teratomas and further testing was performed in immunodeficient mice to ensure lack of tumor formation up to 9 months posttransplant (roughly the life span of these animals) [46]. RPE cell lines derived from the MA09 hESC line were then used for the first clinical trials of embryonic stem cell-derived RPE cell transplantation in humans.

5 Cells Utilized in Human Trials

The embryo used to create the MA09 human embryonic stem cell line was an excess embryo created by in vitro fertilization for reproductive purposes donated for research by an anonymous couple. The MA09-hESC line was derived from the blastomere stage of the embryo and expanded on mitotically inactivated mouse embryonic fibroblasts according to the Good Manufacturing Practices [48]. Because the cells were cultured with animal cells, the derivatives of the MA09-hESC line are classified as a xenotransplantation product. After embryonic body formation and cellular outgrowth, the MA09-hESCs were induced to differentiate into RPE cell patches [41, 48]. These cell patches were harvested at a lighter stage of pigmentation, as lower melanin-containing cells were shown to attach to substrate and grow to confluency with much higher efficiency than more darkly pigmented MA09-hRPE cells [48]. These cell patches were isolated with collagenase, purified, trypsinized, and passaged in culture for a total of three passages (P0, P1, P2). After the second passage MA09-hRPE cells were cryopreserved and stored as the cell product for clinical use [48].

The MA09-hRPE cells were then tested for safety and terminal differentiation into mature RPE cells by gene expression analysis, karyotyping, phagocytosis assay, differentiation and purity evaluation by morphology, quantitative polymerase chain reaction, and quantitative immune staining for RPE and hESC markers [41, 48]. To mitigate the risk of communicable diseases, extensive in vitro and in vivo pathogen testing was performed according to Food and Drug Administration (FDA) and International Conference on Harmonization (ICH) guidelines, just as was performed in the preclinical studies. Each lot of MA09-hRPE bulk product was tested for mycoplasma, retroviruses, microbial contaminants, unacceptable endotoxin levels, and mouse antibody production against 19 viruses.

6 Translational Strategies/Safety Considerations

Even with promising results from preclinical studies, the translation of data from animal models of disease to human patients requires critical additional considerations that may include dosing adjustments, clinical trial design, and most importantly safety considerations.

6.1 Risks of Tumorigenicity and Pathogenicity

For a cell product to be considered safe for clinical use, the potential for abnormal growth, teratoma formation, ectopic tissue proliferation, negative structural effects on the host tissues, immune mediated rejection, inflammation, or deleterious visual outcomes must be assessed. The hESC-derived RPE (hRPE) cells must be free of human and animal pathogens, appropriately differentiated into fully functioning RPE cells, and be free of undifferentiated stem cells. Transplantation of undifferentiated hESCs has been shown to produce teratomas in all injected rodent eyes; however subretinal injection of hESC–hRPE cell suspension which consists of greater than 99% pure differentiated RPE cells leads to no tumors in animals injected with 5–10 × 10^4 RPE cells spiked with either 0.01, 0.1, or 1% undifferentiated hESCs in preclinical rodent models [45, 48].

6.2 Risks Associated with Immunosuppression and "Xenotransplantation"

The subretinal space is rendered immunoprivileged by an intact and healthy RPE layer [49, 50]. Embryonic stem cells possess immunosuppressive qualities, and multiple animal studies have shown that immunosuppression is not necessary for subretinal transplantation of embryonic stem cells as long as the blood-retina barrier is not breached [50]. However the cells being transplanted in the clinical trials discussed are terminally differentiated RPE cells, which do not express stem cell properties. Moreover, subretinal transplantation may activate resident microglia and lead to low graft survival rates [50]. The immunogenicity of hESC-derived RPE cells transplanted into human eyes with severe RPE atrophy is not known; thus transplant rejection and cell death are potential risks, and immunosuppressants may be required at least temporarily. The risks of immunosuppression, including infection, lymphoproliferative disorders, neoplasia, and leukoencephalopathy, should be considered. Furthermore, elderly patients such as those with AMD may be at higher risk for these complications due to their age. Since the human embryonic stem cells are expanded on mouse fibroblasts, as described above, the cell product is considered a "xenotransplantation product." Thus transplantation necessitates enrollment in a registry as well as deferral of future blood, blood component, or tissue donation.

6.3 Risks Associated with the Surgical Procedure

Risks include those associated with a standard three-port pars plana vitrectomy such as endophthalmitis, retinal detachment, vitreous hemorrhage, intraocular inflammation, sympathetic ophthalmia, raised intraocular pressure, and cataract. Other risks associated with subretinal surgery with introduction of a cell product under the retina include possible subretinal abscess, retinal/choroidal/vitreous hemorrhage, intraocular inflammation, retinal detachment due to either the retinotomy or the subretinal transplantation itself, and proliferative vitreoretinopathy.

7 Target Population Selection

AMD is the most common cause of blindness in the elderly population in industrialized countries [51]. The manifestations of AMD include oxidative damage; drusen, lipofuscin, and activated complement deposition; structural changes in the Bruch's membrane; and eventual development of choroidal neovascularization and/or geographic atrophy [51]. In the early stages of AMD, the RPE layer begins to show pleomorphism, cell loss, and accumulation of extracellular debris between the RPE and its basal lamina (basal laminar deposit) or between the RPE basal lamina and Bruch's membrane (basal linear deposit). As the disease progresses and RPE cells begin to die, the adjacent RPE cells accumulate lipofuscin, probably due to phagocytosis of extracellular lipofuscin released by the surrounding dying cells [51]. RPE death leads to disorganization of the overlying photoreceptor outer segments and eventual photoreceptor death with resultant vision loss. However, the area of photoreceptor loss can be either smaller or much greater than the area of RPE atrophy [43, 44]. Rods have been shown to die off earlier than the cones [44]. In addition, the choriocapillaris under the RPE atrophy loses normal vessel fenestrations, and may eventually undergo atrophy [4, 51]. These findings are important in the design of clinical trials, since the area of photoreceptor loss may be much greater than the area of RPE atrophy [44]. Despite the recent advances in the treatment of choroidal neovascularization in AMD, there is still no treatment for the atrophic age-related macular degeneration.

It has been shown that in SMD the initial pathophysiologic insult is abnormal deposition of lipofuscin in the RPE cells, followed by RPE degeneration and finally photoreceptor loss [52]. Because of the central role the RPE plays in the pathophysiology of AMD and SMD and their high prevalence, these patient populations were selected as the target for the first human trial employing hESC-derived RPE cell transplantation.

8 Human Trial Design

The first published human trials of hESC-derived RPE cell transplantation are sponsored by Ocata Therapeutics Inc. (Marlborough, MA, formerly Advanced Cell Technology Inc. and now Astellas Institute of Regenerative Medicine). At the time of writing, the trials are ongoing phase I/II, open-label, multicenter, prospective, nonrandomized trials to determine safety and tolerability of hESC-derived RPE cell suspension (MA09-hESC cell line) delivered into subretinal space of patients with atrophic AMD and SMD. The Stein Eye Institute, Bascom Palmer Eye Institute, Massachusetts's Eye and Ear Infirmary, Wills Eye Hospital in the United States, Moorfields Eye Hospital in the United Kingdom, and CHA University in Korea are participating in the trials.

Transplantation doses were selected based on the safety data from animal studies, and comparison of the human eye size with the animal eyes used in preclinical models. Four dose cohorts in both atrophic AMD (AMD trial) and Stargardt macular degeneration patients (SMD trial) were included in the dose-escalating low-vision (vision 20/400 or less in the study eye) arms of the trials. The doses were 50,000 cells transplanted (3 AMD and 3 SMD patients), 100,000 cells transplanted (3 AMD and 3 SMD patients), 150,000 cells transplanted (3 AMD and 3 SMD patients), and 200,000 cells transplanted (3 AMD and 3 SMD patients). Cohorts were recruited sequentially. Six weeks after the first patient in each cohort was treated, an independent Data Safety Monitoring Board reviewed the clinical data before enrollment was allowed for an additional two patients in that cohort. The Data Safety Monitoring Board performed a full safety review of all patients before the next cohort received a transplant. A second population of four patients with better vision (20/100 or worse in the study eye) was later added to the AMD and SMD US-based trials. Each of these patients received a single subretinal injection of 100,000 MA09-hRPE cells in the study eye.

8.1 Inclusion and Exclusion Criteria

Adults over 18 years of age in the SMD trial and 55 years of age in the AMD trial with the best-corrected Early Treatment Diabetic Retinopathy Study (ETDRS) visual acuity no better than 20/400 (low-vision group) or no better than 20/100 (better-vision group) in the treated eye were included (Table 1). The ETDRS visual acuity of the fellow eye was no worse than 20/400 in the low-vision group and no worse than 20/100 in the better-vision group. Patients are tested to show ERG and visual field testing consistent with atrophic AMD or SMD. In the SMD arm, an abnormal multifocal ERG with a normal or an abnormal full-field ERG is required, and patients are genotyped. Patients are also screened to be free of significant systemic or concurrent ocular disease, have a negative cancer screening, and should be willing to use an effective form of birth control during the study.

8.2 Outcome Measures

The primary endpoint in the phase I/II trial is safety and tolerability of hESC-derived RPE cell product transplantation. Patients are assessed by slit-lamp examination, indirect ophthalmoscopy, visual acuity testing, and reading speed measurements. Ancillary testing includes dilated color fundus photos and autofluorescence images, fluorescein angiography, spectral-domain optical coherence tomography (SD-OCT), Humphrey perimeter or Goldmann visual field testing (if patient is not able to perform automated perimetry reliably), and full-field and multifocal ERG. These techniques were chosen to evaluate the eye for cellular engraftment, unintended proliferation, or immunologic rejection.

Patients underwent periodic physical examinations with vital signs, chest radiographs, electrocardiogram, hematological and serological testing, and blood collection for xenogeneic transplantation archiving.

Table 1 Inclusion and exclusion criteria from the Phase I/II Trial of MA09-hRPE Cell Transplantation for Atrophic Age-related Macular Degeneration and Stargardt Disease

Inclusion criteria	AMD	SMD
Age	>55	>18
Advanced dry AMD with >250 microns of geographic atrophy involving central fovea	X	–
BCVA of study eye	20/400 or worse	20/400 or worse
BCVA of fellow eye = 20/400 or better	20/400 or better	20/400 or better
Electrophysiological findings consistent with disease	X	X
Good general health, expected survival for >4 years after treatment	X	X
Normal SMA-20, CBC, PT/aPTT	X	X
Negative urine screen for drugs of abuse	X	X
Negative human immunodeficiency virus, hepatitis B and C serology	X	X
Negative cancer screening with previous 6 months	X	X
Able to understand and willing to sign the informed consent	X	X
Peripheral visual field constriction	–	X
–	–	–
Exclusion criteria		
Presence of active or inactive CNV in the study eye	X	X
Presence or history of unrelated retinal disease including retinal vascular	X	X
History of optic neuropathy	X	X
Other known cause of macular atrophy	X	X
Presence of glaucomatous optic neuropathy in the study eye, uncontrolled IOP, or use of two or more agents to control IOP	X	X
Cataract likely to need surgery <1 year	X	X
History of retinal detachment repair in the study eye	X	X
Axial myopia of greater than −8 diopters or axial length greater than 28 mm	X	X
History of malignancy (except excised basal cell or squamous cell of the skin)	X	X
History of myocardial infarction in previous 12 months	X	X
History of diabetes mellitus	X	X
History of cognitive impairments or dementia	X	X
Any current immunosuppressive therapy other than intermittent or low-dose corticosteroids	X	X
ALT/AST >1.5 times upper limit of normal or any known liver disease	X	X
Creatinine level ≥1.3 mg/dL	X	X
Hemoglobin <10 gm/dL; platelet count <100 k/mm^3, neutrophil count <1000/mm^3	X	X
Participation within previous 6 months in any clinical trial of a drug by ocular or systemic administration	X	X
Any other sight-threatening ocular disease	X	X
Uveitis or other intraocular inflammatory disease	X	X
Cataract or other eye surgery within previous 3 months	X	X
If female, pregnancy or lactation	X	X
Other medical condition that limits patient compliance and patient safety or alters study results	X	X

Inclusion and exclusion criteria from the Ongoing Phase I/II Trial of MA09-hRPE Cell Transplantation for Atrophic Age-related Macular Degeneration and Stargardt Disease (reproduced with permission from Human embryonic stem cell-derived retinal pigment epithelium in patients with age-related macular degeneration and Stargardt macular dystrophy: follow-up of two open-label phase 1/2 studies. Lancet. 2015; 385:509–16)

8.3 Surgical Procedure and Cell Preparation

On the day of transplantation, a qualified cell processing facility associated with each study site processes the cells locally. The cells are thawed, washed, and suspended to correct concentration under aseptic conditions. Viability and Gram staining are performed to assure sterility of the final product prior to release to the surgical team [48]. The cells are resuspended in Balanced Salt Solution Plus (BSS Plus, Alcon, Hunenberg, Switzerland) immediately prior to transplantation by the surgeon in the operating room [48].

Surgery is performed under either conscious sedation with retrobulbar anesthesia or general anesthesia. Each patient undergoes pars plana vitrectomy with posterior vitreous separation in the worse-seeing eye. Following an optional subretinal injection of balanced salt solution that is used to elevate a small bleb prior to cell injection, 150 microliters of MA09-hRPE cell suspension is injected into the subretinal space through a MedOne PolyTip cannula 23/38 gauge or 25/38 gauge (external/internal gauge, MedOne Surgical, Sarasota, FL, USA) connected to a 1-milliliter syringe [45]. The cannula used to transplant the cells is held in position for an additional minute in the attempt to limit reflux of cells. An optional air-fluid exchange is then performed. The eye is closed in the standard fashion to assure watertight closure. Intraoperative or postoperative steroid administration is not allowed as the effects on graft survival and outcomes are unknown. The patient is maintained in the supine position for at least 6 h to allow cell adhesion onto Bruch's membrane.

8.4 Transplantation Site

The transplantation site is chosen over a "transition zone located between the atrophic macular center and more peripheral, relatively normal tissue" [45]. The transition zone is chosen based on autofluorescence and SD-OCT imaging showing compromised RPE and overlying photoreceptors with preserved Bruch's membrane, which is thought to be important for cell engraftment since prior studies have demonstrated the ability of transplanted RPE cells to repopulate intact Bruch's membrane [42]. Specifically, RPE cells attach and become confluent when plated on the basal lamina of the RPE, the innermost layer of Bruch's membrane, but show extensive loss when plated onto inner collagenous, elastin, or outer collagenous layers [53]. In addition, since the area of photoreceptor loss may be much larger than the area of RPE atrophy [51], rescue of the photoreceptors outside of the geographic atrophy may provide a meaningful visual benefit to the patients.

8.5 Immunosuppressive Regimen

Immunosuppression is induced with low-dose tacrolimus (target blood levels 3–7 ng/mL) and mycophenolate mofetil (ranging 0.5 to 2 g orally twice a day) 1 week prior to implantation and for a period of 6 weeks after transplantation. Monotherapy with mycophenolate is then continued for an additional 6 weeks. Further immunosuppression is determined on an individualized basis based on the ocular findings at follow-up.

8.6 Results Published to Date

Results from the first 18 patients (9 AMD and 9 SMD) who were followed for at least 6 months after surgery were published on line in October 2014 [45]. These patients were enrolled between July 2011 and January 2014. Six eyes (3 AMD and 3 SMD) received 50,000 MA09-hRPE cells, six eyes (3 AMD and 3 SMD) received 100,000 MA09-hRPE cells, and six eyes (3 AMD and 3 SMD) received 150,000 MA09-hRPE cells. Eleven of eighteen patients were female; seventeen were Caucasian and one African-American. The median age was 77 years (range 70–88) in AMD trial and 50 years (range 20–71) in SMD trial. Median follow-up was 22 months (4 patients with <12 months, 12 patients with 12–36 months, and 2 patients with >36 months).

8.7 Reported Structural Outcomes

No eyes developed teratomas, macular edema, contractile epiretinal membrane, or retinal detachment. Subretinal fluid associated with subretinal injection of cells was resorbed in 17 eyes by the first postoperative day and in one eye by day 2. Absence of untoward effects to the host retina was demonstrated by serial SD-OCT, fundus photography, autofluorescence, and fluorescein angiography.

Thirteen of eighteen (72%) patients developed subretinal pigmentation in the area of cell transplantation, and was typically present at the border of the atrophic lesion (Fig. 1). SD-OCT scans of these areas demonstrated increased hyperreflectivity lining the inner aspect of Bruch's membrane in some eyes (Fig. 1d, e). Fundus autofluorescence was unchanged in some eyes and increased in others, and was not always consistently present during the follow-up period (Fig. 2).

Preretinal, pigmented cell growth was detected near the injection site in three eyes (one patient with SMD and two patients with AMD) but did not result in a contractile epiretinal membrane or other adverse effects (Fig. 3).

8.8 Reported Visual Outcomes

Visual field testing, ERG, and reading speed did not show differences between pretransplantation and posttransplantation visits [45]. Preoperative best-corrected visual acuity in the study eyes ranged from 20/200 to hand motion. Visual acuity changes for atrophic AMD and SMD patients improved in the majority of patients at 6 and 12 months after transplantation (Fig. 4). Median best-corrected visual acuity in eyes with atrophic AMD and SMD, excluding one patient with AMD and three patients with SMD who experienced vision loss presumptively due to cataract progression, improved significantly in treated eyes versus fellow eyes in the atrophic AMD arm but did not reach statistical significance, due to a small sample size, in the SMD arm (Fig. 5). Two out of four treated eyes undergoing cataract extraction experienced improvement in visual acuity (9–15-letter gain). The vision returned to baseline in the two other eyes after cataract surgery.

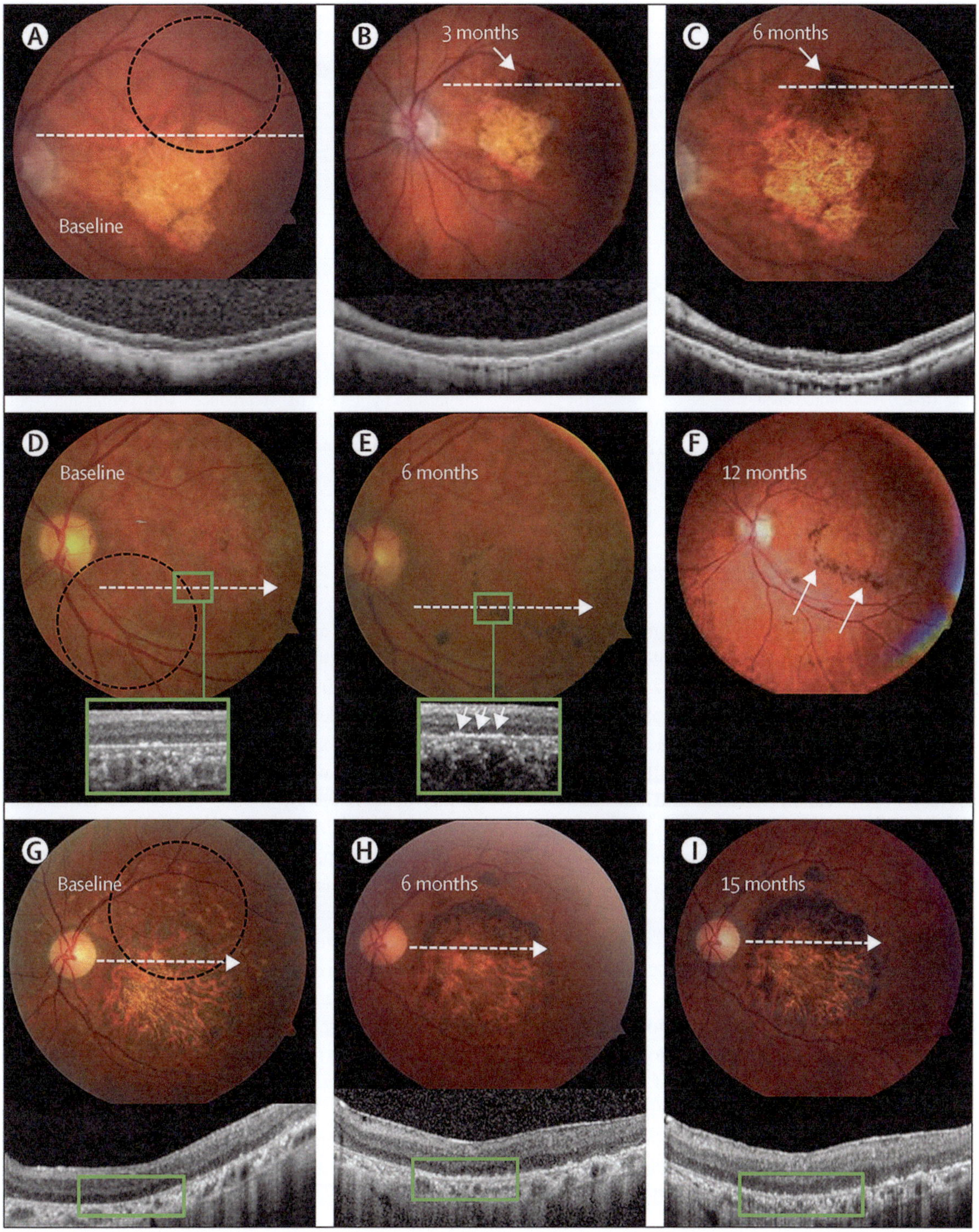

Fig. 1 Fundus images of eyes with pigmentation after transplantation with hESC-derived retinal pigment epithelial cells in patients with atrophic age-related macular degeneration and Stargardt disease. (**a–c**) Color fundus photographs and SD-OCT images at baseline of an eye from a patient with age-related macular degeneration (*dotted circle* shows an outline of the transplanted area), 3 months, and 6 months. Note the presence of a pigmented patch of transplanted cells (**b**, **c**, *arrows*) that becomes larger and more pigmented by 6 months. OCTs (inset) show the presence of cells on the inner aspects of Bruch's membrane at 6 months compared with baseline. (**d–f**) Color fundus photographs and SD-OCT images at baseline of an eye from a patient with Stargardt macular dystrophy (*dotted circle* shows an outline of the transplanted area), 6 months, and 1 year. Note the absence of pigment in the preoperative photograph (**d**). Patches of pigmented cells are evident around the border of baseline atrophy in retinal pigment epithelium (**e**) that becomes more prominent at 1 year (**f**, *arrows*). SD-OCT images at baseline (**d**) and 6 months (**e**) show that increasing pigmentation is at the level of the retinal pigment epithelium, normal monolayer retinal pigment epithelium engraftment, and survival at 6 months (**e**, *arrows*) adjacent to a region of bare Bruch's membrane devoid of native retinal pigment epithelium. (**g–i**) Color fundus photographs of a patient with Stargardt macular dystrophy (*dotted circle* shows an outline of the transplanted area). A large central area of atrophy is visible on the preoperative photograph (**g**). An area of transplanted retinal pigment epithelium cells is visible at the superior half of the atrophic lesion at 6 months (**h**) that becomes larger and more pigmented at 15 months (**i**). *hESC* human embryonic stem cells, *SD-OCT* spectral domain optical coherence tomography, *OCT* optical coherence tomography (reproduced with permission from Human embryonic stem cell-derived retinal pigment epithelium in patients with age-related macular degeneration and Stargardt macular dystrophy: follow-up of two open-label phase 1/2 studies. Lancet. 2015; 385:509–16)

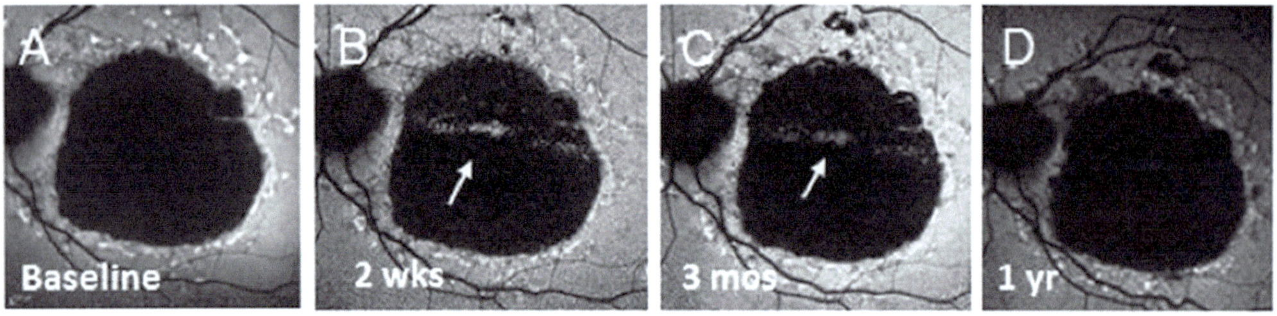

Fig. 2 Autofluorescence images of a Stargardt disease patient with increasing pigmentation consistent with transplanted hESC-derived RPE cells. (**a**–**d**) Autofluorescence images of a Stargardt disease patient, which correspond to the color fundus photographs in Fig. 1g–i. (**a**) Preoperative autofluorescence imaging shows hypo-autofluorescence centrally with small satellite lesions of hypo-autofluorescence, and a surrounding rim of hyper-autofluorescence. (**b**) Two weeks and (**c**) 3 months after surgery. (**b**, **c**, *arrows*) there is speckled hyper-autofluorescence in the *upper half* of the central atrophic lesion that diminishes by 1 year (**d**) (reproduced with permission from Human embryonic stem cell-derived retinal pigment epithelium in patients with age-related macular degeneration and Stargardt macular dystrophy: follow-up of two open-label phase 1/2 studies. Lancet. 2015; 385:509–16)

Despite the multiple imaging techniques employed, it was not possible to distinguish transplanted MA09-hRPE cells from the host RPE cells. There was no correlation between the development or absence of posttransplant pigmentation and visual changes. This lack of correlation may be due to a placebo effect, difficulty in measuring visual acuity in low-vision patients, and our inability to directly correlate structure and function, which might be best captured by other means such as microperimetry or adaptive optics.

8.9 Reported Psychometric Outcomes

The National Eye Institute Visual Function Questionnaire 25, a sensitive and reliable measure of vision-related quality of life, showed a median improvement of 16–25 points at 3–12 months after transplantation in patients with atrophic AMD and a median improvement of 8–20 points at 3–12 months after transplantation in patients with SMD [54]. The areas improved included mental health and vision subscales for general vision, peripheral vision, near activities, and distance activities.

8.10 Reported Complications

Four eyes developed visually significant cataract progression (one AMD and three SMD eyes) as would be expected following vitrectomy. One patient with SMD developed *Staphylococcus epidermidis* endophthalmitis without subretinal abscess 4 days after transplantation. This resolved after intravitreal antibiotic injection, topical antibiotics, and discontinuation of immunosuppression. Cultures and Gram stains of the MA09-hRPE cell lot used in that case were negative. One eye developed vitreous inflammation associated with a transvitreal band at week 3 after transplantation, which resolved without epiretinal membrane or tractional retinal detachment by month 6. No eyes developed tumor formation, ectopic tissue formation, lymphocyte infiltration, signs of rejection, cystoid macular edema, rhegmatogenous or tractional retinal detachment, proliferative vitreoretinopathy, microvascular occlusion, or any other adverse effects. The subretinal bleb was absorbed by the first postoperative day in all but one eye, where the bleb resolved in 2 days. One eye developed a mild epiretinal membrane. One eye showed mild, late fluorescein angiographic leakage at 1 month in an area not imaged preoperatively; however the finding resolved by month 3 without changes in immunosuppressive regimen.

Serious adverse effects likely secondary to the immunosuppression included syncope, chest pain, femoral fracture from a fall, hemiparesis, altered mental status, hospitalization due to urinary tract infection, and altered mental status (Table 2). These were attributed to immunosuppressive medications and underlying health status.

9 Future Directions

The field of stem cell-derived RPE transplantation is still in its infancy. Based on the reported results from the ongoing phase I/II trials of MA09-hRPE cell transplantation for atrophic AMD and SMD, it appears that subretinal injection of the

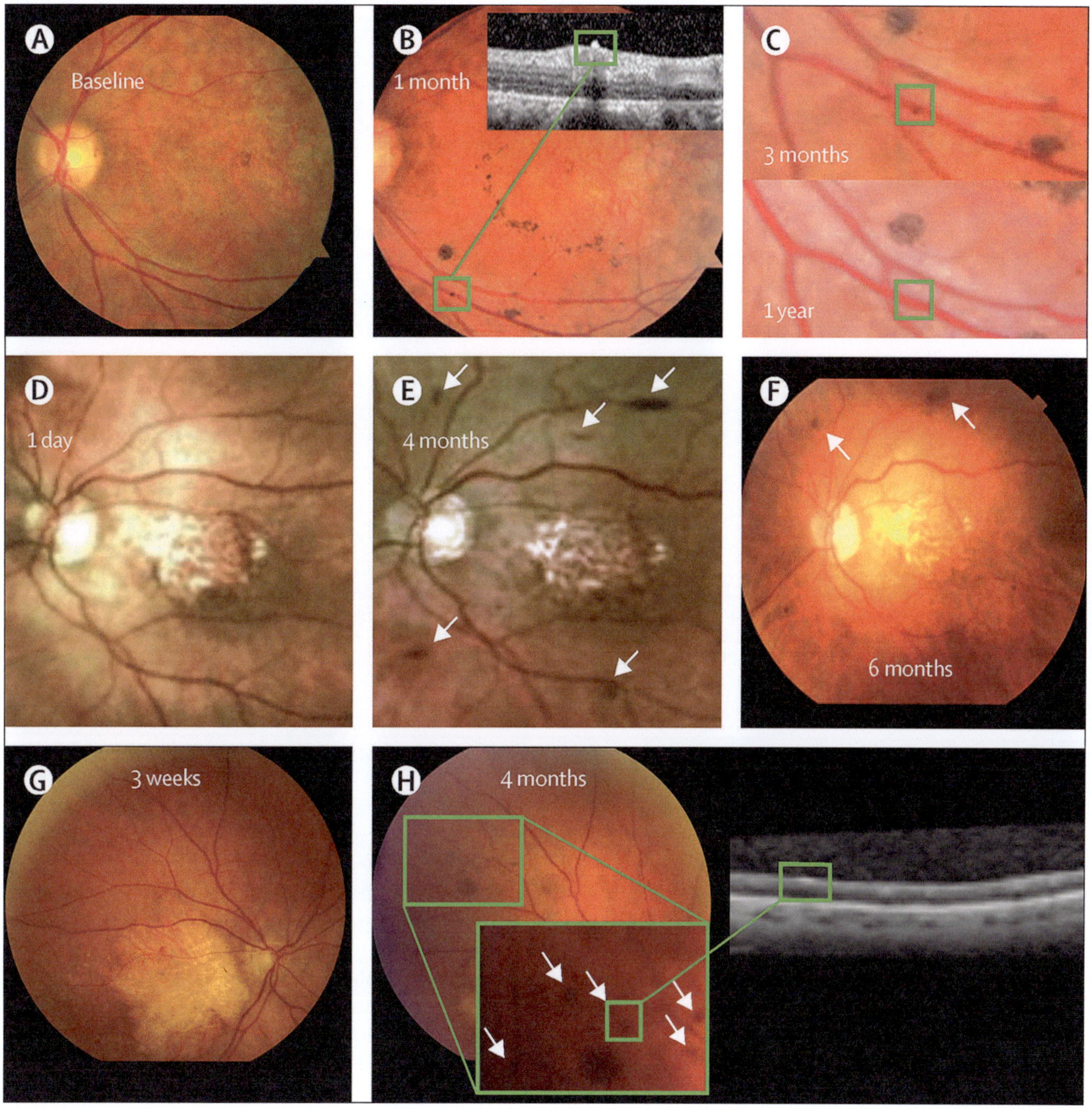

Fig. 3 Preretinal cell growth reported from the Phase I/II Trial of MA09-hRPE Cell Transplantation. Pigmented preretinal growths were seen in the eyes of three patients. The fundus photograph and OCT show a small pigmented clump, perhaps a single cell, in a patient with Stargardt macular dystrophy 1 month after transplantation (**b**) compared with baseline (**a**). (**c**) The pigmented clump resolved and is no longer visible at 12 months. (**d, g**) Fundus images before the development of the preretinal growths. (**e, h**) Two patients with age-related macular degeneration also developed preretinal growths several months after transplantation that were visible on postoperative photographs and OCT. (**f**, *data not shown*) Growth and advancement of the cells subsided by 6–9 months in both patients. No adverse clinical effects were noted from these small epiretinal pigmented patches. *OCT* optical coherence tomography (reproduced with permission from Human embryonic stem cell-derived retinal pigment epithelium in patients with age-related macular degeneration and Stargardt macular dystrophy: follow-up of two open-label phase 1/2 studies. Lancet. 2015; 385:509–16)

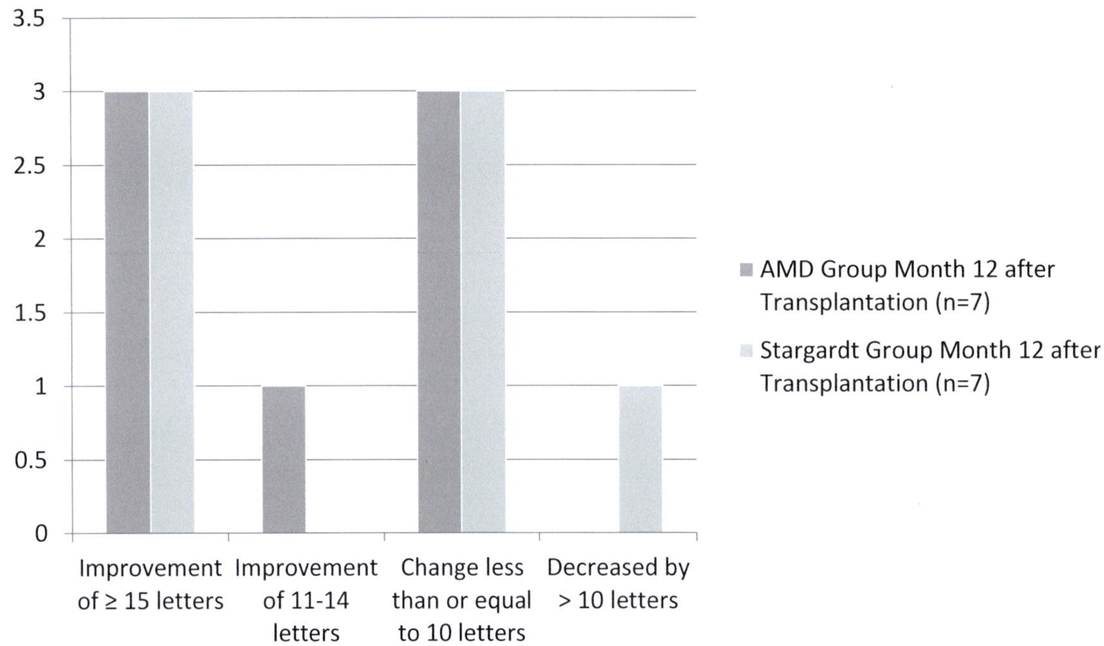

Fig. 4 Graph shows changes in visual acuity 12 months after transplantation in patients with atrophic age-related macular degeneration (*dark gray*) and Stargardt macular dystrophy (*light gray*). Data reported from the Ongoing Phase I/II Trials of MA09-hRPE Cell Transplantation for Atrophic Age-related Macular Degeneration and Stargardt macular dystrophy (reproduced with permission from Human embryonic stem cell-derived retinal pigment epithelium in patients with age-related macular degeneration and Stargardt macular dystrophy: follow-up of two open-label phase 1/2 studies. Lancet. 2015; 385:509–16)

hESC-derived RPE cells does not produce untoward adverse effects in the human eye. Further studies proving sustainability of the cells in subretinal space and enhancement of visual acuity must be performed next. Moreover, given the risks of immunosuppression, future trials should establish if and how immunosuppressive regimens should be administered.

Although further evaluation of the safety and tolerability of the hRPE cells must continue, efficacy of cell transplantation in vision restoration should be studied in a masked and placebo-controlled trial. Once safety is proven, the target population may expand to include patients with earlier stages of disease manifestation, before severe damage to the photoreceptors and RPE cells develops, potentially increasing the likelihood of RPE and photoreceptor rescue and prevention of visual loss.

Refinement of the structural and visual function efficacy endpoints in patients with relatively low vision loss should also be pursued. In addition to visual acuity, the next phase of the clinical development program will evaluate the changes in the area of geographic atrophy, contrast sensitivity, and microperimetry, which have recently been reported as sensitive measures of visual function in patients with AMD [55–57]. These parameters may prove to be complementary or superior for assessing outcomes in these patients compared to the standard ETDRS visual acuity. Finally, adaptive optics imaging should be evaluated as a technique that may play a role in improving the ability to visualize the transplanted cells.

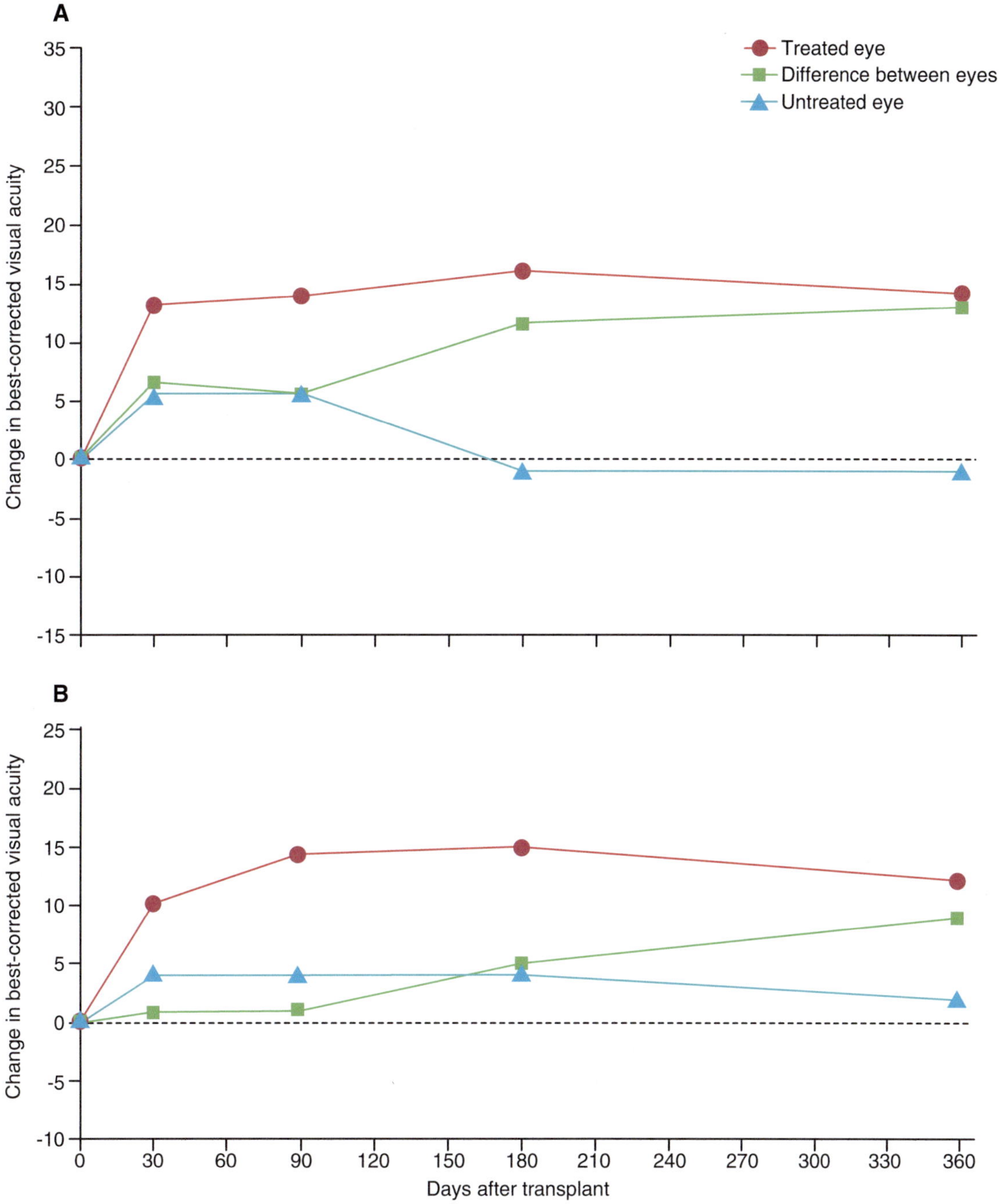

Fig. 5 Change from baseline in best-corrected visual acuity in patients with AMD (**a**) and Stargardt macular dystrophy (**b**) reported from the Phase I/II Trial of MA09-hRPE Cell Transplantation. Median change in best-corrected visual acuity was expressed as number of letters read on the Early Treatment of Diabetic Retinopathy Study visual acuity chart in patients with age-related macular degeneration (**a**) and Stargardt macular dystrophy (**b**). *Red lines* show treated eyes and *blue lines* show untreated eyes of patients during the first year after transplantation of the cells derived from human embryonic stem cells. *Green lines* show the difference between the treated and untreated eyes. Patients who underwent cataract surgery after transplantation are not included in the graph. There was a significant difference in the letters read in transplanted eyes of patients with age-related macular degeneration versus nontransplanted controls at 12 months (median 14 letters vs. −1 letter; p = 0·0117). There was an increase in letters read in transplanted eyes of patients with Stargardt macular dystrophy versus nontransplanted controls at 12 months (median 12 letters vs. 2 letters, although the sample size was too small to allow reliable calculation of the Wilcoxon signed-rank test) (reproduced with permission from Human embryonic stem cell-derived retinal pigment epithelium in patients with age-related macular degeneration and Stargardt macular dystrophy: follow-up of two open-label phase 1/2 studies. Lancet. 2015; 385:509–16)

Table 2 Serious adverse effects reported from the Phase I/II Trial of MA09-hRPE Cell Transplantation For Age-related Macular Degeneration and Stargardt Disease

System organ classification/serious adverse event	SMD	AMD
	n = 9	*n* = 9
	n	*n*
With any serious adverse event	2	4
General disorders	1	0
Chest pain	1	–
Infections	1	1
Endophthalmitis	1	–
Urinary tract infection	–	1
Injury	0	1
Femoral neck fracture	–	1
Neoplasms	0	1
Basal cell carcinoma	–	1
Squamous cell carcinoma	–	1
Nervous system disorders	0	2
Hemiparesis	–	1
Syncope	–	1
Psychiatric disorders	–	1
Mental status change	–	1

Serious adverse effects reported from the Ongoing Phase I/II Trial of MA09-hRPE Cell Transplantation for Age-related Macular Degeneration and Stargardt Disease (reproduced with permission from Human embryonic stem cell-derived retinal pigment epithelium in patients with age-related macular degeneration and Stargardt macular dystrophy: follow-up of two open-label phase 1/2 studies. Lancet. 2015; 385:509–16)

References

1. Friedman DS, O'Colmain BJ, Munoz B, Tomany SC, McCarty C, de Jong PT, et al. Prevalence of age-related macular degeneration in the United States. Arch Ophthalmol. 2004;122(4):564–72.
2. Bourne RR, Stevens GA, White RA, Smith JL, Flaxman SR, Price H, et al. Causes of vision loss worldwide, 1990–2010: a systematic analysis. Lancet Glob Health. 2013;1(6):e339–49.
3. Danis RP, Lavine JA, Domalpally A. Geographic atrophy in patients with advanced dry age-related macular degeneration: current challenges and future prospects. Clin Ophthalmol. 2015;9:2159–74.
4. Bird AC, Phillips RL, Hageman GS. Geographic atrophy: a histopathological assessment. JAMA Ophthalmol. 2014;132(3):338–45.
5. Holz FG, Bellmann C, Margaritidis M, Schutt F, Otto TP, Volcker HE. Patterns of increased in vivo fundus autofluorescence in the junctional zone of geographic atrophy of the retinal pigment epithelium associated with age-related macular degeneration. Graefes Arch Clin Exp Ophthalmol. 1999;237(2):145–52.
6. Walia S, Fishman GA. Natural history of phenotypic changes in Stargardt macular dystrophy. Ophthalmic Genet. 2009;30(2):63–8.
7. Allikmets R, Singh N, Sun H, Shroyer NF, Hutchinson A, Chidambaram A, et al. A photoreceptor cell-specific ATPbinding transporter gene (ABCR) is mutated in recessive Stargardt macular dystrophy. Nat Genet. 1997;15(3):236–46.
8. Logan S, Anderson RE. Dominant Stargardt macular dystrophy (STGD3) and ELOVL4. Adv Exp Med Biol. 2014;801:447–53.
9. Atala A. Human stem cell-derived retinal cells for macular diseases. Lancet. 2015;385(9967):487–8.
10. Yehoshua Z, de Amorim Garcia Filho CA, Nunes RP, Gregori G, Penha FM, Moshfeghi AA, et al. Systemic complement inhibition with eculizumab for geographic atrophy in age-related macular degeneration: the COMPLETE study. Ophthalmology. 2014;121(3):693–701.
11. Ma L, Kaufman Y, Zhang J, Washington I. C20-D3-vitamin A slows lipofuscin accumulation and electrophysiological retinal degeneration in a mouse model of Stargardt disease. J Biol Chem. 2011;286(10):7966–74.
12. Auricchio A, Trapani I, Allikmets R. Gene therapy of ABCA4-associated diseases. Cold Spring Harb Perspect Med. 2015;5(5): a017301.
13. Strauss O. The retinal pigment epithelium in visual function. Physiol Rev. 2005;85(3):845–81.
14. Booij JC, van Soest S, Swagemakers SM, Essing AH, Verkerk AJ, van der Spek PJ, et al. Functional annotation of the human retinal pigment epithelium transcriptome. BMC Genomics. 2009;10:164.
15. Sharma RK, Orr WE, Schmitt AD, Johnson DA. A functional profile of gene expression in ARPE-19 cells. BMC Ophthalmol. 2005;5:25.
16. Wang XF, Cui JZ, Nie W, Prasad SS, Matsubara JA. Differential gene expression of early and late passage retinal pigment epithelial cells. Exp Eye Res. 2004;79(2):209–21.
17. Bressler NM, Bressler SB, Hawkins BS, Marsh MJ, Sternberg Jr P, Thomas MA. Submacular surgery trials randomized pilot trial of laser photocoagulation versus surgery for recurrent choroidal neovascularization secondary to age-related macular degeneration: I. Ophthalmic outcomes submacular surgery trials pilot study report number 1. Am J Ophthalmol. 2000;130(4):387–407.
18. Algvere PV, Berglin L, Gouras P, Sheng Y. Transplantation of fetal retinal pigment epithelium in age-related macular degeneration with subfoveal neovascularization. Graefes Arch Clin Exp Ophthalmol. 1994;232(12):707–16.
19. Algvere PV, Berglin L, Gouras P, Sheng Y, Kopp ED. Transplantation of RPE in age-related macular degeneration: observations in disciform lesions and dry RPE atrophy. Graefes Arch Clin Exp Ophthalmol. 1997;235(3):149–58.
20. Peyman GA, Blinder KJ, Paris CL, Alturki W, Nelson Jr NC, Desai U. A technique for retinal pigment epithelium transplantation for age-related macular degeneration secondary to extensive subfoveal scarring. Ophthalmic Surg. 1991;22(2):102–8.
21. van Zeeburg EJ, Maaijwee KJ, Missotten TO, Heimann H, van Meurs JC. A free retinal pigment epithelium-choroid graft in patients with exudative age-related macular degeneration: results up to 7 years. Am J Ophthalmol. 2012;153(1):120–7. e2
22. van Zeeburg EJ, Cereda MG, Amarakoon S, van Meurs JC. Prospective, Randomized Intervention Study comparing retinal pigment epithelium-choroid graft surgery and anti-VEGF therapy in patients with exudative age-related macular degeneration. Ophthalmologica. 2015;233(3–4):134–45.
23. Thumann G, Aisenbrey S, Schraermeyer U, Lafaut B, Esser P, Walter P, et al. Transplantation of autologous iris pigment epithelium after removal of choroidal neovascular membranes. Arch Ophthalmol. 2000;118(10):1350–5.
24. Aisenbrey S, Lafaut BA, Szurman P, Hilgers RD, Esser P, Walter P, et al. Iris pigment epithelial translocation in the treatment of exudative macular degeneration: a 3-year follow-up. Arch Ophthalmol. 2006;124(2):183–8.
25. Bharti K, Rao M, Hull SC, Stroncek D, Brooks BP, Feigal E, et al. Developing cellular therapies for retinal degenerative diseases. Invest Ophthalmol Vis Sci. 2014;55(2):1191–202.
26. Desai N, Rambhia P, Gishto A. Human embryonic stem cell cultivation: historical perspective and evolution of xenofree culture systems. Reprod Biol Endocrinol. 2015;13:9.
27. Thomson JA, Itskovitz-Eldor J, Shapiro SS, Waknitz MA, Swiergiel JJ, Marshall VS, et al. Embryonic stem cell lines derived from human blastocysts. Science. 1998;282(5391):1145–7.
28. Vaajasaari H, Ilmarinen T, Juuti-Uusitalo K, Rajala K, Onnela N, Narkilahti S, et al. Toward the defined and xeno-free differentiation

of functional human pluripotent stem cell-derived retinal pigment epithelial cells. Mol Vis. 2011;17:558–75.
29. Takahashi K, Tanabe K, Ohnuki M, Narita M, Ichisaka T, Tomoda K, et al. Induction of pluripotent stem cells from adult human fibroblasts by defined factors. Cell. 2007;131(5):861–72.
30. Yu J, Vodyanik MA, Smuga-Otto K, Antosiewicz-Bourget J, Frane JL, Tian S, et al. Induced pluripotent stem cell lines derived from human somatic cells. Science. 2007;318(5858):1917–20.
31. Park IH, Zhao R, West JA, Yabuuchi A, Huo H, Ince TA, et al. Reprogramming of human somatic cells to pluripotency with defined factors. Nature. 2008;451(7175):141–6.
32. Zhao T, Zhang ZN, Rong Z, Xu Y. Immunogenicity of induced pluripotent stem cells. Nature. 2011;474(7350):212–5.
33. Sugita S, Kamao H, Iwasaki Y, Okamoto S, Hashiguchi T, Iseki K, et al. Inhibition of T-cell activation by retinal pigment epithelial cells derived from induced pluripotent stem cells. Invest Ophthalmol Vis Sci. 2015;56(2):1051–62.
34. Hayashi R, Ishikawa Y, Ito M, Kageyama T, Takashiba K, Fujioka T, et al. Generation of corneal epithelial cells from induced pluripotent stem cells derived from human dermal fibroblast and corneal limbal epithelium. PLoS One. 2012;7(9):e45435.
35. Gore A, Li Z, Fung HL, Young JE, Agarwal S, Antosiewicz-Bourget J, et al. Somatic coding mutations in human induced pluripotent stem cells. Nature. 2011;471(7336):63–7.
36. Buchholz DE, Hikita ST, Rowland TJ, Friedrich AM, Hinman CR, Johnson LV, et al. Derivation of functional retinal pigmented epithelium from induced pluripotent stem cells. Stem Cells. 2009;27(10):2427–34.
37. Idelson M, Alper R, Obolensky A, Ben-Shushan E, Hemo I, Yachimovich-Cohen N, et al. Directed differentiation of human embryonic stem cells into functional retinal pigment epithelium cells. Cell Stem Cell. 2009;5(4):396–408.
38. Zahabi A, Shahbazi E, Ahmadieh H, Hassani SN, Totonchi M, Taei A, et al. A new efficient protocol for directed differentiation of retinal pigmented epithelial cells from normal and retinal disease induced pluripotent stem cells. Stem Cells Dev. 2012;21(12):2262–72.
39. Leach LL, Buchholz DE, Nadar VP, Lowenstein SE, Clegg DO. Canonical/beta-catenin Wnt pathway activation improves retinal pigmented epithelium derivation from human embryonic stem cells. Invest Ophthalmol Vis Sci. 2015;56(2):1002–13.
40. Croze RH, Buchholz DE, Radeke MJ, Thi WJ, Hu Q, Coffey PJ, et al. ROCK inhibition extends passage of pluripotent stem cell-derived retinal pigmented epithelium. Stem Cells Transl Med. 2014;3(9):1066–78.
41. Klimanskaya I, Hipp J, Rezai KA, West M, Atala A, Lanza R. Derivation and comparative assessment of retinal pigment epithelium from human embryonic stem cells using transcriptomics. Cloning Stem Cells. 2004;6(3):217–45.
42. Tezel TH, Del Priore LV. Reattachment to a substrate prevents apoptosis of human retinal pigment epithelium. Graefes Arch Clin Exp Ophthalmol. 1997;235(1):41–7.
43. Hynes SR, Lavik EB. A tissue-engineered approach towards retinal repair: scaffolds for cell transplantation to the subretinal space. Graefes Arch Clin Exp Ophthalmol. 2010;248(6):763–78.
44. Grulova I, Slovinska L, Blasko J, Devaux S, Wisztorski M, Salzet M, et al. Delivery of alginate scaffold releasing two trophic factors for spinal cord injury repair. Sci Rep. 2015;5:13702.
45. Schwartz SD, Regillo CD, Lam BL, Eliott D, Rosenfeld PJ, Gregori NZ, et al. Human embryonic stem cell-derived retinal pigment epithelium in patients with age-related macular degeneration and Stargardt's macular dystrophy: follow-up of two open-label phase 1/2 studies. Lancet. 2015;385(9967):509–16.
46. Lu B, Malcuit C, Wang S, Girman S, Francis P, Lemieux L, et al. Long-term safety and function of RPE from human embryonic stem cells in preclinical models of macular degeneration. Stem Cells. 2009;27(9):2126–35.
47. Lund RD, Wang S, Klimanskaya I, Holmes T, Ramos-Kelsey R, Lu B, et al. Human embryonic stem cell-derived cells rescue visual function in dystrophic RCS rats. Cloning Stem Cells. 2006;8(3):189–99.
48. Schwartz SD, Hubschman JP, Heilwell G, Franco-Cardenas V, Pan CK, Ostrick RM, et al. Embryonic stem cell trials for macular degeneration: a preliminary report. Lancet. 2012;379(9817): 713–20.
49. Wenkel H, Streilein JW. Evidence that retinal pigment epithelium functions as an immune-privileged tissue. Invest Ophthalmol Vis Sci. 2000;41(11):3467–73.
50. Xian B, Huang B. The immune response of stem cells in subretinal transplantation. Stem Cell Res Therapy. 2015;6:161.
51. Zarbin MA, Casaroli-Marano RP, Rosenfeld PJ. Age-related macular degeneration: clinical findings, histopathology and imaging techniques. Dev Ophthalmol. 2014;53:1–32.
52. Cideciyan AV, Aleman TS, Swider M, Schwartz SB, Steinberg JD, Brucker AJ, et al. Mutations in ABCA4 result in accumulation of lipofuscin before slowing of the retinoid cycle: a reappraisal of the human disease sequence. Hum Mol Genet. 2004;1(5):525–34.
53. Tezel TH, Del Priore LV. Repopulation of different layers of host human Bruch's membrane by retinal pigment epithelial cell grafts. Invest Ophthalmol Vis Sci. 1999;40(3):767–74.
54. Mangione CM, Lee PP, Gutierrez PR, Spritzer K, Berry S, Hays RD, et al. Development of the 25-item national eye institute visual function questionnaire. Arch Ophthalmol. 2001;119(7):1050–8.
55. Wu Z, Ayton LN, Luu CD, Guymer RH. Longitudinal changes in microperimetry and low luminance visual acuity in age-related macular degeneration. JAMA Ophthalmol. 2015;133(4):442–8.
56. Sunness JS, Rubin GS, Broman A, Applegate CA, Bressler NM, Hawkins BS. Low luminance visual dysfunction as a predictor of subsequent visual acuity loss from geographic atrophy in age-related macular degeneration. Ophthalmology. 2008;115(9):1480–8. 8 e1-2.
57. Sunness JS. Stem cells in age-related macular degeneration and Stargardt's macular dystrophy. Lancet. 2015;386(9988):29.

Stem Cell-Derived RPE Transplantation: The Feasibility and Advantages of Delivery as Monolayers

Odysseas Georgiadis, Lyndon da Cruz, and Peter Coffey

1 Introduction and Background

1.1 The Normal RPE and the Dual Role It Plays in Visual Function

The retinal pigment epithelium (RPE) is a monolayer of hexagonal, pigmented cells, derived from the neuroepithelium and located between the light-sensitive photoreceptors' outer segments and the choriocapillaris (Fig. 1a). The RPE lies on a specialized basement membrane—Bruch's membrane—and together with the photoreceptors constitutes a structural and functional unit that provides the transducing interface for visual perception [1].

The numerous functions of the RPE, by which it contributes to normal retinal function, can be summarized under two broad headings—barrier and support. Firstly, the RPE constitutes a **barrier** (in conjunction with Bruch's membrane) between the neural retina and the choroid. In terms of the barrier function there are two components—one physiological and the other physical. The physiological barrier is formed between RPE cells that are connected with intercellular tight junctions (zonulae occludentes), which establish the outer blood-retinal barrier preventing the passage of large molecules form the choriocapillaris to the subretinal space. The second broad function of the RPE is the provision of physiological **support** to the retina. In terms of this physiological support, the most critical metabolic function of the RPE is the regeneration of bleached opsins, taking place in the cytosol of the RPE cells. The recycling of the visual pigments together with the phagocytosis of the photoreceptors' outer segments, which are being shed in a circadian basis, represents one of the most important parts of the physiological visual cycle. Additionally, RPE participates in retinal and polyunsaturated fatty acid metabolism and cellular and extracellular homeostasis, it secretes growth factors (PEDF, VEGF), it maintains the subretinal space and the extraphotoreceptor matrix, and it absorbs light and protects against photooxidative stress.

1.2 In Order to Meet Its Dual Role, RPE Must Maintain Its Differentiation and Polarity and Its Adhesion to Bruch's Membrane

Two unique features of the normal RPE are essential for its central *barrier* and *support* roles. Firstly, its existence as a monolayer rather than individual separated cells and secondly its orientation (polarization). The cells need to be uniformly polarized in order for the microvilli of their apical cellular membrane to interdigitate with the photoreceptors, thereby facilitating outer-segment phagocytosis (Fig. 1a). The polarization is also important for the regulation of ion and water transport from the apical side to the basolateral surface. This active transport is the way that the RPE implements the removal of water that is generated by the metabolic activity of the photoreceptors and constantly flowing from the vitreous through the retina, from the subretinal space to the choroidal vasculature. Additionally, by removing water, the RPE establishes an adhesive force that contributes to the attachment of the neural retina [1].

The original version of this chapter was revised. An erratum to this chapter can be found at DOI 10.1007/978-3-319-49479-1_11

O. Georgiadis
Vitreoretinal Clinical Research, Moorfields Eye Hospital, NHS Foundation Trust, London, UK
e-mail: odysseas.georgiadis@moorfields.nhs.uk

L. da Cruz
Vitreoretinal Surgery, NIHR Moorfields Biomedical Research Centre, Moorfields Eye Hospital NHS Foundation Trust, London, UK
e-mail: Lyndon.dacruz@moorfields.nhs.uk

P. Coffey (✉)
Institute of Ophthalmology, University College London, 11-43 Bath Street, London EC1V 9EL, UK
e-mail: p.coffey@ucl.ac.uk

S.D. Schwartz et al. (eds.), *Cellular Therapies for Retinal Disease*, DOI 10.1007/978-3-319-49479-1_2

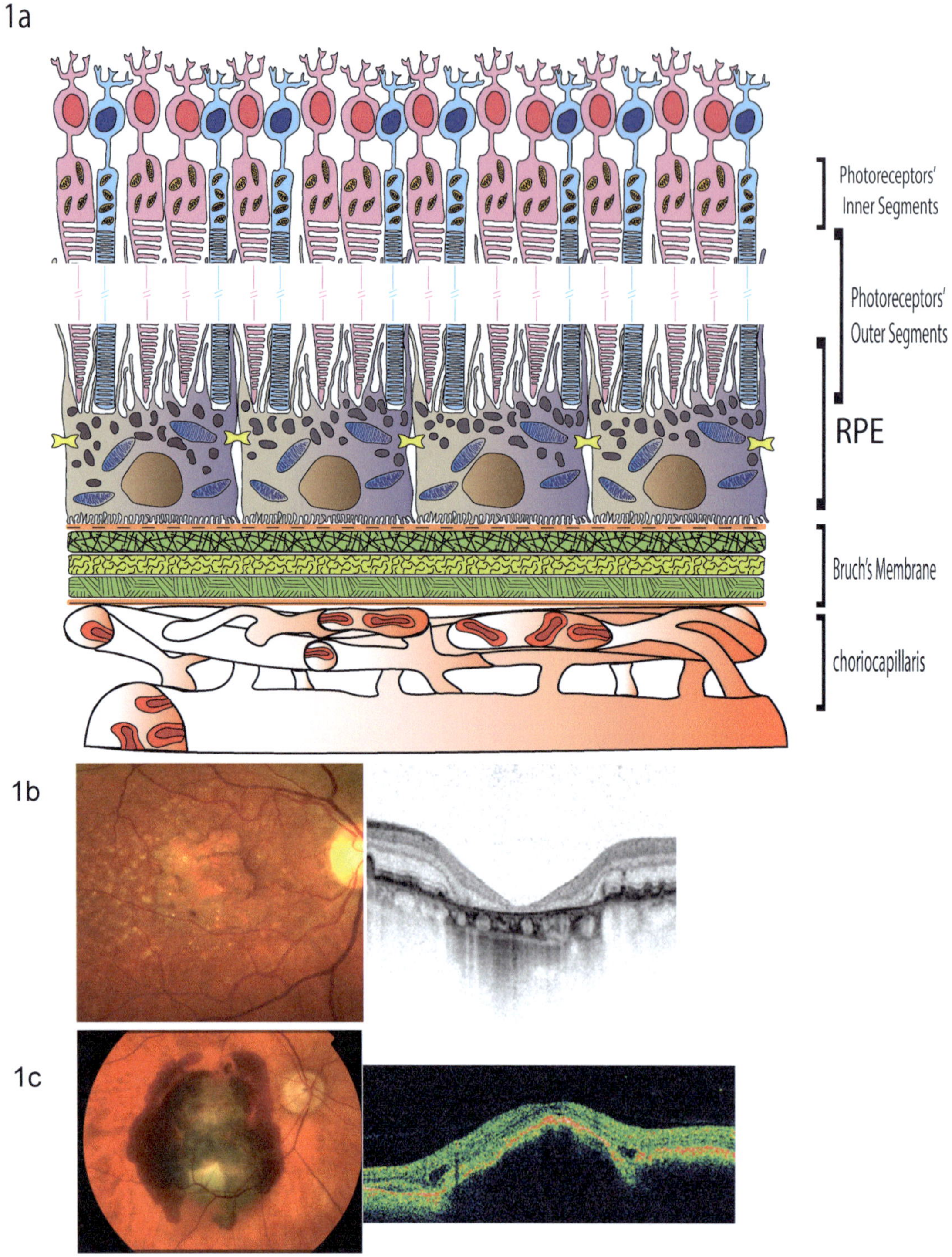

Fig. 1 (**a**) Graphic representation of RPE demonstrating cell polarization and monolayer organization with intracellular tight junctions, interdigitation with the photoreceptors' outer segments, and interface with Bruch's membrane and the choriocapillaris. (**b**) Color fundus photo and macular OCT scan showing geographic atrophy and central loss of retina layers in dry AMD. (**c**) Color fundus photo showing severe submacular hemorrhage and macular OCT scan showing that, despite the severity of the hemorrhage, the layers of central retina are preserved

Both of the critical features of the RPE—physical barrier and physiological support—depend on the active interface with its specialized basement membrane, which is the innermost layer of Bruch's membrane. The consistent adherence between RPE and Bruch's constitutes a physical barrier, which prevents cellular migration, but also allows the regulation of diffusion between retina and choroid.

2 RPE Dysfunction and Disease

2.1 Diseases or Conditions That Affect the RPE Have an Impact on Normal Retinal Function and Thus on Sight

Aging of the RPE, as well as a variety of pathological factors, such as genetic defects, drugs, and dietary (vitamin A) insufficiency can impair its homeostatic functions and affect the photoreceptor renewal process. Subsequently, deterioration or loss of the RPE cells results in corresponding atrophy of the overlying photoreceptors and underlying choriocapillaris.

Macular degenerations are a group of acquired and inherited diseases that affect the central part of the retina that is most critical for vision. Age-related macular degeneration (AMD) is the leading cause of permanent visual impairment in the over 55 age group worldwide [2]. It is a complex degenerative disease caused by a chronic, low-grade inflammation in the central outer retina, which leads to degeneration of the RPE and Bruch's membrane. The RPE and Bruch's membrane complex is the primary target of AMD, which can lead to either of the two pathophysiological pathways of the disease—dry/atrophic, when the supportive role of the complex is compromised (Fig. 1b), and wet/neovascular, when its barrier aspect is defective as a result of progressive cell loss, or by participating in the inflammatory process which leads to neovascularization (Fig. 1c). The common endpoint for both pathways is cell loss, especially within the photoreceptor/RPE/choroid complex [2, 3].

Further to its epidemiological and clinical significance, AMD is also an important paradigm for other similar conditions with wet and dry subtypes such as myopia, trauma, and inherited maculopathies. The latter group can manifest the same loss of RPE with a number of these being RPE specific such as pseudoxanthoma elasticum-associated angioid streaks, choroideremia, vitelliform dystrophy (Best's disease), Stargardt disease, and some subtypes of retinitis pigmentosa such as those associated with MERTK gene abnormalities. Stargardt disease (STGD) is the most prevalent juvenile retinal degenerative disease, estimated to affect 1/8000–10,000 young individuals [4], whom may leave legally blind during the first two decades of their life. Due to these various disorders leading to cell and vision loss, RPE transplantation is potentially beneficial.

3 Scientific Basis for Intervention

3.1 Replacement of RPE Prevents Progression and May Even Reverse These Disease Processes

3.1.1 The RCS Paradigm

Extensive in vivo experiments on the Royal College of Surgeons (RCS) rat have shown the plausibility of photoreceptor rescue after RPE transplantation. This dystrophic strain of rat has a recessive mutation in the MERTK gene, resulting in failure of the RPE to phagocytose shed rod outer segments [5]. This defect leads to accumulation of subretinal debris, photoreceptor death, and, eventually, retinal vascular changes and retinal degeneration. The preliminary studies of RPE transplantation in the RCS rat showed retardation and/or restoration of these changes [6–12]. Furthermore, the transplanted RPE seemed to halt the reduction of central retinal thickness, demonstrating preservation of the outer nuclear layer, outer plexiform layer, and photoreceptors' inner and outer segment [6, 13].

3.2 For the Replacement to Be Effective, the Transplanted RPE Has to Meet the Dual Role of the Normal RPE

For the RPE to function as a barrier, the implanted cells have to organize in a single-layer manner, establish intercellular tight junctions, and maintain adherence with the underlying BM. As well as this, the implanted cells have to polarize in an apical-basal manner and maintain this orientation long-term in order to provide the support function and to maintain the neural retina.

3.3 Current Methods of RPE Replacement Using Cell Suspension-Based Therapy May Not Survive or Develop a Monolayer

The critical importance of the two aforementioned basic characteristics of the RPE (single-layer alignment and apical polarization) could account for the lack of success seen in some attempts to replace the RPE by injecting freshly harvested or cultured cells into the subretinal space in the form of a cell suspension. Although transplantation of an RPE cell suspension may lead to the formation of a monolayer, occasionally it may also lead to uneven distribution of the injected cells and formation of multiple layers of RPE which alternate with bare areas of basal lamina [14, 15]. Crafoord et al. have examined the long-term outcomes of RPE allografts in rabbits and showed that his sectional irregular distribution may result to increased risk of graft failure, as well as damage of the adjacent photoreceptors [14].

Other problems that have been identified in delivering suspensions are the shearing forces that are applied on the suspension during the injection, leading to cell death, the need to isolate the cells and therefore deliver them in a less differentiated form, and the possibility of cell loss due to the potential reflux of the cells from the injection site. For the cells that survive the delivery, a defective or diseased basal lamina may constitute a hostile environment for the new cells and affect their survival rate.

Transplantation trials using cell suspensions have highlighted the substantial role of a healthy Bruch's membrane for the accomplishment of the process. Several groups have demonstrated that RPE cell suspensions applied on an aged or a damaged Bruch's membrane are prone to unsuccessful integration with the host tissue and may fail to survive and remain functional for prolonged periods [16–18]. This inability of the RPE cells to grow well on defective Bruch's membranes is considered as a major cause of possible failure for such approaches to support retina in the long term and restore photoreceptor function and vision in human eyes [19, 18].

Although the importance of Bruch's membrane for the success of the RPE transplantation had initially been overlooked, most recent developments in this field have invested efforts to support RPE grafts with BM-mimicking artificial membranes. Based on these findings transplantation of a polarized RPE monolayer as a sheet on an artificial Bruch's membrane has been proposed as a potentially promising approach [15, 20].

4 Hypothesis for Therapeutic Intervention

4.1 The Hypothesis

RPE transplantation in the form of a patch/sheet holds potential benefits in terms of cell survival, differentiation, polarity, and integration with host, as well as treating a sufficient area for function.

4.2 The "Cataract" Paradigm

Modern cataract treatment offers an ideal paradigm of efficiently and safely treating large numbers of patient with a surgical procedure. At present, phacoemulsification and artificial lens insertion provides the most adequate way to deal with cataract in a single procedure. Using a straightforward surgical technique with proven efficacy, it utilizes a premade and readily available implant (IOL) and it has a low complication rate and a very rapid recovery. To meet this paradigm, a cell delivery system would require a premade and readily available patch of healthy RPE, ideally preloaded to a device similar to an IOL injector and implanted with a straightforward procedure, with low risk for complication and quick recovery.

4.3 The Patch Transplant

The most prevalent stem cell-based RPE patch transplant system consists of a biocompatible substrate or scaffold, seeded with stem cell-derived RPE cells. The aim of the scaffold is to provide a supportive surface for cells to attach, proliferate, differentiate, and perform their normal functions after transplantation while providing structural rigidity that aids delivery [21]. Preclinical studies designed to compare suspension method against sheet transplantation in rats showed superiority of the latter, in terms of longer survival and integration with host [20]. Another recent study by Thomas et al. has demonstrated 87% survivability of RPE cells derived from human embryonic stem cells, when cultured as a monolayer on a parylene membrane and transplanted as a patch into the subretinal space of Royal College of Surgeons rats (RCS). This study also showed a significant rescue of outer nuclear layer cells as well as higher response to light stimulus in the patch-transplantation RCS group, when compared to control groups [22].

In contrast to single-cell injections, cellular sheets supported by biomedical scaffolds can form a solid transplantable graft, enable to deliver an already mature monolayer of cells to a specific destination. Therapeutic stem cell-derived RPE cells demonstrate the ability to adhere and grow on various candidate substrates and retain the critical characteristics of in situ RPE, such as apical polarization, tight junctions, pigmentation, and structural and metabolic integration with overlying photoreceptors and underlying choriocapillaris [23]. These retained characteristics make the RPE sheets on the transplanted patch more likely to meet the dual role of host RPE—namely to provide support and act as a barrier.

4.4 Proof of Principle: Human Transplantation Studies

At present, there is no definite curative treatment for either the wet or the dry types of macular degeneration. In the case of dry AMD, although specific dietary supplementation may slow down progression, there is no treatment able to reverse or even stop the disease. In the case of wet AMD, intravitreal anti-VEGF injections may stabilize the disease; however this treatment is expensive and carries a high treatment burden in the long term.

Given that RPE loss and dysfunction play an important part in many retinal degenerative conditions it is not surprising that RPE replacement has been proposed as a potential treatment in these settings. Various techniques have been proposed with a view to reconstructing the RPE-neural retina interface. These include retinal translocation and autologous RPE choroidal free grafting. The former aims to move the macula neuroretina to overlay healthy extramacular RPE, or, in the case of the latter, to achieve intraoperative harvesting and grafting of healthy RPE under the macula.

4.4.1 Macular Translocation for AMD

Macular translocation surgery is a well-described operation that aims in repositioning of the macula onto a healthier RPE and choroid, usually in the paramacular region [24]. Although it is not strictly a RPE transplantation procedure, it can be considered as such, since it results in a "new" healthy RPE area lying under the relocated fovea (Fig. 2a, b) [13]. The long-term postoperative outcomes of this approach demonstrate improvement in both the visual acuity and the quality of life of a significant proportion of patients [25–27]. However, the complexity of the technique, necessity for further surgery for ocular muscle repositioning to avoid torsional diplopia, and significant rate of complications, including retinal detachment, proliferative vitreoretinopathy, epiretinal membranes, and macular hole, have resulted in considering this approach unviable for the disease group it aims to treat [25].

4.4.2 Macular Autologous Transplantation

The technique of autologous RPE transplantation has evolved through many stages from submacular RPE/choroid pedicle flap rotation [28] to submacular RPE/choroids free graft transposition {Stanga:2001hm}, submacular injection of suspension of RPE cells from the peripheral fundus [29], and submacular insertion of RPE/choroid patch graft from the peripheral fundus (Fig. 2c, d) [30]. The latter group have published long-term outcomes for 133 patients who underwent autologous RPE-choroid grafting after vitrectomy and CNV excision [31]. They demonstrated modest overall outcomes, in terms of visual acuity; nevertheless 5% of patients were showed to have best corrected visual acuity better than 20/40, after 4 years of follow-up [31].

Both macular translocation and autologous RPE transplantation have supported the rationale for RPE replacement and showed it to be beneficial. However, these procedures are extremely demanding in terms of surgical skills and operation times and in addition they are associated with a significant risk of potentially serious complications. Furthermore, the quality of the RPE replacement is limited due to the age of the patient and the damage that may occur to the tissue during the transplantation procedure. Due to these limitations a treatment using an allogenic cell-derived RPE immobilized on a membrane and delivered with a simpler surgical technique may offer a safer and more effective treatment.

5 Feasibility and Requirements for Cellular Therapy

5.1 Potential Sources of RPE Cells Have Been Identified

An ideal cell-based therapy for RPE regeneration would be efficient, safe, reproducible, as well as inexpensive and simple to execute. Transplanted cells may be primary cells that

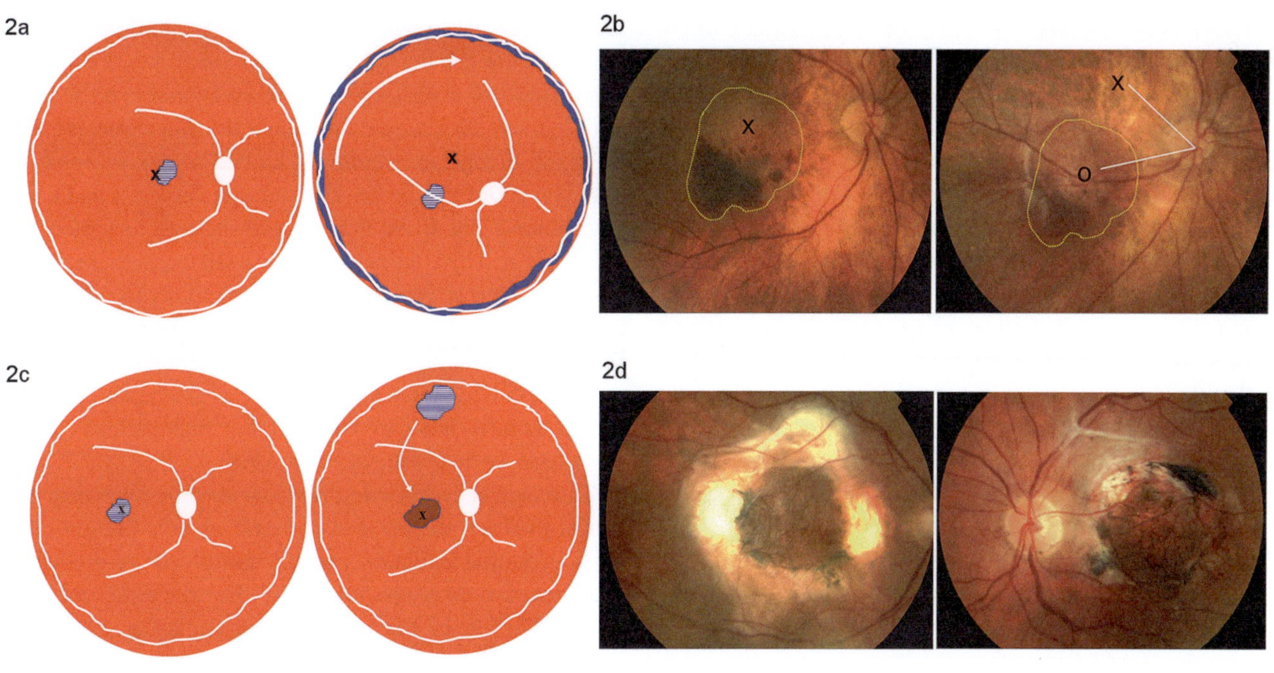

Fig. 2 (**a**, **b**) Graphic representation and color fundus photos of macular translocation surgery: rotation of retina after 360° retinotomy and reposition of the macula (x) over a paramacular area of healthy RPE (o). (**c**) Graphic representation of RPE transplantation surgery: harvesting autologous RPE/choroid graft from the peripheral (superior) fundus and transplantation underneath central macula (x). (**d**) Postoperative color fundus photos of two cases of RPE autologous transplantation

have been freshly harvested or have instead been cultured in vitro, both prior to transplantation. The various published attempts to restore RPE function and prevent photoreceptor loss with cell transplantation have used both RPE and non-RPE cells, from autologous (harvested from and transplanted to the patient themselves), homologous (harvested from donor eye and transplanted to patient), and heterologous (xenograft—harvested from a different species) sources.

5.1.1 RPE Cells

Primary RPE autologous transplantation trials have involved both multilayer patch grafts and cell suspension techniques. Peyman et al. described the technique of using peripheral healthy RPE, harvested together with a bed of Bruch's membrane and partial-thickness choroid and relocated into the submacular space [28]. The authors described superior functional results compared to a homologous RPE-Bruch's graft from an enucleated donor eye [28]. Numerous investigators have used similar techniques of autologous grafting, showing variable outcomes especially regarding the safety of the procedures {Stanga:2001hm} [30, 31].

5.1.2 Non-RPE Cells

Transplantation studies have also used genetically modified or spontaneously transformed RPE cells as well as non-RPE cells as a source for the transplants [9, 32, 33]. Non-RPE cells utilized for RPE replacement include iris pigment epithelial cells [34], Schwann cells [33], Retinal progenitor cells [35], and stem cells (SCs) of various origin, such as bone marrow SCs [36], human embryonic stem cells (hESCs) [11, 37], umbilical/placental SCs [38], and induced pluripotent stem cells (iPSCs) [39, 40].

Stem cells manifest two basic features that make them ideal sources for cell therapies: firstly they are able to self-renew indefinitely in the undifferentiated state which is essential for producing a sufficient volume of tissue for transplant. Secondly, as pluripotent cells, they can differentiate into any cell type [41]. The advent of stem cell technology has allowed RPE to be derived independently of direct harvesting from a donor. This has created a potential for a limitless supply of young and disease-free transplantable cells. As such, stem cell technology holds invaluable promise as a regenerative source of cells and tissues for transplantation in AMD and other retinal degenerative diseases.

5.2 Scaffold/Substrate Technology: Replacing Bruch's

Scaffolds are biocompatible structures specifically engineered to guide the regeneration and restoration of the functional attributes of a damaged and diseased tissue. They can be manufactured from both natural and synthetic materials and their scope of development is to support and maintain the three-dimensional architecture of tissues and thus to optimize cell delivery [42]. Transplantation of SC-derived RPE with a scaffold has been described as an efficient strategy to increase the survival of implanted cells [43].

The ideal scaffold for an RPE-based retinal therapy needs to be biocompatible and nonimmunogenic and provide sufficient signals for proper cell attachment and alignment. Additionally, it has to be extremely thin and implantable by minimally invasive techniques into the subretinal space, flexible so as not to traumatize surrounding tissues, but also mechanically strong, to withstand the necessary surgical manipulation. Ideally, it should also be manageable with the existing vitreoretinal surgery tools for delivery.

Various polymeric scaffolds have been used with the most common among them being poly(L-lactic acid)(PLLA), poly(lactic-co-glycolic acid)(PLGA), matrix metalloproteinase 2 (MMP2), polymethylmethacrylate (PMMA), polycaprolactone (PCL), and poly(glycerol-sebacate) (PGS). A more detailed analysis of each would lie outside of the scope of this chapter.

5.3 Delivery Tool Technology

In addition to the RPE cells and scaffold the third significant issue for RPE transplantation research is the design and manufacture of delivery devices that would allow atraumatic insertion and controlled orientation of these ultrathin cell grafts. An optimal device for SC sheet delivery in the subretinal space has to meet some technical criteria, in order to be eligible for cell transplantation in human eyes. First of all, in terms of size, it has to provide enough space to enclose the graft without damaging it, though in the same time, it has to be small enough to require as limited access incisions as possible, in order to be inserted through the sclera and under the retina. It has to be flexible and delicate to protect the extremely sensitive retinal and surrounding tissues, but along with that, it has to demonstrate enough rigidity to support the cell sheet and allow the necessary intraocular manipulation. Finally, it should be compatible with currently used ophthalmic surgery equipment and also have a reasonable manufacturing cost, if it is to be used for high-patient-volume operations, such as subjects with AMD.

Thumann et al. have described an injector consisting of a carrier platform mounted in a plastic cannula (Teflon tube) for both animal and human retinal transplantation studies. It uses negative pressure to engulf the transplant into the device. The tip of the instrument is angled at 45° which makes the subretinal manipulation and release of the graft easier. Once the carrier platform has been inserted into the subretinal space it is released by exerting a small amount of pressure with the syringe [44]. The London-based group *The*

London Project have also constructed a delivery tool consisting of a protected chamber for the graft that is only ejected from the device once the chamber opening is placed through a retinotomy and into the subretinal space. The graft is slightly rolled to reduce the width of the required sclerotomy and to increase the rigidity that aids the mechanism that pushes it out of the tool.

As mentioned previously, an excellent paradigm to be followed by engineers working on design and development of cell delivery systems is that of current cataract extraction—lens insertion operation. A delivery tool analogous to an IOL injector, ideally loaded with a premade sheet transplant, could serve the therapeutic needs of high-patient-volume diseases in a feasible, efficient, and safe way.

6 Preclinical Data for Differentiation and Safety of Cells for Transplantation

6.1 SC-Derived RPE Appears to Replicate Very Closely the Endogenous Tissue Morphology and Function (Polarity, Apical Orientation, Gene and Protein Expression)

6.1.1 Human Embryonic Stem Cells (hESCs)

hESCs are pluripotent stem cells firstly isolated from human blastocyst and cultured by Thomson et al. in 1998 [41]. The first report of differentiating RPE-like cells from ESCs was by Kawasaki et al. in 2002. This group used monkey ESCs cultured in a differentiating medium in the presence of stromal cells derived from mouse bone marrow. They reported that their ES-derived RPE-like cells expressed typical RPE markers, and they had extensive and functional apical microvilli and were able to show phagocytosis of latex beads [45]. The next important step was made in 2004 by Klimanskaya et al., when they developed an original protocol for human ESC-derived RPE cells without coculture with animal cells or factors [46].

6.1.2 Induced Pluripotent Stem Cells (iPSCs)

Another revolutionary development, initiated by Takahashi and Yamanaka, was the generation of ESC-like cells from adult somatic cells, by reprogramming the latter with multiple factors and transforming them to iPSCs [47]. Using this method, RPE has subsequently been derived from iPSCs [40].

6.1.3 Proof of Differentiation

During the last decade, numerous researchers have explored several differentiation strategies to derive RPE cells from stem cells and several methods have been developed that can verify the structural and functional characteristics of the derived RPE. These methods include the following:

- Histology: microscopic examination of the histological aspects of the generated RPE cells, e.g., hexagonal shape, pigmentation, organization in a monolayer, polarization, and the presence of tight junctions [48] (Fig. 3a–d).
- Physiology: assessment of RPE cell functioning as a barrier: measurement of water and ion transport resistance through the epithelial monolayer [48] and as support: phagocytosis potency demonstrated with intracellular tracing of polystyrene beads [49] or labeled photoreceptor outer segments, either bovine or human [50] (Fig. 3b).
- Molecular biology: differential distribution of surface molecules among basal and apical cell membrane. Analysis of the sectional concentration of proteins, such as receptors, growth factors, channels, and transporters [51].
- Genetics: demonstration of RPE "signature genes" involved in melanogenesis, intercellular tight junctions, visual cycle, phagocytosis, sensory perception, water and ion transport, oxidoreductase metabolism [52]. Exclusion of gene and protein expression associated with early development and absence of pluripotency or nonepithelial differentiation [50].
- Immunology: investigation of polarized secretion of cytokines, such as VEGF, TGF-β2, PEDF, PDGF [51] (Fig. 3c).

7 Translational Strategies (e.g., Patient Selection, Safety Considerations, Outcomes)

7.1 Retinal Disease Candidates*:* AMD, Stargardt, RP

The aim of RPE transplantation is to retard or halt the loss of photoreceptors caused by a defective RPE-Bruch's membrane-choriocapillaris complex and support their regeneration. There are numerous diseases that can affect this complex and lead to retinal degeneration and hence loss of vision. Since a detailed study of all of the retinal degenerative diseases would lie outside the scope of this chapter, we give a short description here of the ones that most commonly have been or are likely to be targeted for cell therapy approaches, namely AMD, RP, and Stargardt disease, and for which there are still no alternative curative therapies available.

7.1.1 AMD

The first attempts to replace RPE in humans were performed as part of the effort to reconstruct the RPE-Bruch's-choriocapillaris defect created by surgical removal of the CNV in wet AMD [13]. At that time the only available treatments were laser photocoagulation and PDT. Today despite the proven effectiveness and extensive use of anti-VEGF agents that have revolutionized the clinical management of

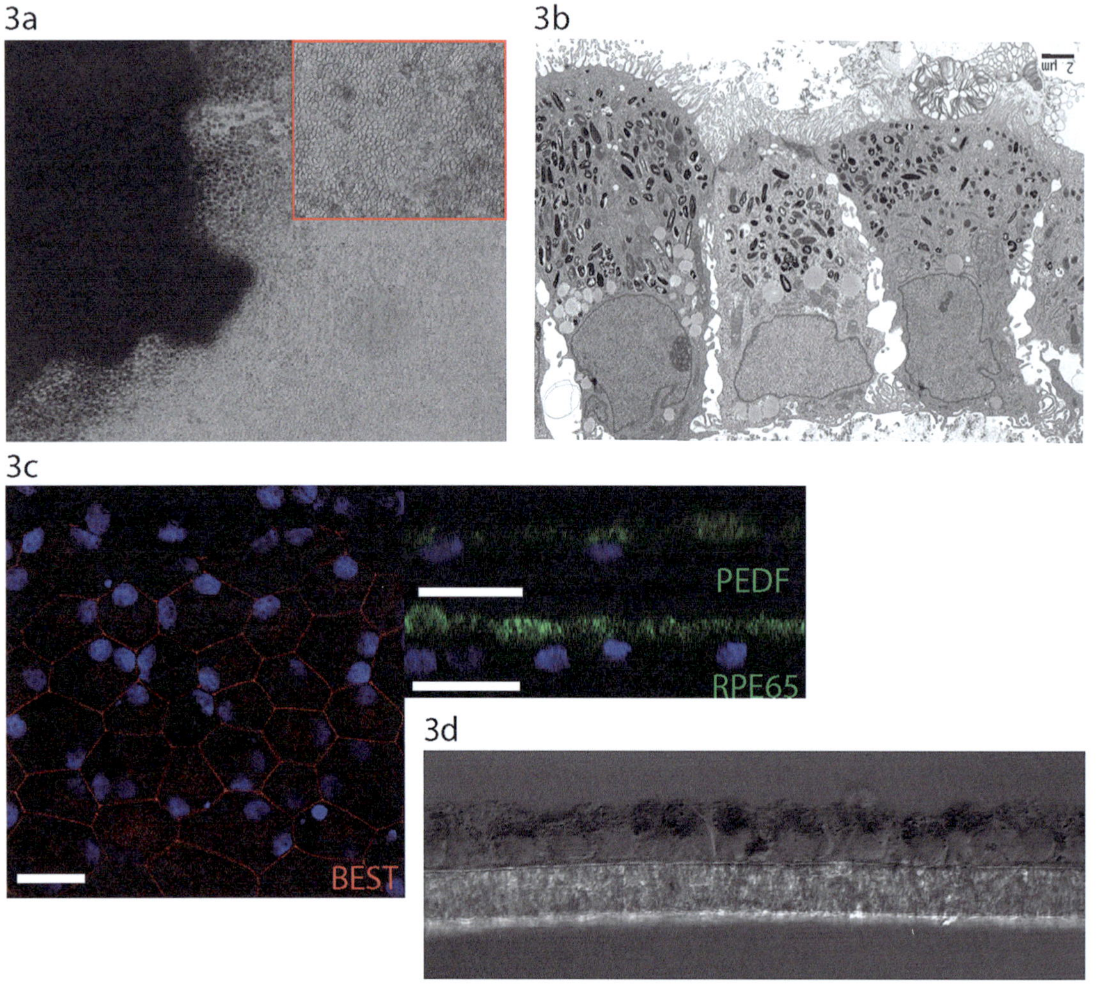

Fig. 3 (**a**) Differentiation of RPE cells from pluripotent stem cells. Microscopy photo showing transformation of cells from the undifferentiated state (*left side* of picture) to mature hexagonal RPE cells (*right side* of picture and further magnification in *top-right* corner). (**b**) Electronic microscopy showing polarization and phagocytosis function of stem cell-derived RPE cells. (**c**) Immunohistochemistry pictures, equally scaled by 10 µm, of stem cell-derived RPE cells showing expression of *Bestrophin, PEDF*, and *RPE65* genes, correspondingly to natural RPE cells. (**d**) Stem cell-derived RPE organized as a polarized cellular monolayer (*top*) on an artificial membrane (*bottom*)

neovascular AMD [53], there is evidence that the underlying atrophic component of the disease still progresses even under anti-VEGF therapy [54]. In contrast to anti-VEGF treatment that aims to control a complication of the disease, RPE replacement aims to restore a normative architecture and thereby normal visual function. Additionally, for dry AMD, lifestyle and dietary changes have been suggested to delay progression in early stages of the disease [55]; however there is still no definite treatment for the advanced stages of geographic atrophy. So, many patients with advanced wet AMD and almost all patients with advanced dry AMD continue to lose vision and thus could be considered as candidates for innovative cellular therapies.

7.1.2 Stargardt Disease

Stargardt disease is an inherited retinal dystrophy in which genetically altered photoreceptor outer segments are thought to cause secondary degeneration of the RPE. It is the most common cause of macular disease in children and it classically presents in the two first decades of life. Although the molecular genetics and pathophysiology of STGD have been broadly studied, there are currently no recognized treatments for this degeneration that deprives from their central vision and leaves legally blind approximately 1:10,000 young adults. The first attempt to treat STGD with SC-derived RPE was carried out by *Ocata Therapeutics, Inc.* (formerly ACT (Advanced Cell Technology)).

7.1.3 RP

Retinitis pigmentosa is most commonly associated with primary photoreceptor degeneration due to gene defects affecting the photoreceptors. There are however subtypes that appear to begin with primary failure of the RPE due to defects in RPE-specific genes, such as MERTK [56] and RPE65 [57]. These subtypes could be potential targets for stem cell-derived RPE replacement strategies in the future.

7.2 Immunology, Immunosuppression, and Graft Rejection

As with all transplantation-based treatments, the viability of SC-derived RPE replacement therapy can be limited by the threat imposed by the recipient's immune system. However, in the case of subretinal delivery of RPE we have to take into consideration two major factors that are peculiar to this particular therapy. Firstly, the special immunological microenvironment of the subretinal space and secondly the individual immune properties of the implanted SCs.

7.2.1 Subretinal Space Immunology

It has been well described that there are particular physiological features of the eye that render it an "immune-privileged" site and make it an attractive target for transplantation approaches. The biological aspects of this property are not fully understood, but among the major components are the tight junction-based blood-ocular barrier and the capacity of the healthy RPE to regulate the subretinal and retinal immune environment. Unfortunately for RPE replacement strategies, the immunological advantage of the subretinal space can be compromised by diseases that damage the RPE. Wenkel and Streilein demonstrated that grafted defective RPE cells from CD95-deficient mice can promote immune rejection, even when implanted into an immune-privileged site like the anterior segment of the eye [58, 59]. Furthermore, the inevitable surgical trauma that occurs during the implantation can compromise the blood-retinal barrier and the regulatory properties of the host RPE, as well as induce inflammation and thus expose implant to an increased level of immune recognition and reaction.

7.2.2 Transplant Immunology

The immune properties of the implanted SC-derived tissue depend on the source of the SCs.

- Mesenchymal stem cells (MSCs) appear to have the ability to downregulate the immune response by suppressing B-cell and T-cell activation, by inducing regulatory T-cells (Tregs) and by affecting the dendritic cell maturation [60].
- Human embryonic stem cells (hESCs), although they are subject to immune recognition and rejection, seem to induce low immunogenicity compared to fetal retinal progenitors [61].
- Induced pluripotent stem cells (iPSCs) are theoretically identical to the antigenic contents of the recipient, and thus are not expected to induce immune reaction. However, Zhao et al. have reported that posttranslational modifications in iPSCs during the dedifferentiation/differentiation process can lead to slightly different antigenic properties that could increase the risk of immune rejection [62]. More recent studies though have demonstrated only insignificant immune reaction to iPSC-derived tissue after transplantation [63].

7.3 Uncontrolled Proliferation/Tumorigenesis: Long-Term Safety

Laboratory experience has shown that hESCs and iPSCs can differentiate relatively easily into RPE cells, possibly because they appear during the early stages of embryogenesis. The differentiation process in vitro does not support survival of stem cells. However, the possibility of a small number of stem cells to remain undifferentiated and proliferate into the recipient tissue leading to tumor growth remains a major concern. Moreover, the possibility for a number of the implanted cells to dedifferentiate within the subretinal space, giving rise to teratoma formation, has to be considered.

Although preclinical safety tests, such as toxicology and tumorigenicity analyses, are required to isolate populations of undifferentiated cells [64, 65], the long-term behavior of the implanted cells cannot be definitively predicted and the risk of uncontrolled proliferation cannot be ruled out. Arnhold et al. demonstrated a 50% ratio of neoplasia formation 8 weeks after subretinal engraftment of embryonic stem cells in rhodopsin (−/−) mice [66]. Conversely, Lund et al. showed no evidence of tumor formation, after using 18 different hESC lines [67]. Nonetheless, the average duration of most animal studies (6–9 months) is not enough to clarify the long-term proliferation potentials of these cells. That said, the possibility of tumor formation, as a complication of the cell therapy, should always be a major key point of discussion with candidate patients, before obtaining the informed consent.

8 Human Data

8.1 The ACT Trial: First RPE Suspension Approach

The first phase I/II study to examine the safety and tolerability of subretinal transplantation of hESC-RPE in humans has been sponsored by *Ocata Therapeutics, Inc.* In this open label, multicenter prospective study patients with advanced dry AMD and Stargardt disease received subretinal cell suspension of hESC-RPE (specifically line MA09-hRPE). The trial has been conducted in four centers in the USA. Schwartz et al. have published the study methods and the 18-month outcomes for 18 patients, 9 with advanced dry AMD (age >55 years) and 9 with advanced STGD (age >18 years) [68]. Both groups were divided into three dose cohorts each (50,000, 100,000, and 150,000 cells) and transplanted patients were followed up for a median of 22 months, by use of systemic, ophthalmic, and imaging examinations. Tacrolimus and mycophenolate mofetil were used for immunosuppression for about 12 weeks. Authors have reported no serious ocular or systemic safety issues, including evidence of adverse proliferation or rejection. Any adverse events were associated with either immunosuppression or the vitreoretinal surgery itself. 72% of patients (13 of 18) showed areas with increasing subretinal pigmentation, consistent with transplanted RPE. In terms of functional outcomes, visual acuity and vision-related quality of life demonstrated improvement in the majority of subjects. These results offer the first evidence of medium- to long-term safety, transplant survival, and possible function of pluripotent stem cell progeny in degenerative retinal disease. Consequently, they suggest hESCs as a potentially safe source of cells for therapy of medical conditions that require tissue replacement.

8.2 The London Project: hESC/iPSC RPE Sheets

The London Project to Cure Blindness (LPCB) and University College London, supported by Pfizer, have commenced a trial to reconstruct the anatomy of the RPE in a phase I study. They have developed a lozenge-shaped patch that consists of a thin biomedical polyester membrane on which they have cultured hESC-RPE cell line (approximately 100,000 cells) as a confluent polarized monolayer [69]. They utilize subretinal microsurgery techniques and a purpose-designed injector to transplant the patch (approximately 6 mm × 3 mm in size) into the submacular space. The first two patients have been transplanted and a further eight will be completed. Patients are selected with sudden severe vision loss from subretinal and sub-RPE hemorrhage or RPE tears. Transient perioperative systemic and intraocular depot corticosteroids are being used for immunosuppression. Subjects will be followed up for 52 weeks, during which they will be examined with systemic and ophthalmic, clinical, laboratory, and functional tests. The overall goal of this trial is to demonstrate whether hESC-derived RPE immobilized on a membrane and implanted as a sheet can offer a safe and effective treatment approach for degenerative maculopathies.

LPCB is also planning another trial to treat AMD patients using iPSC-derived RPE monolayers on a similar polyester scaffold. The first step of this trial is already running at present, and is planned to involve 12 patients with dry AMD. Cells will be isolated from a skin biopsy, which will be used to generate iPSCs and subsequently these pluripotent cells will be differentiated into RPE cells for transplantation.

8.3 RIKEN-Takahashi: iPSC RPE Sheet

The first clinical trial using iPSC-derived RPE has been approved from July 2013 to run in Japan. Preclinical data has shown that the phagocytic ability and gene expression characteristics of the iPSC-RPE are analogous to the native RPE [37]. The researchers plan to enroll six patients with exudative AMD, with fibrotic scar, subfoveal CNV, or RPE tear and BCVA between hand motion and 0.3. The iPSCs will be generated from a patient's skin, differentiated to iPSC-RPE, and cultured to form a monolayer on a temporary scaffold made by collagen gel. The approximate size of the RPE patch transplant will be 1.3 mm × 3 mm. Follow-up period is planned to last 4 years. On September 2014 the first patient was operated in the study, but on November 2015 RIKEN announced the termination of subject enrollment without having enrolled more subjects. This cessation was due to the *Act on the Safety of Regenerative Medicine* legislation, which had come into effect on November 2014, giving a 1-year grace period to the researches to complete the enrollment. Primary monitoring results of the single operated patient have been announced in the company's website (http://www.riken-ibri.jp/AMD/img/20151009en.pdf).

8.4 The California Project: hESC RPE Sheet

Regenerative Patch Technologies (RPT), in collaboration with the California Institute for Regenerative Medicine (CIRM) and the University of Southern California Eye Institute, have been planning a phase I/IIa clinical trial to explore the safety and tolerability of subretinal hESC-RPE implants in patients with geographic atrophy involving the central fovea. The graft consists of an H9 line hESC-derived RPE monolayer on an ultrathin parylene membrane. Patients will be recruited in two cohorts with BCVA of ≤20/400 and ≤20/100 to 20/400. Immunosuppression will last initially for about 2 months and gradually tapered. Subjects will be monitored with both clinical and imaging examinations. The study was approved in October 2015 and the first recruited patient was recently operated. No results have been announced yet.

9 Future Directions

Due to the parallel investigation of sheet and suspension RPE transplantation, information about the advantages and disadvantages of each will become more apparent. As safety and efficacy data becomes more available it will be possible to look at applications of the patch or suspension treatments to earlier stages of degeneration where a greater number of viable photoreceptors are present. This will be particularly important in slowly progressive disease such as dry AMD and IMD.

Moreover, as greater safety and control over cell transplantation is achieved, the possibility of transplanting neural cells will become feasible. Treatment of a broader spectrum of target diseases will be possible with development of neuroretinal cell transplantation. Whether the cells be dissociated or delivered as a stratified neural retina and RPE sheet as a single complex has yet to be demonstrated. The possibility of developing PR/RPE dual grafts for diseases with severe neural retina and photoreceptor loss has been restricted by problems with sourcing photoreceptor progenitor cells. With the recent demonstrations that it is possible to create optic cups in vitro and create photoreceptors and other retinal elements at any stage of development, neuroretinal replacement has become a very real possibility [70, 71].

The transplantation of cells into the retina and subretinal tissues may well be enhanced in the future by gene insertion into the cells and ex vivo modification of the cell phenotype. Gene therapy using vectors expressing an anti-Hif has been published [72] and similar expression could enhance RPE transplantation in wet and dry AMD. Growth factors such as CNTF and BDNF could also be supportive for the retinal milieu.

Finally, it is likely that the current operating and delivery mechanisms associated with conventional vitreoretinal surgery will be insufficient to give the necessary precision for future cellular delivery of both sheets and cell suspensions. As such future cellular therapy will be increasingly dependent on bioengineering developments and the development of new biocompatible materials. At present however the outcomes of hand-delivered sheet RPE transplants will begin to inform the safety and efficacy of this method and allow for developments in delivery, materials, and cell biology that will become available in the near future.

References

1. Strauss O. The retinal pigment epithelium in visual function. Physiol Rev. 2005;85:845–81.
2. de Jong PTVM. Age-related macular degeneration. N Engl J Med. 2006;355:1474–85.
3. Jager RD, Mieler WF, Miller JW. Age-related macular degeneration. N Engl J Med. 2008;358:2606–17.
4. North V, Gelman R, Tsang SH. Juvenile-onset macular degeneration and allied disorders. In: Cell-based therapy for retinal degenerative disease. Basel: S. KARGER AG; 2014. p. 44–52.
5. D'Cruz PM, Yasumura D, Weir J, Matthes MT, Abderrahim H, LaVail MM, Vollrath D. Mutation of the receptor tyrosine kinase gene Mertk in the retinal dystrophic RCS rat. Hum Mol Genet. 2000;9:645–51.
6. Li LX, Turner JE. Inherited retinal dystrophy in the RCS rat: prevention of photoreceptor degeneration by pigment epithelial cell transplantation. Exp Eye Res. 1988;47:911–7.
7. Lopez R, Gouras P, Kjeldbye H, Sullivan B, Reppucci V, Brittis M, Wapner F, Goluboff E. Transplanted retinal pigment epithelium modifies the retinal degeneration in the RCS rat. Invest Ophthalmol Vis Sci. 1989;30:586–8.
8. Seaton AD, Turner JE. RPE transplants stabilize retinal vasculature and prevent neovascularization in the RCS rat. Invest Ophthalmol Vis Sci. 1992;33:83–91.
9. Lund RD, Adamson P, Sauvé Y, et al. Subretinal transplantation of genetically modified human cell lines attenuates loss of visual function in dystrophic rats. Proc Natl Acad Sci. 2001;98:9942–7.
10. Wang S, Lu B, Lund RD. Morphological changes in the Royal College of surgeons rat retina during photoreceptor degeneration and after cell-based therapy. J Comp Neurol. 2005;491:400–17.
11. Vugler A, Carr A-J, Lawrence J, et al. Elucidating the phenomenon of HESC-derived RPE: anatomy of cell genesis, expansion and retinal transplantation. Exp Neurol. 2008;214:347–61.
12. Carr A-J, Vugler AA, Hikita ST, et al. Protective effects of human iPS-derived retinal pigment epithelium cell transplantation in the retinal dystrophic rat. PLoS One. 2009;4:e8152.
13. da Cruz L, Chen FK, Ahmado A, Greenwood J, Coffey P. RPE transplantation and its role in retinal disease. Prog Retin Eye Res. 2007;26:598–635.
14. Crafoord S, Algvere PV, Seregard S, Kopp ED. Long-term outcome of RPE allografts to the subretinal space of rabbits. Acta Ophthalmol Scand. 1999;77:247–54.
15. Binder S. Scaffolds for retinal pigment epithelium (RPE) replacement therapy. Br J Ophthalmol. 2011;95:441–2.
16. Tezel TH, Kaplan HJ, Del Priore LV. Fate of human retinal pigment epithelial cells seeded onto layers of human Bruch's membrane. Invest Ophthalmol Vis Sci. 1999;40:467–76.
17. Tsukahara I, Ninomiya S, Castellarin A, Yagi F, Sugino IK, Zarbin MA. Early attachment of uncultured retinal pigment epithelium from aged donors onto Bruch's membrane explants. Exp Eye Res. 2002;74:255–66.
18. Gullapalli VK, Sugino IK, Van Patten Y, Shah S, Zarbin MA. Impaired RPE survival on aged submacular human Bruch's membrane. Exp Eye Res. 2005;80:235–48.
19. Binder S, Krebs I, Hilgers R-D, Abri A, Stolba U, Assadoulina A, Kellner L, Stanzel BV, Jahn C, Feichtinger H. Outcome of transplantation of autologous retinal pigment epithelium in age-related macular degeneration: a prospective trial. Invest Ophthalmol Vis Sci. 2004;45:4151–60.
20. Diniz B, Thomas P, Thomas B, et al. Subretinal implantation of retinal pigment epithelial cells derived from human embryonic stem cells: improved survival when implanted as a Monolayer Subretinal implantation of RPE cells from hESC. Invest Ophthalmol Vis Sci. 2013;54:5087–96.
21. Hynes SR, Lavik EB. A tissue-engineered approach towards retinal repair: scaffolds for cell transplantation to the subretinal space. Graefes Arch Clin Exp Ophthalmol. 2010;248:763–78.
22. Thomas BB, Zhu D, Zhang L, et al. Survival and functionality of hESC-derived retinal pigment epithelium cells cultured as a monolayer on polymer substrates transplanted in RCS RatsRetinal pigment epithelium transplantation. Invest Ophthalmol Vis Sci. 2016;57:2877–87.

23. Pennington BO, Clegg DO. Pluripotent stem cell-based therapies in combination with substrate for the treatment of age-related macular degeneration. J Ocul Pharmacol Ther. 2016;32(5):261–71.
24. Machemer R, Steinhorst UH. Retinal separation, retinotomy, and macular relocation II. A surgical approach for age-related macular degeneration? Graefes Arch Clin Exp Ophthalmol. 1993;231: 635–41.
25. Chen FK, Patel PJ, Uppal GS, Tufail A, Coffey PJ, da Cruz L. Long-term outcomes following full macular translocation surgery in neovascular age-related macular degeneration. Br J Ophthalmol. 2010;94:1337–43.
26. Chen FK, Patel PJ, Uppal GS, Rubin GS, Coffey PJ, Aylward GW, da Cruz L. A comparison of macular translocation with patch graft in neovascular age-related macular degeneration. Invest Ophthalmol Vis Sci. 2009;50:1848–55.
27. Cahill MT, Stinnett SS, Banks AD, Freedman SF, Toth CA. Quality of life after macular translocation with 360 degrees peripheral retinectomy for age-related macular degeneration. Ophthalmology. 2005;112:144–51.
28. Peyman GA, Blinder KJ, Paris CL, Alturki W, Nelson NC, Desai U. A technique for retinal pigment epithelium transplantation for age-related macular degeneration secondary to extensive subfoveal scarring. Ophthalmic Surg. 1991;22:102–8.
29. Binder S, Stolba U, Krebs I, et al. Transplantation of autologous retinal pigment epithelium in eyes with foveal neovascularization resulting from age-related macular degeneration: a pilot study. Am J Ophthalmol. 2002;133:215–25.
30. van Meurs JC, Van Den Biesen PR. Autologous retinal pigment epithelium and choroid translocation in patients with exudative age-related macular degeneration: short-term follow-up. Am J Ophthalmol. 2003;136:688–95.
31. van Zeeburg EJT, Maaijwee KJM, Missotten TOAR, Heimann H, van Meurs JC. A free retinal pigment epithelium–choroid graft in patients with exudative age-related macular degeneration: results up to 7 Years. Am J Ophthalmol. 2012;153:120–7.e2.
32. Coffey PJ, Girman S, Wang SM, Hetherington L, Keegan DJ, Adamson P, Greenwood J, Lund RD. Long-term preservation of cortically dependent visual function in RCS rats by transplantation. Nat Neurosci. 2002;5:53–6.
33. Wang S, Lu B, Wood P, Lund RD. Grafting of ARPE-19 and Schwann cells to the subretinal space in RCS rats. Invest Ophthalmol Vis Sci. 2005;46:2552–60.
34. Thumann G, Bartz-Schmidt KU, Heimann K, Schraermeyer U. Phagocytosis of rod outer segments by human iris pigment epithelial cells in vitro. Graefes Arch Clin Exp Ophthalmol. 1998; 236:753–7.
35. Kumar R, Dutt K. Enhanced neurotrophin synthesis and molecular differentiation in non-transformed human retinal progenitor cells cultured in a rotating bioreactor. Tissue Eng. 2006;12:141–58.
36. Arnhold S, Heiduschka P, Klein H, et al. Adenovirally transduced bone marrow stromal cells differentiate into pigment epithelial cells and induce rescue effects in RCS rats. Invest Ophthalmol Vis Sci. 2006;47:4121–9.
37. Kamao H, Mandai M, Okamoto S, Sakai N, Suga A, Sugita S, Kiryu J, Takahashi M. Characterization of human induced pluripotent stem cell-derived retinal pigment epithelium cell sheets aiming for clinical application. Stem Cell Reports. 2014;2:205–18.
38. Lund RD, Wang S, Lu B, et al. Cells isolated from umbilical cord tissue rescue photoreceptors and visual functions in a rodent model of retinal disease. Stem Cells. 2009;25:602–11.
39. Silva M, Daheron L, Hurley H, et al. Generating iPSCs: translating cell reprogramming science into scalable and robust biomanufacturing strategies. Cell Stem Cell. 2015;16:13–7.
40. Buchholz DE, Hikita ST, Rowland TJ, Friedrich AM, Hinman CR, Johnson LV, Clegg DO. Derivation of functional retinal pigmented epithelium from induced pluripotent stem cells. Stem Cells. 2009;27:2427–34.
41. Thomson JA, Itskovitz-Eldor J, Shapiro SS, Waknitz MA, Swiergiel JJ, Marshall VS, Jones JM. Embryonic stem cell lines derived from human blastocysts. Science. 1998;282:1145–7.
42. Garg T, Singh O, Arora S, Murthy R. Scaffold: a novel carrier for cell and drug delivery. Crit Rev Ther Drug Carrier Syst. 2012;29:1–63.
43. Thomson HAJ, Treharne AJ, Walker P, Grossel MC, Lotery AJ. Optimisation of polymer scaffolds for retinal pigment epithelium (RPE) cell transplantation. Br J Ophthalmol. 2011;95:563–8.
44. Thumann G, Aisenbrey S, Schaefer F, Bartz-Schmidt KU. Instrumentation and technique for delivery of tissue explants to the subretinal space. Ophthalmologica. 2006;220:170–3.
45. Kawasaki H, Suemori H, Mizuseki K, et al. Generation of dopaminergic neurons and pigmented epithelia from primate ES cells by stromal cell-derived inducing activity. Proc Natl Acad Sci. 2002;99:1580–5.
46. Klimanskaya I, Hipp J, Rezai KA, West M, Atala A, Lanza R. Derivation and comparative assessment of retinal pigment epithelium from human embryonic stem cells using transcriptomics. Cloning Stem Cells. 2004;6:217–45.
47. Takahashi K, Yamanaka S. Induction of pluripotent stem cells from mouse embryonic and adult fibroblast cultures by defined factors. Cell. 2006;126:663–76.
48. Sonoda S, Spee C, Barron E, Ryan SJ, Kannan R, Hinton DR. A protocol for the culture and differentiation of highly polarized human retinal pigment epithelial cells. Nat Protoc. 2009;4:662–73.
49. Kevany BM, Palczewski K. Phagocytosis of retinal rod and cone photoreceptors. Physiology. 2010;25:8–15.
50. Carr A-J, Vugler A, Lawrence J, et al. Molecular characterization and functional analysis of phagocytosis by human embryonic stem cell-derived RPE cells using a novel human retinal assay. Mol Vis. 2009;15:283–95.
51. Zhu D, Deng X, Spee C, Sonoda S, Hsieh C-L, Barron E, Pera M, Hinton DR. Polarized secretion of PEDF from human embryonic stem cell–derived RPE promotes retinal progenitor cell survival. Invest Ophthalmol Vis Sci. 2011;52:1573–85.
52. Strunnikova NV, Maminishkis A, Barb JJ, et al. Transcriptome analysis and molecular signature of human retinal pigment epithelium. Hum Mol Genet. 2010;19:2468–86.
53. Gragoudas ES, Adamis AP, Cunningham Jr ET, Feinsod M, Guyer DR. Pegaptanib for neovascular age-related macular degeneration. N Engl J Med. 2004;351:2805–16.
54. Young M, Chui L, Fallah N, Or C, Merkur AB, Kirker AW, Albiani DA, Forooghian F. Exacerbation of choroidal and retinal pigment epithelial atrophy after anti-vascular endothelial growth factor treatment in neovascular age-related macular degeneration. Retina. 2014;34:1308–15.
55. Age-Related Eye Disease Study Research Group, JP SG, Chew EY, Clemons TE, Ferris FL, Gensler G, Lindblad AS, Milton RC, Seddon JM, Sperduto RD. The relationship of dietary carotenoid and vitamin A, E, and C intake with age-related macular degeneration in a case-control study. Arch Ophthalmol. 2007;125:1225–32.
56. Vollrath D, Gal A, Li Y, Thompson DA, Weir J, Orth U, Jacobson SG, Apfelstedt-Sylla E. Mutations in MERTK, the human orthologue of the RCS rat retinal dystrophy gene, cause retinitis pigmentosa. Nat Genet. 2000;26:270–1.
57. Hamel CP, Jenkins NA, Gilbert DJ, Copeland NG, Redmond TM. The gene for the retinal pigment epithelium-specific protein RPE65 is localized to human 1p31 and mouse 3. Genomics. 1994;20:509–12.
58. Wenkel H, Streilein JW. Analysis of immune deviation elicited by antigens injected into the subretinal space. Invest Ophthalmol Vis Sci. 1998;39:1823–34.
59. Wenkel H, Streilein JW. Evidence that retinal pigment epithelium functions as an immune-privileged tissue. Invest Ophthalmol Vis Sci. 2000;41:3467–73.
60. Griffin MD, Ritter T, Mahon BP. Immunological aspects of allogeneic mesenchymal stem cell therapies. Hum Gene Ther. 2010;21: 1641–55.

61. Luo J, Baranov P, Patel S, et al. Human retinal progenitor cell transplantation preserves vision. J Biol Chem. 2014;289:6362–71.
62. Zhao T, Zhang Z-N, Rong Z, Xu Y. Immunogenicity of induced pluripotent stem cells. Nature. 2011;474:212–5.
63. Tan Y, Ooi S, Wang L. Immunogenicity and tumorigenicity of pluripotent stem cells and their derivatives: genetic and epigenetic perspectives. Curr Stem Cell Res Ther. 2014;9:63–72.
64. Kuroda T, Yasuda S, Kusakawa S, Hirata N, Kanda Y, Suzuki K, Takahashi M, Nishikawa S-I, Kawamata S, Sato Y. Highly sensitive in vitro methods for detection of residual undifferentiated cells in retinal pigment epithelial cells derived from human iPS cells. PLoS One. 2012;7:e37342.
65. Kanemura H, Go MJ, Shikamura M, Nishishita N, Sakai N, Kamao H, Mandai M, Morinaga C, Takahashi M, Kawamata S. Tumorigenicity studies of induced pluripotent stem cell (iPSC)-derived retinal pigment epithelium (RPE) for the treatment of age-related macular degeneration. PLoS One. 2014;9:e85336.
66. Arnhold S, Klein H, Semkova I, Addicks K, Schraermeyer U. Neurally selected embryonic stem cells induce tumor formation after long-term survival following engraftment into the subretinal space. Invest Ophthalmol Vis Sci. 2004;45:4251.
67. Lund RD, Wang S, Klimanskaya I, Holmes T, Ramos-Kelsey R, Lu B, Girman S, Bischoff N, Sauvé Y, Lanza R. Human embryonic stem cell-derived cells rescue visual function in dystrophic RCS rats. Cloning Stem Cells. 2006;8:189–99.
68. Schwartz SD, Regillo CD, Lam BL, et al. Human embryonic stem cell-derived retinal pigment epithelium in patients with age-related macular degeneration and Stargardt's macular dystrophy: follow-up of two open-label phase 1/2 studies. Lancet. 2015;385:509–16.
69. Carr A-JF, Smart MJK, Ramsden CM, Powner MB, da Cruz L, Coffey PJ. Development of human embryonic stem cell therapies for age-related macular degeneration. Trends Neurosci. 2013;36:385–95.
70. Nakano T, Ando S, Takata N, Kawada M, Muguruma K, Sekiguchi K, Saito K, Yonemura S, Eiraku M, Sasai Y. Self-formation of optic cups and storable stratified neural retina from human ESCs. Cell Stem Cell. 2012;10:771–85.
71. Parfitt DA, Lane A, Ramsden CM, et al. Identification and correction of mechanisms underlying inherited blindness in human iPSC-derived optic cups. Cell Stem Cell. 2016;18:769–81.
72. Rakoczy EP, Lai C-M, Magno AL, et al. Gene therapy with recombinant adeno-associated vectors for neovascular age-related macular degeneration: 1 year follow-up of a phase 1 randomised clinical trial. Lancet. 2015;386:2395–403.

Induced Pluripotent Stem Cell-Derived Autologous Cell Therapy for Age-Related Macular Degeneration

Vladimir Khristov, Balendu Shekhar Jha, Aaron Rising, Yichao Li, Haohua Qian, Arvydas Maminishkis, Juan Amaral, Maria Campos, and Kapil Bharti

1 Introduction

Retinal pigment epithelium (RPE) is a polarized monolayer of cells located adjacent to retinal photoreceptors. The two tissues are intimately connected such that the outer segments of photoreceptors are in direct contact with the apical processes of RPE cells. This anatomical proximity is developmentally driven since both cell types originate from the same layer of optic neuroepithelium during ocular morphogenesis. This proximity helps RPE perform several functions that are critical for the health and functional integrity of photoreceptors. RPE functions include the following: (1) *Facilitate nutrient and fluid flow between the photoreceptors and the blood supply.* Tight junctions formed by membrane protein interactions between neighboring RPE cells provide impedance to fluid and nutrient flow to maintain a barrier between photoreceptors and blood supply. Thus, all nutrients and metabolites to and from photoreceptors and the blood supply must pass through individual RPE cells [1–3]. (2) *Replenishment of visual pigment.* During the visual cycle, the opsin-bound visual pigment 11-cis retinal is isomerized to all-trans-retinal. This *trans* form of visual pigment must be reisomerized back to the *cis* form so it can function again as a visual pigment. For reisomerization, all-trans-retinal must be transported to RPE cells [4]. (3) *Phagocytosis of photoreceptor outer segments.* RPE cells actively phagocytose outer segments of photoreceptors on a diurnal basis such that new outer segments are continuously added towards the proximal end of cells. It is estimated that a single RPE cell in human phagocytoses hundreds of thousands of outer segment discs over its lifetime [5]. (4) *Maintenance of ionic homeostasis in the subretinal space.* Photoreceptors are in a depolarized state under dark conditions and maintain a passive efflux of K^+ balanced by an equal influx of K^+ through the Na^+/K^+ pump. Following dark-to-light transition, gated Na^+ channels close causing photoreceptors to hyperpolarize, and Na^+/K^+ pump influx of K^+ increases to balance the charge inside photoreceptors. As a result, the subretinal potassium concentration drops from 5 to 2 mM leading to hyperpolarization of RPE apical membrane by 10–20 mV. This opens up RPE potassium channels to correct the subretinal space potassium concentration so that photoreceptor polarization state can be maintained in the subsequent dark-to-light transition [6, 7]. (5) *Maintenance of cytokine balance in the retina/RPE/choroid.* RPE cells secrete cytokines in a polarized fashion towards their apical and basal sides. For example, to maintain the health of both the photoreceptors and the choroid, RPE secretes more PEDF apically and more VEGF basally. Furthermore, RPE cells secrete immunomodulatory cytokines like such as TGF-beta and IL-6 to locally control adaptive and the innate immune responses, extending its role beyond a metabolic support for photoreceptors, the retina, and the choroid [8].

Defects in several of the above-mentioned RPE functions have been linked to congenital or late-onset retinal degenerative diseases. Retinitis pigmentosa (RP) and Leber's congenital amaurosis (LCA) are types of early-onset retinal degenerations caused by mutations in genes that affect RPE function. For instance CRALBP, MFRP, MERTK, and LCHAD mutations lead to RP, and RPE65, KCNJ13, and GUCY2D mutations lead to LCA [9]. Other examples of RPE-associated retinal degenerations include polygenic disease like age-related macular degeneration (AMD), one of the leading causes of blindness among elderly. The disease has two advanced stages: "dry" AMD or geographic atrophy

V. Khristov, B.S. • A. Rising, Ph.D. • A. Maminishkis, Ph.D.
Section on Epithelial and Retinal Physiology and Disease, National Eye Institute, 10 Center Drive, Room 10B04, Bethesda, MD 20892, USA

B.S. Jha, Ph.D. • K. Bharti, Ph.D. (✉)
Unit on Ocular and Stem Cell Translational Research, National Eye Institute, 10 Center Drive, Room 10B10, Bethesda, MD 20892, USA
e-mail: kapilbharti@nei.nih.gov

Y. Li, M.S. • H. Qian, Ph.D. • J. Amaral, M.D. • M. Campos, M.D.
National Eye Institute, National Institutes of Health, 9000 Rockville Pike, Bethesda, MD 20892, USA

S.D. Schwartz et al. (eds.), *Cellular Therapies for Retinal Disease*, DOI 10.1007/978-3-319-49479-1_3

(GA) [10] that is characterized by localized atrophy of RPE in the macula, and "wet" AMD with choroidal neovascularization (CNV), where capillaries of the choroid breach through Bruch's membrane and the RPE monolayer enter the retina and leak fluid or blood in the eye leading to the loss of central vision [11, 12]. While CNV treatment can be effectively managed by regular intravitreal injections of anti-VEGF antibodies that cause regression of neovascular tissue, there is currently no treatment for RPE atrophy seen in dry AMD [13]. Approximately 90% of advanced AMD cases in the United States are of the dry AMD type which amounts up to two million individuals with varying degrees of visual dysfunction [14]. Since AMD mostly affects elderly people over the age of 65, as the US population ages the incidence of this disease is likely to double in the coming decade [14], posing a serious health care burden on our society.

Because advanced AMD stage is triggered by RPE loss, it is thought that replacing the atrophied RPE layer with a "younger" monolayer might slow down or stop disease progression in areas of the retina where photoreceptors are still present. In this chapter, we provide an overview on RPE replacement as a potential therapy for AMD and summarize our most recent efforts on developing an autologous RPE replacement therapy for AMD patients using patient-specific induced pluripotent stem cells.

2 RPE Replacement Therapy: The Logic and the History

Early AMD gradually progresses to the GA stage with a localized death of photoreceptor cells in the macula. Histological analysis of cadaver eyes from AMD patients has demonstrated that RPE atrophy precedes photoreceptor cell death. The border of the GA lesion has a transition zone with atrophied RPE cells but relatively healthy photoreceptors. This transition zone can extend up to a couple of millimeters with variable photoreceptor loss, but a consistent RPE loss [15]. It is thought that transplantation of a healthy RPE monolayer in this region can protect the overlying photoreceptors and stop disease progression.

Several sources of RPE graft have been tested. Initial attempts in the 1980s used xenografts from cadaver human eyes tested in monkey and rabbit eyes to optimize the surgical procedures for RPE graft transplantation. These procedures were relatively invasive and required detaching and folding over a large portion of the retina and provided minimal engraftment of transplanted cells [16]. This work was followed by testing of allo-RPE grafts isolated from fetal or cadaver human eyes transplanted into AMD patients. These allo-RPE sheets also did not provide a significant long-term engraftment likely due to surgical complications and immune rejection of the allo-tissue [17]. A breakthrough in the field of RPE transplantation occurred with the use of autologous RPE sheets isolated from nasal or temporal regions of the same patient's eye [18, 19]. Surgeons harvested RPE/choroid from peripheral retina and transplanted this graft to the macula of the same eye. Initially this procedure was tested only on patients with wet AMD, where it also required removing the neovascular tissue from the subretinal space [18, 20]. The procedure has also been tested on patients with GA [21]. Long-term follow-up (up to 7 years and more) has demonstrated varying degrees of success with this procedure. This was likely because the harvesting of autologous grafts of RPE/choroid often led to postsurgical complications involving proliferative vitreoretinopathy (PVR) and fibrosis in the periphery of the eye. Only in a small number of cases was the surgery successful with minimal postoperative complications, the graft was found to integrate in to the host Bruch's membrane, and patient's visual acuity improved over a short-term follow-up [18–20]. Despite its complications and low success rate, this approach of autologous RPE transplantation provided a proof of concept that an RPE graft transplantation therapy for AMD is feasible. Furthermore, these attempts laid the groundwork for surgical techniques that are currently being combined with more advanced manufacturing methods for developing autologous RPE grafts for AMD patients.

3 Pluripotent Stem Cells as Source for RPE Transplant

The original attempts to obtain an RPE graft for transplantation studies in AMD patients used either allogeneic cadaver, fetal tissue sources, or autologous RPE tissue from peripheral retina. All these tissue sources are extremely difficult to obtain, limited in supply, lack functional validation, and are immune incompatible (at least in the case of allogeneic sources). Recent advances in pluripotent stem cell technologies and methods to differentiate pluripotent stem cells into desired cell types have made it possible to obtain copious amounts of well-characterized and functionally validated RPE for testing safety and toxicology in animals and for use in human trials [22–26]. Both human embryonic stem (ES) cell and induced pluripotent stem (iPS) cells have been used as a source of manufacturing the RPE transplant [27–29]. Obviously the transplant derived from ES cells is an allogeneic product and requires immune suppression whereas iPS cell-derived transplant can be autologous, thus potentially eliminating the need for systemic immune suppression [30]. This increases the possibility of longer term graft survival and integration, and reduces the possibility of systemic immune suppression-associated complications in older patients. Nevertheless, it is too early to tell which approach will be effective in patients and both approaches are currently being tested [17]. Here, we summarize previous work and ongoing efforts in developing an autologous cell therapy for AMD using patient-specific iPS cells.

Allogeneic and autologous cell therapies require fundamentally different approaches towards obtaining the regulatory approval from the Food and Drug Administration (FDA). In the case of allogeneic therapy, the approval is obtained for the final clinical product. As part of Investigational New Drug (IND)-enabling studies, current Good Manufacturing Practices (cGMP)-grade clinical product needs to be scaled up and manufactured to form a big lot. Samples from this cGMP lot of clinical product are used for preclinical animal toxicity and efficacy studies. In contrast, for an autologous therapy the regulatory approval is obtained for the clinical product manufacturing process and not the product. The reason is that the clinical product needs to be manufactured individually for each patient after the IND approval is obtained. Therefore, as part of IND-enabling studies one needs to demonstrate that the clinical-grade product can be manufactured consistently from different patients and the final clinical-grade product from different patients meets the release criteria set for validation of the product. This necessitates a robust manufacturing process that is efficient and reproducible, and generates pure cells that form a functional RPE transplant. Here we summarize such a cGMP-compatible manufacturing process, functionally validated cGMP-compatible RPE graft, and preclinical studies that demonstrate the safety and efficacy of such a graft (Fig. 1).

4 Developing an Autologous Cell Therapy Using Patient-Specific iPS Cells

4.1 Making Clinical-Grade iPS Cells

Development of an autologous cell therapy requires generation of patient-specific clinical grade iPS cell banks that do not carry exogenous sequences, integrated in the genome during the reprogramming process, and thus are considered "footprint" free. Several previous publications have focused on such footprint-free technologies to generate iPS cells [31]. Therefore, we will focus here mainly on manufacturing of clinical grade iPS cell banks for autologous cell therapy and also describe some critical quality attributes for autologous iPS cell banks. We have used peripheral blood-purified CD34+ cells for generation of clinical grade iPS cells. CD34+ cells were chosen because of the following advantages: (1) minimally invasive procedure to obtain cells from peripheral blood; (2) non-senescent nature of cells under nutrient-limiting clinical-grade manufacturing conditions [32, 33]; (3) previous protocols to obtain high-quality iPS cells from CD34+ cells [34, 35]; and (4) availability of regulatory compliant protocols for isolation of CD34+ cells from the hematopoietic stem cell field [36, 37]. Using a footprint-free, xeno-free protocol [38], we manufactured one Working Cell Bank (WCB) per clone of GA patients' iPS cells. The WCB provides several advantages for the manufacturing process: (1) it provides an intermediate stage

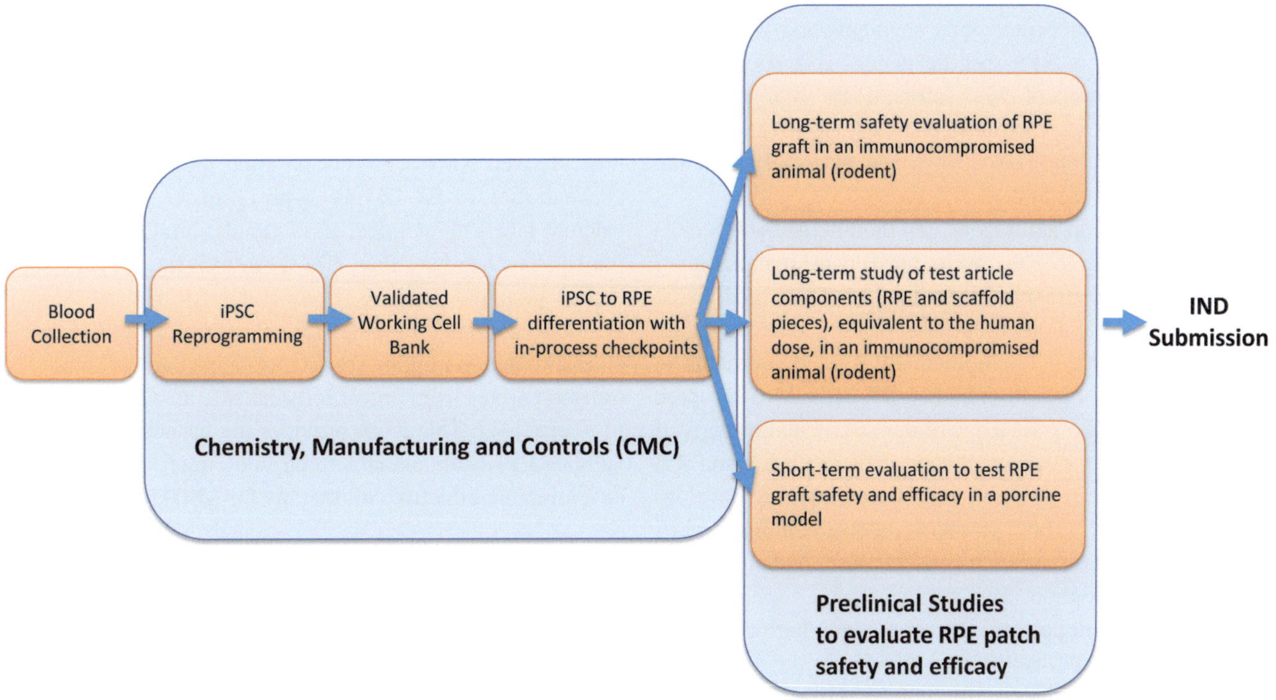

Fig. 1 Schematic of study design

where cells can be cryopreserved and the manufacturing process can be halted, if required; (2) it provides a stage for in-process quality controls for critical quality attributes of iPS cells: for instance, normal pluripotency, normal G-band karyotyping, loss of reprogramming plasmids, identity match to the patient, sterility, and absence of potentially oncogenic mutation in the iPS cell oncogene exome. This in-process quality control is critical to confirm the success of the reprogramming process for generating clinical-grade iPS cells that can be efficiently differentiated into an autologous RPE transplant.

4.2 Making Clinical-Grade RPE Cells

4.2.1 RPE Differentiation Protocols

Multiple protocols have been developed to differentiate RPE cells from human pluripotent stem cells and they vary significantly in the efficiency of differentiation. One of the first reports on differentiation of human ES cells into RPE-like cells was published in 2004 [39]. The protocol used in this report was rather simple. Basic fibroblast growth factor was withdrawn from ES cells and the cells were allowed to spontaneously differentiate. Within 4–6 weeks, pigmented polygonal RPE cell clusters appeared in those cultures. It is, however, worth mentioning that pluripotent stem cells spontaneously differentiate into cells of all three germ layers. Because RPE is pigmented, it was relatively easy to identify in those cultures. Pigmented RPE cultures are manually and enzymatically purified to obtain almost pure cultures of RPE-like cells. Klimanskaya et al. further demonstrated several RPE-like features in those cells including their ability to phagocytose photoreceptor outer segments and the expression of several key RPE-specific genes. This spontaneous differentiation protocol was subsequently reproduced in other labs for both ES and iPS cells [22]. However, the efficiency of differentiation has always been extremely poor and variable with this method.

In the case of an autologous cell therapy using iPS cells where individualized RPE grafts need to be manufactured for each patient, the manufacturing process needs to be robust, efficient, and reproducible. Several labs including ours have optimized developmentally guided protocols to manufacture RPE cells from human iPS cells. Most of these protocols utilize the knowledge of developmental biology of the RPE and coax pluripotent stem cells (both ES and iPS cells) in vitro through various stages of embryonic development to make RPE. These protocols involve combining various growth factors at different concentrations and at different stages of differentiation. We have developed a triphasic stepwise differentiation protocol that works with high efficiency on several human iPS cells tested thus far. In the first phase, cells express eye-field transcription factors like PAX6, RAX, OTX1, and SIX3. In the second optic neuroectoderm phase, cells are differentiated into committed RPE cells by activation of the TGF pathway. Previous work has shown that the TGF pathway specifically increases the expression of RPE-specific combination of PAX6, MITF, and OTX2 transcription factors [28, 40]. In the third phase RPE cells are matured by specific activation of the primary cilium in RPE cells. Our recent work has shown that pharmacological activation of primary cilium in confluent RPE cells causes downregulation of RPE progenitor markers like PAX6 and specifically increases the expression of RPE maturity markers like RPE65 and BEST1. Fully mature RPE cells present with extensive apical processes and electrical responses that mimic native human RPE cells (see below for details). Overall, our stepwise differentiation protocol is reproducible and generates RPE at a very high efficiency.

Another critical aspect of an autologous cell therapy is the ability to functionally authenticate RPE cells generated from genetically diverse individuals. Functional responses of RPE cells obtained from different individuals are different likely due to genetic and experimental variations. Therefore, it is critical to understand the range in which RPE functional responses can be considered "acceptable" for clinical application. To address this concern, we generated 15 different iPS cell lines from epigenetically distinct tissue sources obtained from genetically diverse individuals [41]. RPE cells derived from all these iPS cell lines were validated for structural (immunostaining and transmission electron microscopy), molecular (mRNA and miRNA expression), and functional (electrophysiology, intracellular calcium imaging, and the ability to transport water from apical to basal side) features. Our work confirms the notion that distinct iPSC-RPE show variability in their functional and molecular features. Furthermore, the interdonor variability exceeds the intradonor clonal variability or starting tissue epigenetic variability. We provide the following functional criteria to authenticate iPSC-derived RPE cells: (1) baseline intracellular calcium concentration of 80–120 nM with up to 50 nM changes in calcium levels upon apical ATP stimulation; (2) transepithelial resistance greater than 300 Ω cm^2; (3) baseline transepithelial potential ranging from 2 to 10 mV and apical K^+-induced change ranging from 1 to 8 mV, whereas apical ATP stimulation ranging from 0.5 to 5 mV; and (4) ability to transport water from apical to basal side at steady-state 5–7 μL/h/cm^2 [41]. This work provides the basis to functionally validate RPE cells manufactured in a clinical setting and to develop more effective cell therapy for AMD patients.

4.2.2 Clinical-Grade RPE Differentiation Protocols

Our lab has converted the lab-grade iPSC-RPE protocol into a clinical-grade manufacturing process. This protocol takes 40 days to differentiate iPS cells into RPE cells that can be used to manufacture an RPE transplant. We have also established two in-process quality control checks at day 25 when the iPS cells have differentiated to RPE progenitors and at day 40 when RPE progenitors are in their maturation phase. At both these time points, RPE differentiation is controlled for RPE markers like PAX6, MITF, PMEL17, and TYRP1, and the absence of pluripotency markers like OCT4 and TRA1–60. Our differentiation protocol generates pure RPE cells and has been tested using clinical-grade iPS cells from three different AMD patients.

4.2.3 Making an RPE Transplant

Polarization in epithelial cell types is induced by the formation of tight junctions between neighboring cells leading to differential expression of different receptors, ion channels, and other membrane proteins on the apical and the basolateral membranes. This polarization is also reflected in the intracellular signaling machinery, which is able to distinguish between apical and basal membrane-associated signal transduction. In vivo RPE polarization is intimately associated with the maturation process. The "ideal" RPE transplant is the one that is able to reproduce the native-like maturation and polarization state of RPE cells. Furthermore, for a cell therapeutic application it is also required that the scaffold used for making the transplant is biocompatible, preferably FDA approved, and easier to manufacture consistently at commercial scale in a regulated, cell therapy-compatible environment (for a detailed discussion of various substrate materials and their industrial applications, see [42]). It is also desired that the scaffold material used for creating the RPE transplant is biodegradable such that when it degrades, the RPE monolayer has the possibility of resting upon host Bruch's membrane.

Previous studies have tested various scaffold materials for RPE transplant manufacturing including biological membranes, nondegradable plastics, and biodegradable polymers (see [42] or summary). The logical choice for a scaffold from a biological perspective is the native Bruch's membrane. This penta-laminar proteinaceous membrane is found between the RPE cell basement membrane and the basement membrane of choroidal endothelial cells, and contains various collagen proteins, laminins, fibronectin, elastin, and chondroitin sulfates [43]. Specific layers of Bruch's membrane were found to support RPE growth better than others. For instance, RPE survival depended on the presence of the basal layer consisting mostly of collagen IV α3-5 [44–46], whereas other layers of Bruch's membrane did not support RPE growth as well. Human amniotic membrane has been used as another alternate substrate for RPE transplant [47–50]. RPE grown on amniotic membranes express higher levels of the RPE genes RPE65, CRALBP, BESTROPHIN, and TYROSINASE RELATED PROTEIN (TRP)-2 compared to cells cultured in plastic dishes [47]. Apart from these limited gene expression and morphological analysis, RPE grown on these biological membranes were not functionally validated. Therefore, it remains unclear if those cells truly behaved like native RPE cells. By far the biggest challenge of using either an explant of Bruch's membrane or an amniotic membrane is their limited availability and the difficulty in maintaining clinically compatible conditions during manufacturing.

Artificial substrates have the advantage that they can be produced in a controlled, xeno-free environment. Parylene-C is poly-xylene-based biologically compatible polymer that is routinely used in implants [51, 52]. Submicron thickness layer of parylene supports RPE growth and monolayer formation that attains structural features similar to native RPE cells [24]. Bare parylene membrane supported by a mesh structure has diffusion properties similar to the Bruch's membrane, but RPE grown on these membranes was not validated functionally [53]. Also, parylene is not a degradable polymer and it is expected that the polymer will likely have to integrate into the host Bruch's membrane for a long-term survival of the implant.

In comparison, biodegradable polymers provide the advantage that once the polymer substrate degrades away, the cellular monolayer with its own natural extracellular matrix is able to integrate into the host Bruch's membrane. Biodegradable polymers can be made into scaffolds either as a sheet or a nanofibrous structure with fibers (several nanometers in diameter) electrospun to obtain a three-dimensional architecture. Nanofibrous structure is preferred for RPE transplantation because high-resolution imaging of the Bruch's membrane suggests that it's a nanofibrous structure. Nanofiber scaffolds have also been fabricated from natural materials such as collagen [54, 55] and bacterial cellulose [56]. As compared to bacterial cellulose, collagen-derived scaffolds were much better in supporting RPE cell growth and function. RPE cells grown on collagen scaffolds demonstrated improved gene expression, cytokine secretion, and phagocytic activity as compared to RPE grown on polystyrene plastics. The main concern with the use of collagen-based scaffold is the lack of control over the manufacturing process in order to generate a material that is xeno-free and can be manufactured using FDA-approved industry standards for clinical purposes.

Synthetic nanofiber scaffolds have multiple advantages over natural scaffolds. These include direct control over the manufacturing process that allows production of a scaffold composed of defined xeno-free constituents, defined fiber diameter, porosity, and thickness. Additionally, the exact fiber composition, surface characteristics, and degradation rates can be precisely controlled to yield a reproducible surface for RPE maturation. Electrospun scaffolds composed of multiple materials have been reported [57]. Many synthetic monomers have been used as the basis to produce synthetic nanofiber scaffolds. Some examples include poly(L-lactide-*co*-e-caprolactone) (PLCL) [54], polycaprolactone (PCL) [58], poly-(lactic-*co*-glycolic) acid (PLGA) [59], polyethylene terephthalate (PET) [60], and polytetrafluoroethylene (PTFE) [61]. The surface of these synthetic scaffolds can be further modified by oxidation, or by applying a coating such as vitronectin, fibronectin, laminins, and other human ECM proteins. The aim of these modifications is to enhance RPE cell attachment and maturation.

After considering biocompatibility, biodegradability, and regulatory compliance, our team elected to use an electrospun PLGA nanofibrous scaffold for the production of the RPE transplant. To generate a confluent and functionally mature iPSC-RPE cell monolayer, we have optimized the use of electrospun 400 nm diameter fibers of PLGA. An important consideration for fiber diameter is to avoid RPE cells from migrating through the voids between neighboring fibers and form multiple layers within the scaffold. Typically, the void diameter is two times the fiber diameter. Thus, a scaffold with 400 nm fiber diameter provides voids smaller than 1 μm and significantly smaller than a typical RPE cell. Our results are supported by a recent publication that compared RPE cell growth on nanofiber scaffolds of different diameter and determined the optimal fiber diameter between 250 and 400 nm for RPE cell growth [62]. The scaffolds are mounted on a semipermeable membrane such that medium can be added both on the apical and basal sides of the cells. Our protocol generates fully polarized cells within 4–6 weeks of culture (Fig. 2a, b).

4.2.4 RPE Transplant Release Criteria

One of the chief regulatory requirements of any cell therapy is to develop "release criteria" that define critical quality attributes of the final clinical product. This requirement is particularly relevant for an autologous cell therapy where individualized clinical products will be manufactured for each patient. In the case of autologous iPSC-RPE transplant these criteria include (1) absence of fungal or bacterial contaminants; (2) absence of mycoplasma; (3) endotoxins below the accepted range; (4) identity of RPE transplant as confirmed by short tandem repeat analysis and comparison to patient material; (5) purity analysis for the expression of RPE markers and absence of pluripotent cell makers; (6) morphometric analysis for RPE cell shape, size, and area; (7) ability of cells to secrete cytokines such as VEGF and PEDF in a polarized fashion; and (8) transepithelial resistance of several hundred $\Omega \bullet cm^2$. Our results demonstrate that clinical-grade autologous iPSC-RPE transplant generated from AMD patients in an cGMP-compatible manner meets all of the above release criteria.

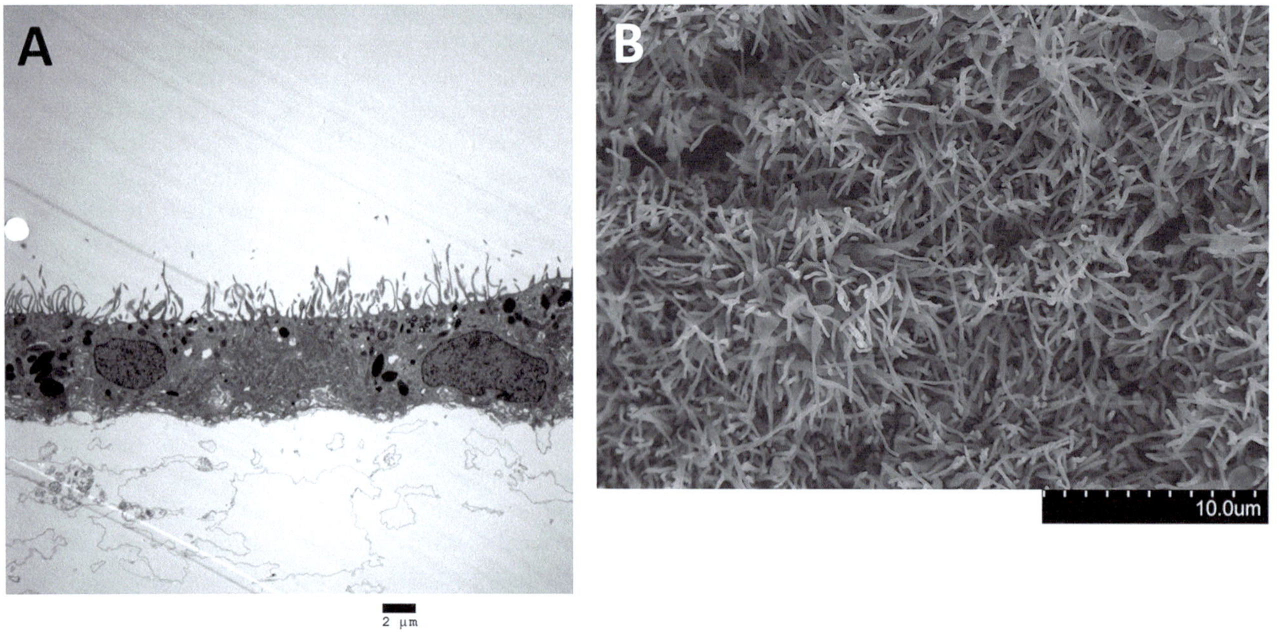

Fig. 2 Electron microscopy (EM) of RPE on the scaffold. (**a**) Transmission EM of iPSC-RPE on a partially degraded scaffold shows the monolayer structure and polarization. (**b**) Scanning EM of iPSC-RPE on scaffold shows abundant apical processes crucial for the interface with photoreceptor outer segments

4.3 Preclinical Animal Testing

4.3.1 A Pig Model of Selective RPE Ablation to Induce Secondary Retinal Degeneration

To test the safety and efficacy of an AMD patient-derived iPSC-RPE transplant, we used the porcine animal model. The size, anatomy, vasculature, and photoreceptor distribution of the porcine eye are similar to the human eye making it an ideal model for our preclinical studies [63, 64]. Since no genetic porcine model of RPE degeneration is available, we developed an acute injury model with RPE ablation. RPE injury and ablation induce a secondary retinal degeneration that mimics the retinal damage and RPE loss seen in dry AMD patients. To ablate the RPE, we use a micropulse laser [65, 66]. The micropulse technology precisely controls the thermal elevation by disrupting a continuous-wave beam into multiple repetitive short pulses at fixed intervals with intermittent breaks. As compared to the continuous wave that delivers a steady stream of energy and causes widespread thermal damage to the tissue, micropulse laser allows tissue to cool in between pulses, thus reducing the thermal damage [67]. Because of selective thermal energy absorbance by melanin granules and the intermittent cooling effect, the thermal damage stays contained within the RPE monolayer and only mildly disperses to photoreceptor outer segments (Fig. 3). Using a scanning laser delivery system we are able to perform confluent high-density threshold applications in the designated area, creating an extensive area of ablated RPE (Fig. 3). The porcine laser injury model was validated using histological and immunohistochemical evaluations.

4.3.2 Transplantation Tool

A surgical tool to transplant the RPE graft need to: (1) maintain RPE transplant integrity and viability; (2) safely deliver the RPE transplant into subretinal space at the correct location without damaging the overlying retina or Bruch's membrane; (3) have an ergonomic design; (4) reliably deliver the implant; (5) be composed of biocompatible material; (6) be small enough to avoid a large sclerotomy and retinotomy; (7) avoid rolling or folding of the RPE graft to maintain its polarization; and (8) be disposable. We have developed a tool that meets all these criteria, and we are using it currently to test safety and efficacy of the RPE transplant (Fig. 4).

Figure 4 shows schematics of the tool design and its use for transplantation into the subretinal space. This tool was designed for delivery of flexible RPE implants. It features a flattened cannula attached to an ergonomic handle and tubing connected with a Luer connector. To ensure ease of use and reliable delivery of RPE graft the tool has two distinct curves that are very different from the majority of vitreoretinal instrument shapes. The first bend close to the tool handle (top-left image in diagram) is needed to allow for surgeon to insert this tool through a sclerotomy such that the angle of the tool does not obstruct the surgeons view. The second bend (closer to the tool tip) follows the curvature of the eye and is critical for precise delivery of the RPE transplant in the subretinal space. In the absence of this bend sclera will have to be folded to reach the subretinal space. This creates the potential danger of intraocular pressure fluctuations and distortion of the surgeon's view. The tool handle design is based on previous work describing surgeons' comfort in holding instruments where a consistent distance between instrument tip to grip location proves to be helpful in ensuring successful surgeries [68] [69]. Distance from instrument tip to grip was set to match previously used surgical instruments [70]. This allows the surgeon to better control the tip location in relation to other eye structures.

The tool is disposable and can be sterilized. For transplantation procedures the tool is connected to the viscous fluid injection port (VFI) of a vitrectomy device (e.g., Alcon Constellation), allowing the surgeon foot pedal control of the implant delivery. It also provides physical and mechanical support to maintain implant viability. It is made of hydrophobic,

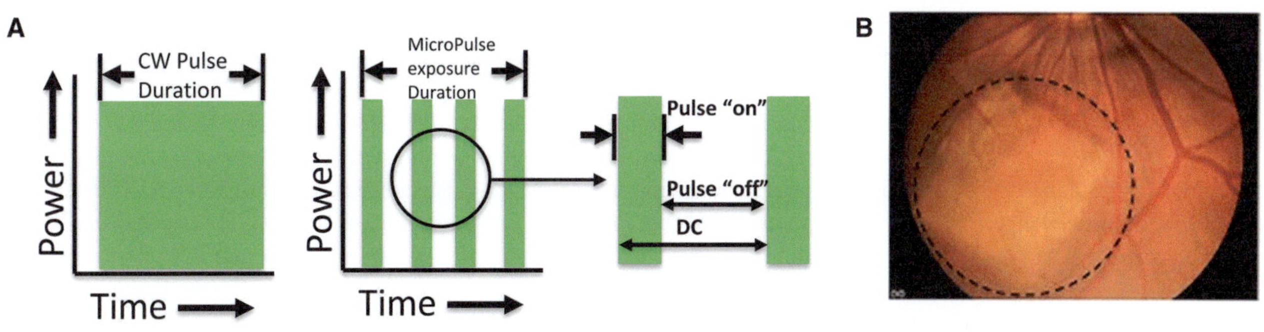

Fig. 3 Laser-induced RPE injury model. (**a**) The *left* graph shows a continuous-wave laser where laser power is delivered continuously for a period of time. In comparison, micropulse laser (*middle* and *right* graphs) delivers the same amount of power in smaller time intervals. Micropulse laser allows the tissue to cool down in between two laser pulses, thus confining the damage to the outer retina. (**b**) *Dotted circle* marks the area of laser-induced RPE damage as seen in a fundus photograph of the pig eye. *CW* continues wave, *DC* duty cycle

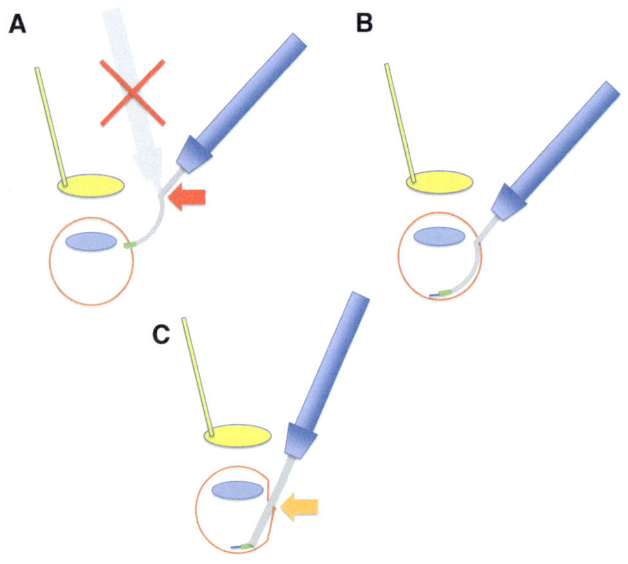

Fig. 4 Implantation tool design. (**a**) Shows the implantation tool (*blue*) at the time of globe entry. Note that the bend of the implantation tool (*red arrow*) prevents interference with the fundus visualization system (*yellow*). (**b**) Shows how the bend in the implantation tool allows advancing the instrument deeper inside the eye in the subretinal space without distorting the sclera. (**c**) Shows scenario without bend in implantation tool, which would induce folds in the sclera (*orange arrow*) as the tool is angled to deliver the implant

flexible materials that minimize damage to the retina while accessing the subretinal space and can be used to manipulate the position of the implant under the retina.

4.3.3 Use of Porcine Model to Test the Efficacy of iPSC-RPE Transplants

We are using the laser-induced RPE ablation pig model and our transplantation tool to test the safety and efficacy of RPE transplant derived from AMD patients' iPS cells. AMD iPSC-RPE grafts are transplanted in laser-injured pig eyes 2 days after the injury. Our imaging analysis suggests that the photoreceptor nuclear layer is well preserved at this time point (Fig. 5). A standard 25-gauge pars plana vitrectomy is performed to deliver the RPE transplant into the subretinal space. After inducing a posterior vitreous detachment, a subretinal fluid bleb is created in the lasered area. A retinotomy is performed to allow the delivery of the scaffold from the tool into the subretinal space. After delivery of the RPE transplant in the subretinal space, the retina is reattached using a fluid air exchange. This surgical procedure and the transplantation tool will also be used for human surgeries.

4.3.4 Functional Assessment of Transplanted iPSC-Derived RPE Cells

Optical Coherence Tomography (OCT)

Optical coherence tomography is an imaging technique that uses long-wavelength light to visualize the retinal layers at high resolution [71, 72]. We utilized OCT to perform morphological assessments of the retina prior to and after all surgical manipulations of the eye. The baseline assessments help to determine if the retina has the appropriate morphological characteristics [72]. We assess the retina after laser ablation to determine the extent of the damage; this also helps with post hoc analysis of recovery. Laser conditions used in our protocol cause minimal damage to photoreceptors and the Bruch's membrane with localized damage to the RPE. After transplantation of the RPE graft, we obtain a weekly OCT to assess its integration and to determine any morphological changes. Our work shows that in successful surgeries the RPE graft lays down on barren Bruch's membrane and over time integrates into this surrounding tissue. These longitudinal measurements help to monitor regeneration of the photoreceptor nuclear layer and the photoreceptor outer segments. Additionally, the OCT helps to determine when the scaffold of the RPE transplant has degraded. Our analysis shows that the scaffold degrades in 6–8 weeks. Photoreceptor regeneration is also evident after several weeks of transplantation. Similarly, OCT scans will be used in human studies as well to assess RPE and photoreceptor health. Figure 5 shows a porcine eye with the implant engrafted under the retina.

Multifocal Electroretinography (mfERG)

Electroretinography (ERG) is a standard technique for determining retinal health in clinical and in laboratory settings. Light stimulation produces an electrical signal in the form of a waveform that represents the function of certain retinal cell types [73]. The full-field ERG (ffERG) utilizes flashes of varying light intensities and produces different waveforms depending on the light stimulus provided [73, 74]. While the ffERG test has been the most widely used diagnostic measurement for overall retinal health and for determining phenotypical characteristic of specific diseases [75], it cannot distinguish between specific regions in the retina within the same eye. This limitation of ffERG is particularly crucial in our laser-induced RPE ablation pig model where the area of damage to the retina and the transplanted area are relatively small as compared to the whole eye. ffERG does not have sufficient resolution to generate distinct signals from healthy regions and the RPE transplant area. In contrast to ffERG, mfERG relies on repeated light stimulation of a small area of the retina [74]. The advantage of mfERG over ffERG is that multiple smaller individual regions, usually between 39 and 103, over a specific region of the retina can be measured. Because of this advantage mfERG can better determine changes in the lasered or transplanted regions as compared to the healthy parts of the same retina. Roughly half of these subregions are being stimulated at a given time and by using a mathematical equation (called m-sequence) a waveform is produced for each of the stimulated subregions [74]. Figure 5 depicts an example of the localized areas that can be stimu-

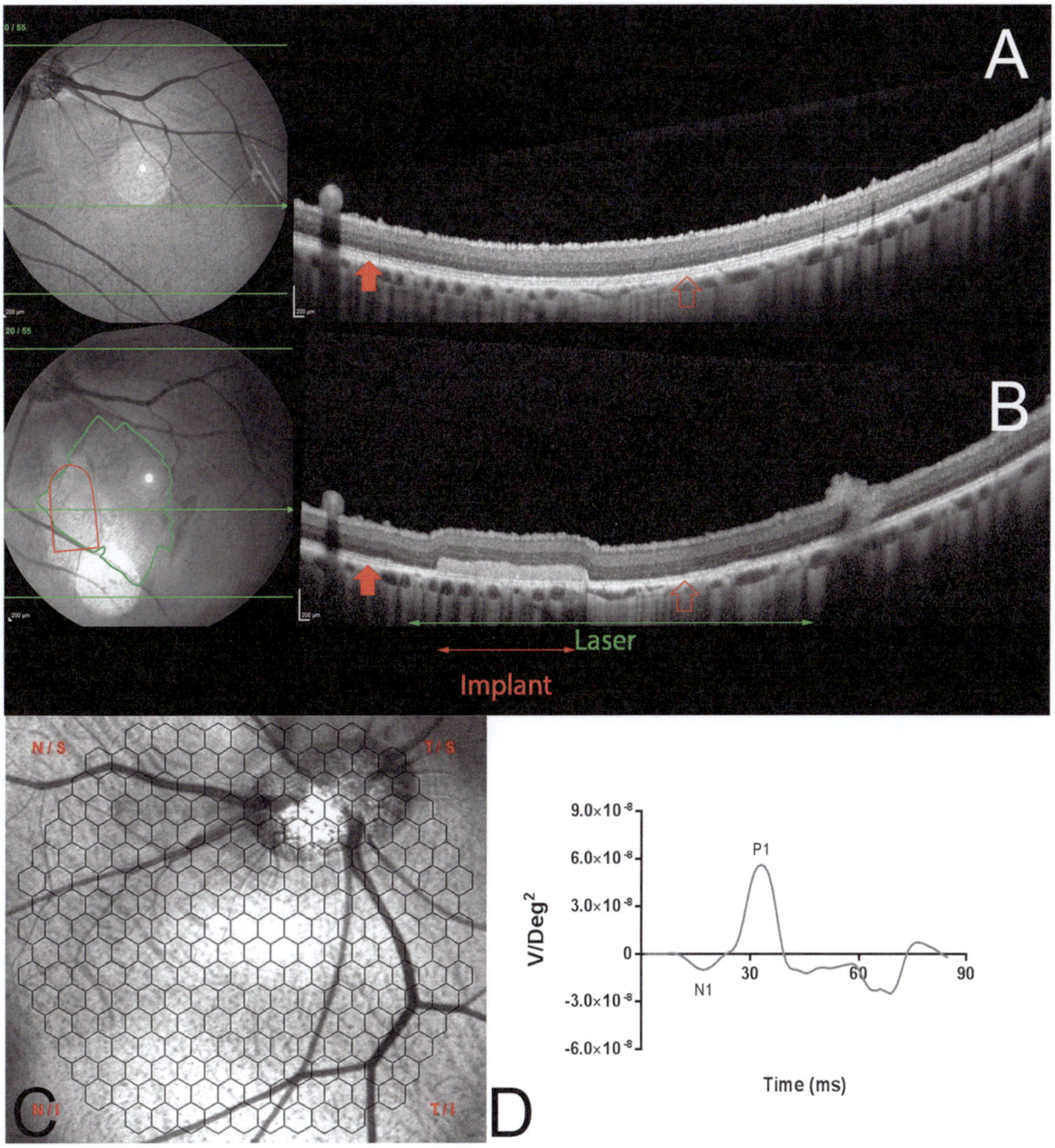

Fig. 5 Functional assessment of transplanted iPSC-derived RPE cells by OCT and mfERG. (**a** and **b**) Show the same section of retina (as depicted by the *green arrow* on the *left*) from a single pig prelaser (**a**) and postlaser approximately 10 days after the initial laser (1 week post-implant) (**b**). The implant containing iPSC-derived RPE cells is outlined in *red* and the lasered area in *green* on the *left*. The *red solid arrows* show healthy RPE with the *thin double line* represent the inner segments of the photoreceptors (the *upper band*) and the RPE layer (the *lower band*). The *red open arrow* shows laser-damaged RPE where only one band remains 10 days after the initial laser (**b**). (**c**) Fundus image of a pig eye that has not had laser ablation or RPE transplant. (**d**) An example of the wave form from a representative mfERG. *Y*-axis is the average voltage over a given area (one degree squared) and the *x*-axis shows the time since the visual stimulus

lated and the waveform generated by this technique. Our results show that mfERG is able to clearly show a difference between the lasered area and the healthy area of the same eye (not shown). mfERG responses clearly decrease in the lasered area as compared to the healthy area likely because of RPE ablation and secondary damage caused to the photoreceptors. Furthermore, transplantation of an empty scaffold in the subretinal space also decreases mfERG signals, likely because the empty graft interferes with photoreceptor-RPE interactions. However, as the scaffold degrades, the retina starts to recover and eventually fully recovers in 6–8 weeks when the scaffold degrades (data not shown). A successful transplantation of an RPE graft in the lasered area can also be distinguished from healthy area as mfERG responses begin to recover with regeneration of the photoreceptor cells. Overall, the combination of imaging and functional assessments allows the possibility of determining RPE transplant safety and efficacy in pig models of RPE injury.

5 Conclusion

RPE transplantation provides a potential therapy for the treatment of RPE-associated retinal degenerative diseases. Multiple different approaches are currently being pursued. Here we present the outline of an autologous approach that uses patient-specific iPSC-derived RPE grown on a biodegradable scaffold. We have developed a streamlined GMP-grade manufacturing process for AMD patient-specific iPSC-RPE. This RPE transplant was tested in a pig model of RPE injury to confirm its safety and efficacy. Our work suggests that a human iPSC-RPE transplant is able to rescue dying photoreceptors in this pig model of RPE injury. Autologous approach provides potentially fewer immune challenges as compared to allogeneic stem cell-derived RPE-based approaches. However, autologous iPS cell therapy provides several manufacturing challenges. The manufacturing process is fairly long and not currently cost effective. Ongoing efforts to develop HLA-matched iPS cell lines and to automate the manufacturing process will help make this therapy commercially successful.

References

1. Edelman JL, Miller SS. Epinephrine stimulates fluid absorption across bovine retinal pigment epithelium. Invest Ophthalmol Vis Sci. 1991;32:3033–40.
2. Maminishkis A, Miller SS. Experimental models for study of retinal pigment epithelial physiology and pathophysiology. J Vis Exp. 2010; doi:10.3791/2032.
3. Strauss O. The retinal pigment epithelium in visual function. Physiol Rev. 2005;85:845–81.
4. Tang PH, Kono M, Koutalos Y, Ablonczy Z, Crouch RK. New insights into retinoid metabolism and cycling within the retina. Prog Retin Eye Res. 2013;32:48–63.
5. Kevany BM, Palczewski K. Phagocytosis of retinal rod and cone photoreceptors. Physiology (Bethesda). 2010;25:8–15
6. Oakley 2nd B, Flaming DG, Brown KT. Effects of the rod receptor potential upon retinal extracellular potassium concentration. J Gen Physiol. 1979;74:713–37.
7. Oakley 2nd B, Miller SS, Steinberg RH. Effect of intracellular potassium upon the electrogenic pump of frog retinal pigment epithelium. J Membr Biol. 1978;44:281–307.
8. Holtkamp GM, Kijlstra A, Peek R, de Vos AF. Retinal pigment epithelium-immune system interactions: cytokine production and cytokine-induced changes. Prog Retin Eye Res. 2001;20:29–48.
9. Chacon-Camacho OF, Zenteno JC. Review and update on the molecular basis of Leber congenital amaurosis. World J Clin Cases. 2015;3:112–24.
10. Zajac-Pytrus HM, Pilecka A, Turno-Krecicka A, Adamiec-Mroczek J, Misiuk-Hojlo M. The dry form of Age-Related Macular Degeneration (AMD): the current concepts of pathogenesis and prospects for treatment. Adv Clin Exp Med. 2015;24:1099–104.
11. Ambati J, Fowler BJ. Mechanisms of age-related macular degeneration. Neuron. 2012;75:26–39.
12. Zarbin MA. Current concepts in the pathogenesis of age-related macular degeneration. Arch Ophthalmol. 2004;122:598–614.
13. Iroku-Malize T, Kirsch S. Eye conditions in older adults: age-related macular degeneration. FP Essent. 2016;445:24–8.
14. Chew EY. Nutrition effects on ocular diseases in the aging eye. Invest Ophthalmol Vis Sci. 2013;54:ORSF42–7.
15. Bird AC, Phillips RL, Hageman GS. Geographic atrophy: a histopathological assessment. JAMA Ophthalmol. 2014;132:338–45.
16. Gouras P, Lopez R. Transplantation of retinal epithelial cells. Invest Ophthalmol Vis Sci. 1989;30:1681–3.
17. Jha BS, Bharti K. Regenerating retinal pigment epithelial cells to cure blindness: a road towards personalized artificial tissue. Curr Stem Cell Rep. 2015;1:79–91.
18. Binder S, Krebs I, Hilgers RD, Abri A, Stolba U, Assadoulina A, Kellner L, Stanzel BV, Jahn C, Feichtinger H. Outcome of transplantation of autologous retinal pigment epithelium in age-related macular degeneration: a prospective trial. Invest Ophthalmol Vis Sci. 2004;45:4151–60.
19. Binder S, Stolba U, Krebs I, Kellner L, Jahn C, Feichtinger H, Povelka M, Frohner U, Kruger A, Hilgers RD, et al. Transplantation of autologous retinal pigment epithelium in eyes with foveal neovascularization resulting from age-related macular degeneration: a pilot study. Am J Ophthalmol. 2002;133:215–25.
20. Joussen AM, Heussen FM, Joeres S, Llacer H, Prinz B, Rohrschneider K, Maaijwee KJ, van Meurs J, Kirchhof B. Autologous translocation of the choroid and retinal pigment epithelium in age-related macular degeneration. Am J Ophthalmol. 2006;142:17–30.
21. Joussen AM, Joeres S, Fawzy N, Heussen FM, Llacer H, van Meurs JC, Kirchhof B. Autologous translocation of the choroid and retinal pigment epithelium in patients with geographic atrophy. Ophthalmology. 2007;114:551–60.
22. Carr AJ, Vugler AA, Hikita ST, Lawrence JM, Gias C, Chen LL, Buchholz DE, Ahmado A, Semo M, Smart MJ, et al. Protective effects of human iPS-derived retinal pigment epithelium cell transplantation in the retinal dystrophic rat. PLoS One. 2009;4:e8152.
23. Diniz B, Thomas P, Thomas B, Ribeiro R, Hu Y, Brant R, Ahuja A, Zhu D, Liu L, Koss M, et al. Subretinal implantation of retinal pigment epithelial cells derived from human embryonic stem cells: improved survival when implanted as a monolayer. Invest Ophthalmol Vis Sci. 2013;54:5087–96.
24. Hu Y, Liu L, Lu B, Zhu D, Ribeiro R, Diniz B, Thomas PB, Ahuja AK, Hinton DR, Tai YC, et al. A novel approach for subretinal implantation of ultrathin substrates containing stem cell-derived retinal pigment epithelium monolayer. Ophthalmic Res. 2012;48:186–91.
25. Lu B, Malcuit C, Wang S, Girman S, Francis P, Lemieux L, Lanza R, Lund R. Long-term safety and function of RPE from human embryonic stem cells in preclinical models of macular degeneration. Stem Cells. 2009;27:2126–35.
26. Schwartz SD, Hubschman JP, Heilwell G, Franco-Cardenas V, Pan CK, Ostrick RM, Mickunas E, Gay R, Klimanskaya I, Lanza R. Embryonic stem cell trials for macular degeneration: a preliminary report. Lancet. 2012;379:713–20.
27. Buchholz DE, Pennington BO, Croze RH, Hinman CR, Coffey PJ, Clegg DO. Rapid and efficient directed differentiation of human pluripotent stem cells into retinal pigmented epithelium. Stem Cells Transl Med. 2013;2:384–93.
28. Idelson M, Alper R, Obolensky A, Ben-Shushan E, Hemo I, Yachimovich-Cohen N, Khaner H, Smith Y, Wiser O, Gropp M, et al. Directed differentiation of human embryonic stem cells into functional retinal pigment epithelium cells. Cell Stem Cell. 2009;5:396–408.
29. Reh TA, Lamba D, Gust J. Directing human embryonic stem cells to a retinal fate. Methods Mol Biol. 2010;636:139–53.

30. Kamao H, Mandai M, Okamoto S, Sakai N, Suga A, Sugita S, Kiryu J, Takahashi M. Characterization of human induced pluripotent stem cell-derived retinal pigment epithelium cell sheets aiming for clinical application. Stem Cell Reports. 2014;2: 205–18.
31. Warren L, Ni Y, Wang J, Guo X. Feeder-free derivation of human induced pluripotent stem cells with messenger RNA. Sci Rep. 2012;2:657.
32. Sidney LE, Branch MJ, Dunphy SE, Dua HS, Hopkinson A. Concise review: evidence for CD34 as a common marker for diverse progenitors. Stem Cells. 2014;32:1380–9.
33. Silvestri F, Banavali S, Baccarani M, Preisler HD. The CD34 hemopoietic progenitor cell associated antigen: biology and clinical applications. Haematologica. 1992;77: 265–73.
34. Hayes M, Zavazava N. Strategies to generate induced pluripotent stem cells. Methods Mol Biol. 2013;1029:77–92.
35. Su RJ, Baylink DJ, Neises A, Kiroyan JB, Meng X, Payne KJ, Tschudy-Seney B, Duan Y, Appleby N, Kearns-Jonker M, et al. Efficient generation of integration-free ips cells from human adult peripheral blood using BCL-XL together with Yamanaka factors. PLoS One. 2013;8:e64496.
36. Munoz J, Shah N, Rezvani K, Hosing C, Bollard CM, Oran B, Olson A, Popat U, Molldrem J, McNiece IK, et al. Concise review: umbilical cord blood transplantation: past, present, and future. Stem Cells Transl Med. 2014;3:1435–43.
37. Solves P, Mirabet V, Perales A, Carbonell-Uberos F, Roig R. Banking strategies for improving the hematopoietic stem cell content of umbilical cord blood units for transplantation. Curr Stem Cell Res Ther. 2008;3:79–84.
38. Mack AA, Kroboth S, Rajesh D, Wang WB. Generation of induced pluripotent stem cells from CD34+ cells across blood drawn from multiple donors with non-integrating episomal vectors. PLoS One. 2011;6:e27956.
39. Klimanskaya I, Hipp J, Rezai KA, West M, Atala A, Lanza R. Derivation and comparative assessment of retinal pigment epithelium from human embryonic stem cells using transcriptomics. Cloning Stem Cells. 2004;6:217–45.
40. Fuhrmann S, Levine EM, Reh TA. Extraocular mesenchyme patterns the optic vesicle during early eye development in the embryonic chick. Development. 2000;127:4599–609.
41. Miyagishima KJ, Wan Q, Corneo B, Sharma R, Lotfi MR, Boles NC, Hua F, Maminishkis A, Zhang C, Blenkinsop T, et al. In pursuit of authenticity: induced pluripotent stem cell-derived retinal pigment epithelium for clinical applications. Stem Cells Transl Med. 2016;5(11):1562–74.
42. Hotaling NA, Khristov V, Wan Q, Sharma R, Jha BS, Lotfi M, Maminishkis A, Simon Jr CG, Bharti K. Nanofiber scaffold-based tissue-engineered retinal pigment epithelium to treat degenerative eye diseases. J Ocul Pharmacol Ther. 2016;32:272–85.
43. Booij JC, Baas DC, Beisekeeva J, Gorgels TG, Bergen AA. The dynamic nature of Bruch's membrane. Prog Retin Eye Res. 2010; 29:1–18.
44. Curcio CA, Johnson M. Chapter 20 – Structure, function, and pathology of Bruch's membrane. In: Sadda SR, Hinton DR, Schachat AP, Sadda SR, Wilkinson CP, Wiedemann P, Schachat AP, editors. A2 – Ryan, Stephen J. in Retina. 5th ed. London: W.B. Saunders; 2013. p. 465–81.
45. Tezel TH, Del Priore LV. Repopulation of different layers of host human Bruch's membrane by retinal pigment epithelial cell grafts. Invest Ophthalmol Vis Sci. 1999;40:767–74.
46. Tezel TH, Kaplan HJ, Del Priore LV. Fate of human retinal pigment epithelial cells seeded onto layers of human Bruch's membrane. Invest Ophthalmol Vis Sci. 1999;40:467–76.
47. Akrami H, Soheili ZS, Sadeghizadeh M, Khalooghi K, Ahmadieh H, Kanavi MR, Samiei S, Pakravesh J. Evaluation of RPE65, CRALBP, VEGF, CD68, and tyrosinase gene expression in human retinal pigment epithelial cells cultured on amniotic membrane. Biochem Genet. 2011;49:313–22.
48. Capeans C, Pineiro A, Pardo M, Sueiro-Lopez C, Blanco MJ, Dominguez F, Sanchez-Salorio M. Amniotic membrane as support for human retinal pigment epithelium (RPE) cell growth. Acta Ophthalmol Scand. 2003;81:271–7.
49. Ohno-Matsui K, Ichinose S, Nakahama K, Yoshida T, Kojima A, Mochizuki M, Morita I. The effects of amniotic membrane on retinal pigment epithelial cell differentiation. Mol Vis. 2005;11:1–10.
50. Singhal S, Vemuganti GK. Primary adult human retinal pigment epithelial cell cultures on human amniotic membranes. Indian J Ophthalmol. 2005;53:109–13.
51. Marei I, Chester A, Carubelli I, Prodromakis T, Trantidou T, Yacoub MH. Assessment of Parylene C thin films for heart valve tissue engineering. Tissue Eng Part A. 2015;21:2504–14.
52. Thompson CH, Zoratti MJ, Langhals NB, Purcell EK. Regenerative electrode interfaces for neural prostheses. Tissue Eng Part B Rev. 2016;22:125–35.
53. Lu B, Zhu D, Hinton D, Humayun MS, Tai YC. Mesh-supported submicron parylene-C membranes for culturing retinal pigment epithelial cells. Biomed Microdevices. 2012;14:659–67.
54. Sorkio A, Porter PJ, Juuti-Uusitalo K, Meenan BJ, Skottman H, Burke GA. Surface modified biodegradable electrospun membranes as a carrier for human embryonic stem cell-derived retinal pigment epithelial cells. Tissue Eng Part A. 2015;21:2301–14.
55. Sorkio AE, Vuorimaa-Laukkanen EP, Hakola HM, Liang H, Ujula TA, Valle-Delgado JJ, Osterberg M, Yliperttula ML, Skottman H. Biomimetic collagen I and IV double layer Langmuir-Schaefer films as microenvironment for human pluripotent stem cell derived retinal pigment epithelial cells. Biomaterials. 2015;51:257–69.
56. Goncalves S, Padrao J, Rodrigues IP, Silva JP, Sencadas V, Lanceros-Mendez S, Girao H, Dourado F, Rodrigues LR. Bacterial cellulose as a support for the growth of retinal pigment epithelium. Biomacromolecules. 2015;16:1341–51.
57. Warnke PH, Alamein M, Skabo S, Stephens S, Bourke R, Heiner P, Liu Q. Primordium of an artificial Bruch's membrane made of nanofibers for engineering of retinal pigment epithelium cell monolayers. Acta Biomater. 2013;9:9414–22.
58. McHugh KJ, Tao SL, Saint-Geniez M. Porous poly(epsilon-caprolactone) scaffolds for retinal pigment epithelium transplantation. Invest Ophthalmol Vis Sci. 2014;55:1754–62.
59. Thomson HA, Treharne AJ, Walker P, Grossel MC, Lotery AJ. Optimisation of polymer scaffolds for retinal pigment epithelium (RPE) cell transplantation. Br J Ophthalmol. 2011;95:563–8.
60. Liu Z, Yu N, Holz FG, Yang F, Stanzel BV. Enhancement of retinal pigment epithelial culture characteristics and subretinal space tolerance of scaffolds with 200 nm fiber topography. Biomaterials. 2014;35:2837–50.
61. Krishna Y, Sheridan C, Kent D, Kearns V, Grierson I, Williams R. Expanded polytetrafluoroethylene as a substrate for retinal pigment epithelial cell growth and transplantation in age-related macular degeneration. Br J Ophthalmol. 2011;95:569–73.
62. Thieltges F, Stanzel BV, Liu Z, Holz FG. A nanofibrillar surface promotes superior growth characteristics in cultured human retinal pigment epithelium. Ophthalmic Res. 2011;46:133–40.
63. Middleton S. Porcine ophthalmology. Vet Clin North Am Food Anim Pract. 2010;26:557–72.
64. Sanchez I, Martin R, Ussa F, Fernandez-Bueno I. The parameters of the porcine eyeball. Graefes Arch Clin Exp Ophthalmol. 2011;249: 475–82.
65. Roider J, Michaud NA, Flotte TJ, Birngruber R. Response of the retinal pigment epithelium to selective photocoagulation. Arch Ophthalmol. 1992;110:1786–92.
66. Ruskovic D, Boulton M, Ulbig MW, Watt M, McHugh DA, Marshall J. Effect of micropulsed diode laser on human RPE in vivo and in vitro. Invest Ophthalmol Vis Sci. 1997;38:3483.

67. Sivaprasad S, Elagouz M, McHugh D, Shona O, Dorin G. Micropulsed diode laser therapy: evolution and clinical applications. Surv Ophthalmol. 2010;55:516–30.
68. Santos-Carreras L, Hagen M, Gassert R, Bleuler H. Survey on surgical instrument handle design: ergonomics and acceptance. Surg Innov. 2012;19:50–9.
69. Furnish G. Surgical instrument handle (Google Patents). (1996).
70. Ryan SJ, Schachat AP, Wilkinson CP, Hinton DR, Sadda SVR, Wiedemann P. Retina. London: W.B. Saunders; 2012.
71. Hassenstein A, Meyer CH. Clinical use and research applications of Heidelberg retinal angiography and spectral-domain optical coherence tomography – a review. Clin Exp Ophthalmol. 2009;37:130–43.
72. Spaide RF, Curcio CA. Anatomical correlates to the bands seen in the outer retina by optical coherence tomography: literature review and model. Retina. 2011;31:1609–19.
73. Hood DC, Bach M, Brigell M, Keating D, Kondo M, Lyons JS, Marmor MF, McCulloch DL, Palmowski-Wolfe AM, International Society for Clinical Electrophysiology of Vision. ISCEV standard for clinical multifocal electroretinography (mfERG) (2011 edition). Doc Ophthalmol. 2012;124:1–13.
74. Poloschek CM, Sutter EE. The fine structure of multifocal ERG topographies. J Vis. 2002;2:577–87.
75. McCulloch DL, Marmor MF, Brigell MG, Hamilton R, Holder GE, Tzekov R, Bach M. ISCEV Standard for full-field clinical electroretinography (2015 update). Doc Ophthalmol. 2015;130:1–12.

Scaffolds for Cell Transplantation

Meena S. George, Hossein Nazari, Debbie Mitra, Dennis Clegg, David R. Hinton, and Mark S. Humayun

1 Introduction and Background

Many degenerative retinal diseases, whether they have genetic or environmental causes, result in damage to the RPE, Bruch's membrane, and/or the choriocapillaris, ultimately resulting in photoreceptor dysfunction or death and vision loss. Replacement of the damaged RPE or photoreceptors with healthy cells is being investigated for patients with AMD, retinitis pigmentosa, and other related conditions [1]. Stem cells represent an unlimited source for such cell replacement therapies. Transplanted stem cell-derived-RPE (SC-RPE) has been recently shown to interface with the host retina and rescue dying photoreceptors, both in animal models and possibly in human studies [2–4]. Also, great progress has been made in the generation of stem cell-derived-neural retinal cells, which are getting closer to clinical trials [1, 5, 6].

Vision is highly dependent on the stratified organization of photoreceptors and inner retinal neural elements. In addition, in order for photoreceptors to maintain their function, photoreceptor outer segments must engage with the apical microvilli of underlying RPE cells. The RPE is a monolayer of polarized, pigmented cells with an apical domain in contact with photoreceptors and a basal side that attaches to the Bruch's membrane. RPE function is dependent on the RPE's polarized architecture and thus it is likely important for RPE replacement to reestablish functional cell orientation.

Two main approaches to deliver SC-RPE cells to the subretinal space include injection of a suspension of SC-RPE cells under the macula and subretinal implantation of a single-layer SC-RPE on a scaffold. Subretinal delivery of SC-RPE cells as a suspension has been promising in clinical trials, establishing the safety of delivering such cells under the retina and demonstrating promising results showing stabilization or improvement of vision in some patients [2]. However, this method of SC-RPE transplantation and its outcomes may be limited by the irregular distribution and clumping of the cells in the subretinal space, the inability of these cells to establish a uniform monolayer, the possible reflux of the implanted cells into the vitreous cavity, and the compromised life span of the transplanted cells [1]. In this chapter, we review the current status of scaffolds for transplantation of retinal cells, with an emphasis on clinical translation. There have been many excellent reviews in this area recently [4–8].

Meena S. George and Hossein Nazari contributed equally to this work.

M.S. George, M.D., Ph.D.
Keck Medicine of USC, Ophthalmology, USC Roski Eye Institute, 1450 San Pablo Street, 4th floor, Los Angeles, CA 90033, USA
e-mail: meena.george@gmail.com

H. Nazari, M.D.
University of Texas Medical Branch, Ophthalmology, 700 University Blvd, Galveston, Texas 77555, USA
e-mail: honazari@utmb.edu

D. Mitra, Ph.D.
Keck Medicine of USC, Ophthalmology, USC Roski Eye Institute, USC Institute for Biomedical Therapeutics, 1450 San Pablo Street, Room 6525, Los Angeles, CA 90033, USA
e-mail: debbiemi@usc.edu

D. Clegg, Ph.D.
University of California, Santa Barbara, Molecular, Cellular and Developmental biology, Center for Stem Cell Biology and Engineering, University of California, Santa Barbara, CA 93105, USA
e-mail: clegg@lifesci.ucsb.edu

D.R. Hinton, M.D.
Keck Medicine of USC, Pathology, Ophthalmology, USC Roski Eye Institute, 2011 Zonal Ave, HMR 209, Los Angeles, CA 90089, USA
e-mail: dhinton@usc.edu

M.S. Humayun, M.D., Ph.D. (✉)
Keck Medicine of USC, Ophthalmology, USC Roski Eye Institute, USC Institute for Biomedical Therapeutics, 1450 San Pablo Street, Room 6534, Los Angeles, CA 90033, USA
e-mail: humayun@med.usc.edu

S.D. Schwartz et al. (eds.), *Cellular Therapies for Retinal Disease*, DOI 10.1007/978-3-319-49479-1_4

2 Scientific Basis for Intervention

2.1 RPE Structure and Function

The RPE is a seemingly simple epithelial monolayer, but it is actually highly organized in ways that are critical for its function. RPE cells establish the outer blood-retinal barrier; they actively transport small molecules such as oxygen, water, and ions, as well as macromolecules to and from the retina and choriocapillaris. Polarization of the RPE cells into their apical-basal domains is necessary for RPE function. Such polarization requires the RPE monolayer to establish zonulae occludentes (tight junctions) between adjacent cells. Also, RPE cells phagocytose shed photoreceptor outer segments on a diurnal cycle, which is critical for photoreceptor survival. RPE cells play a crucial role in the retinoid cycle by regenerating 11-*cis*-retinal, a cofactor required for light signal transduction. The RPE layer also secretes key growth factors both apically and basally, and its pigmented melanosomes absorb stray light [9]. Without a functional RPE monolayer, the adjoining photoreceptors rapidly perish, and vision loss results.

2.2 Biophysical Characteristic of Bruch's Membrane

Bruch's membrane (BM) is a unique acellular elastin- and collagen-rich extracellular matrix sheet situated between choriocapillaris and RPE. This 2–4 μm thick multilayered matrix consists of (1) the basal lamina of RPE, (2) an inner collagenous zone, (3) a thick porous band of elastic fibers, (4) an outer collagenous zone, and (5) the basal lamina of the choriocapillaris. BM is highly permeable to small molecules and acts as part of the regulatory system for the reciprocal movement of biomolecules, nutrients, oxygen, fluids, and metabolic waste products between the retina and the systemic circulation [10]. Macromolecules such as retinal binding protein, transferrin, and ceruloplasmin diffuse from the choriocapillaris across BM and are then transported by the RPE to the neural retina. Similarly, photoreceptor and RPE metabolic by-products are transferred through BM and reenter the systemic circulation via the choriocapillaris.

Increasing evidence suggests that the molecular, structural, and functional properties of BM can be influenced by aging, environmental factors, and genetics [10]. In general, with age, lipid products accumulate progressively in BM and become calcified. Hydraulic conductivity of BM decreases leading to an exponential decrease in the capacity of BM to facilitate the exchange of molecules between the choriocapillaris, RPE, and neural retina [11–15]. When seeded with healthy RPE cells, such aged BM does not readily support RPE cell differentiation and survival. In addition, localized RPE defects, cell death, and abnormal cellular morphology were frequently seen on aged Bruch's membrane seeded with RPE cells [16]. In contrast, SC-RPE bioimplants composed of a "prefabricated" monolayer of RPE grown on a natural or synthetic basement membrane-like structure may better imitate native RPE and survive, integrate, and ultimately preserve photoreceptors.

3 Scaffolds for SC-RPE Monolayer Bioimplants

It has been shown that SC-RPE cells implanted in rats as a polarized monolayer on a scaffold survive longer than SC-RPE injected as a suspension [1, 3]. A supporting scaffold allows for cell attachment, differentiation, polarization, survival, and maintenance of function [17–19]. Ideally, this scaffold should mimic Bruch's membrane in permeability and be durable enough to allow for manipulation and handling during cell culture and subsequent surgical implantation.

Various natural or synthetic, either biodegradable or non-biodegradable, materials (highlighted in Table 1) have been utilized as scaffolds. While natural materials may seem appealing, they can have drawbacks, such as nonpredictability, limited availability, and risks of disease transmission that may limit their acceptance for clinical use. Synthetic materials also have both positive and negative characteristics.

3.1 Natural Materials

Natural materials such as donor Descemet's membrane, BM, lens capsule, as well as cross-linked and non-cross-linked collagen [30, 31], gelatin [32], and fibrinogen [33, 34] are generally biocompatible. Their biomechanical properties mimic native BM and make them suitable for the subretinal environment. However, the thickness, stiffness, and flexibility of these materials are less predictable. In addition, these materials may be allogenic, resulting in unwanted immune-reactions from both the natural material and the stem cell-derived RPE.

3.1.1 Cross-Linked Gelatin

Gelatin is biodegradable and is used as an embedding matrix to protect cells, in tissue engineering, and in cell replacement therapies [21, 35]. Non-cross-linked gelatin dissolves rapidly in the aqueous environment and is thus not suitable as a basement membrane-like substrate. In contrast, noninflammatory formulations of cross-linked gelatin have shown good structural solidity and biologic compatibility for ocular

Table 1 Synthetic substrates for SC-RPE and neural retinal cell scaffolds

Substrate material	Biodegradability	Potential advantage	Potential disadvantage
Parylene [20]	Nonbiodegradable	Good mechanical strength, biostability, chemical inertness, supports RPE growth and polarization. Submicron ultrathin areas are permeable to water, nutrients, and RPE waste products. Little to no inflammatory response	Need to be micromachined precisely to allow diffusion
Polyester [21]	Nonbiodegradable	Biostable with good mechanical strength and machinability	Is not permeable and needs micropores for water, nutrient, and waste transport. Cells may clog pores and compromise transport
Ultrathin poly(methyl methacrylate) (PMMA) [22]	Nonbiodegradable	Contain pores that help to retain adherent RPE to a considerably greater extent than unmachined versions during the transplantation process, and these can serve as a biocompatible substrate for cell delivery in vivo	Pores are required to ensure permeability to nutrients and waste material. However, cells may migrate and grow into the pores making the scaffold surface uneven
Poly(glycerol sebacate) (PGS) [23]	Biodegradable	Supports neural retinal cell survival and expresses mature retinal proteins in host retina	No long-term in vivo data available. Potential inflammation as substance degrades
Poly(ε-caprolactone) (PCL) nanowire [24]	Biodegradable	Supports neural retinal cell survival and expresses mature retinal proteins in a host retina	Variable proliferation and pigmentation pattern observed. No long-term data available
Hyaluronic acid (HA)-based hydrogels [25]	Biodegradable	Degradable material for neural retinal cell transplants	No long-term data available
Poly(L-lactide-*co*-ε-caprolactone) (PLCL) [26]	Biodegradable	Degradable substances that may allow better transport of water and molecules. Proposed for delivering RPE monolayers to the subretinal space	Optimal fiber diameter and surface form is not known. No long-term data available
Poly(D,L-lactic-*co*-glycolic acid) (PLGA)/poly(L-lactic acid) (PLLA) polymer [27–29]	Biodegradable	Degradable substances that may allow better transport of water and molecules across the RPE layer. Proposed as a better means for delivering neural retinal cells to the subretinal space	Compared to retinal tissue, PLGA/PLLA polymers are harder and more brittle limiting the ability to manipulate the scaffolds. Mild local tissue reactions may occur at the site of PLGA implantation

use [36]. In a preparation of gelatin admixed with a specific spider protein and polycaprolactone (PCL), named RWSF/PCL/Gt membrane, RPE cells grew and attained appropriate polarity upon implantation in rabbit eyes. The RPE cells also expressed a full range of mature RPE genes without induction of inflammation in short-term studies [37]. No long-term observational study of gelatin scaffolds for RPE cells has been reported yet.

3.1.2 Collagen Film

Collagen films, prepared as thin as 2.4 μm thick sheets, have been tested for RPE transplantation. These thin collagen film membranes allowed stable RPE monolayer development and permitted the flow of nutrients and waste to and from the RPE [38]. Although there have been concerns over inherent inability of cross-linked collagen to degrade [31], no deleterious effect on overlying RPE was observed in preclinical studies [38]. Material engineering hurdles may limit feasibility of using collagen films. For example, the method described for preparation of thin collagen films involves fabrication of collagen film over a Teflon layer that needs to be removed before implantation and this adds a significant amount of complexity to the process. A fairly similar product, cross-linked fibrinogen sheets, did not properly support RPE cell cultures in one study [33]. Fetal RPE cells lost their monolayer status and formed microspheres around these fibrinogen sheets [33]. In an alternate method, a collagen gel was used as a temporary scaffold to facilitate seeding of induced pluripotent stem cells (iPSC)-RPE as a confluent layer [39]. Following collagenase treatment, the collagen can be removed to afford an iPSC-RPE cell sheet. Challenges in titrating the enzymatic digestion and damage to the cells, scaling up the product, surgical handling and delivery of the iPSC-RPE cell sheet, and implanting large sheets could be potential limitations of this approach. Furthermore, this method did not address the concerns of implanting cells on aged BM with altered cell adhesion and diffusion characteristics as discussed previously. The first transplant was successfully conducted in 2014 and we await further results from this effort (Clinical trial ID: **JPRN-UMIN000011929**) [40].

3.1.3 Hyaluronic Acid (HA)-Based Hydrogels

Hyaluronic acid (HA)-based hydrogels are used to encapsulate neural retinal cells for delivery as an injectable engineered tissue in the subretinal space [25]. HA may also serve as a feeder layer for stem cell cultures. Mechanical and biologic parameters including three-dimensional architecture, thickness, and viscosity and degradation of HA can be easily controlled. The viscous HA gel containing neural retinal cells causes very little disruption to the host retinal architecture and degrades completely in 3 weeks leaving an even distribution of neural retinal cells in the subretinal space [25]. The system would not support RPE monolayer growth and implantation, however, because of its unique viscous characteristics.

3.1.4 Lens Capsule and Other Natural Membranes

Iris pigmented epithelium and RPE cells cultured on bovine, porcine, and human anterior lens capsule have been shown to exhibit epithelial characteristics, such as formation of monolayers with intercellular tight junctions [41–44]. However, availability and risk of infection transmission among other reasons have limited their clinical use.

3.2 Synthetic Materials

Synthetic polymers are advantageous over natural materials as a scaffold substrate as they are micromachinable and have predictable mechanical and biological properties. In addition, some synthetic biodegradable substances can be designed to degrade in a desired time course [45]. Biodegradable polymers, poly-L-lactic (PLLA), poly(L-lactide-*co*-ε-caprolactone) (PLCL), and poly-DL-lactic-*co*-glycolic acid (85:15) (PLGA) (both of molecular weight 105 kD), are commonly studied for RPE monolayer culture and implantation in animal models [26, 28, 45–48]. RPE cells have been shown to maintain their differentiated cell phenotype and function on PLLA, PLCL, and PLGA sheets (Fig. 1) [26, 28]. Human stem cell-derived RPE and photoreceptor cells grow effectively on thin biodegradable PLCL and PLGA films [26, 49]. The degradation rate of PLLA and PLGA can be manipulated by changing the molecular weight and the ratio of lactic to glycolic units in order to attain an appropriate degradation timescale. Parylene-C, discussed below, is a nonbiodegradable substance that is being used as scaffold in a SC-RPE monolayer bioimplant trial. In a unique approach, nonscaffold RPE sheets are obtained when RPE cells are harvested through nonenzymatic methods using a temperature-responsive culture dish that is comprised of a poly(*N*-isopropylacrylamide) graft. Through reduction of temperature from 37 to 20 °C, the culture surface is modified from hydrophilic to hydrophobic giving the ability of the RPE sheet to detach thermally from the polymer graft surface. The in vitro and in vivo studies gave promising results [50–53].

3.2.1 Parylene-C

Parylene-C has been extensively used in medical devices because of its machinability and biocompatibility. Micromachined ultrathin parylene-C sheets have been shown to serve as an excellent artificial BM for SC-RPE cells. The permeability of ultrathin parylene-C membranes for various submicron thicknesses was evaluated by Tai and Humayun's group [54]. They showed that parylene-C membranes thinner than 0.3 μm possess similar molecular weight exclusion limits and nutrient diffusion flux to that of healthy human BM. This conclusion was further validated by in vitro perfusion cell viability tests. The group designed a mesh-supported submicron parylene membrane (MSPM) to provide sufficient

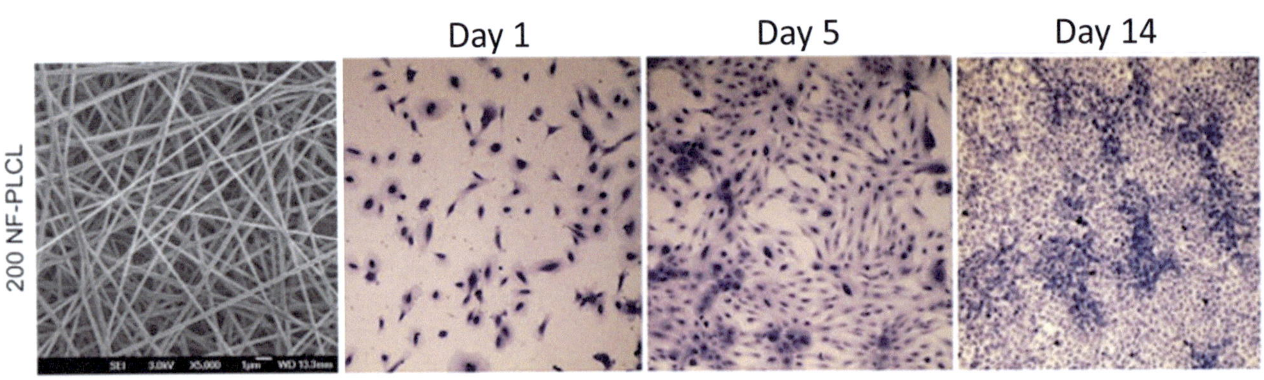

Fig. 1 Scanning electron microscopy and light microscopy images of nanofibrous poly(L-lactide-*co*-ε-caprolactone) (NF-PLCL) substrates show the topography of substrates with a random fibrous morphology and RPE cells seeded on the membrane (scale bar indicates 1 μm). Fiber diameter is 200 nm. Human fetal RPE cells attached and proliferated on PLCL substrates over 14 days (reproduced with permission from Liu Z, Yu N, Holz FG, Yang F, Stanzel BV. Enhancement of retinal pigment epithelial culture characteristics and subretinal space tolerance of scaffolds with 200 nm fiber topography. Biomaterials. 2014;35(9):2837–50)

mechanical support because of the difficulty handling a submicron membrane during production, cell culture, and surgical transplantation (Fig. 2) [54]. An MSPM covered by a monolayer of SC-RPE cells was then implanted in the subretinal space of rat models (Fig. 3) [20, 55]. The implanted SC-RPE cell monolayer was found to be intact with less than 2% cell loss from the substrate up to 6 months postimplantation [3, 55]. A clinical trial led by Regenerative Patch Technologies at the University of Southern California and sponsored by California Institute for Regenerative Medicine (CIRM) is currently testing the implantation of SC-RPE monolayers cultured on such membranes in human subjects (NCT02590692). This trial has just started enrolling patients and the first implant was recently placed in the first human subject [56]. Implantation has been successful and proper placement of the implant was verified via optical coherence tomography (OCT) imaging on patient follow-up. Short- and long-term outcomes over a larger cohort of patients are still needed in order to establish the safety and efficacy of this approach.

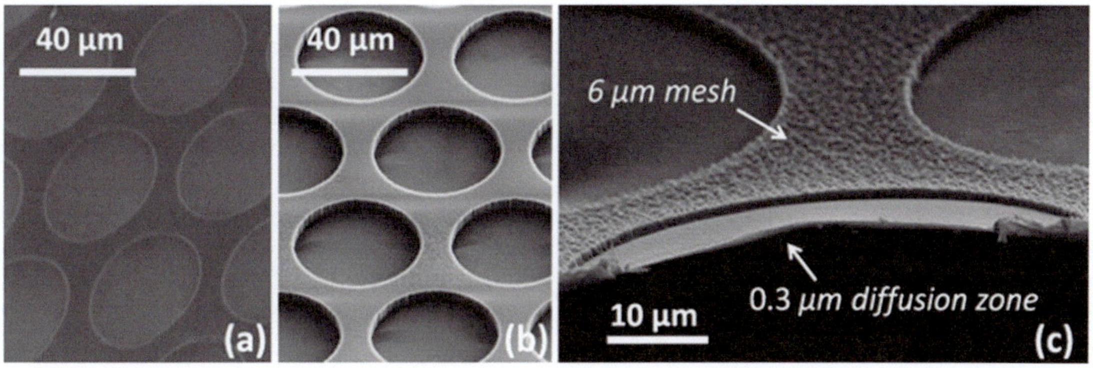

Fig. 2 (**a**) The front side, (**b**) the backside, and (**c**) the cross section of a microfabricated 0.30 µm ultrathin parylene-C as seen in SEM images (adapted with permission From Lu B, Zhu D, Hinton D, Humayun MS, Tai YC. Mesh-supported submicron parylene-C membranes for culturing retinal pigment epithelial cells. Biomed Microdevices. 2012;14(4): 659–67)

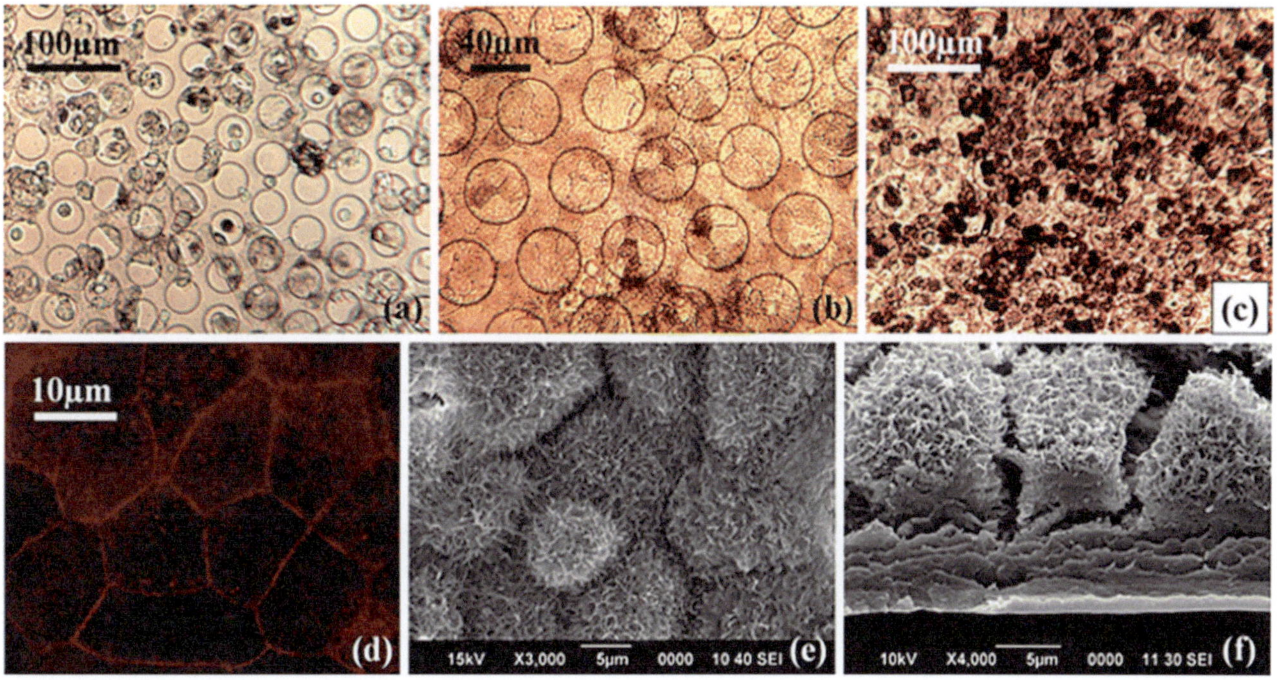

Fig. 3 Human embryonic stem cell derived-RPE cell culture on ultrathin parylene-C membrane. (**a**) 1 day, (**b**) 1 week, and (**c**) 4 weeks after seeding. Cells started to become pigmented after 4 weeks. (**d**) The tight junctions among cells and their hexagonal shapes confirmed with anti-Z0-1 staining. (**e**, **f**) The top and cross-sectional SEM images show the formation of microvilli on well-polarized cells (adapted with permission From Lu B, Zhu D, Hinton D, Humayun MS, Tai YC. Mesh-supported submicron parylene-C membranes for culturing retinal pigment epithelial cells. Biomed Microdevices. 2012;14(4):659–67)

3.2.2 Polyester

Polyester is extensively utilized in cell culture labware production. Polyester surfaces support cell growth and maintain a stable chemical composition over a long period of time after implantation in the body. Subretinal implantation of *blank* polyester membranes as proof of concept was performed successfully in rabbits and was atraumatic; as expected, these blank membranes caused significant toxicity to the overlying retinal cells. RPE cell monolayers on porous polyester membranes have also been implanted in the subretinal space of rabbits [21]. However, clogging of the polyester membrane pores with cell debris after 4–6 weeks of implantation and potential inflammatory response to material will need to be addressed [21]. In another model, SC-RPE on a porous polyester scaffold using both human embryonic-SC-RPE and induced pluripotent-RPE have been studied [57, 58]. The first transplant in a human occurred in 2015 with preliminary results indicating survival of RPE cells [59].

3.2.3 Methyl Methacrylate (PMMA)

Nonporous or micromachined porous PMMA scaffolds (6 μm thickness and 11 μm diameter) have been used for implantation of neural retinal cells in the subretinal space of rodent models [22, 60]. Porous PMMA membranes demonstrated enhanced neural retinal cell retention and allowed for greater process outgrowth and cell migration into the host retinal layers [22]. These results suggest that porous ultrathin PMMA membranes can serve as a biocompatible scaffold for optimum differentiation, survival, and delivery of retinal progenitor cells.

3.2.4 Poly ε-Caprolactone (PCL)

Polycaprolactone/PCL is a biodegeradable synthetic material frequently used in cell delivery studies. Subretinal implantation of mouse neural retinal cells cultured on a laminin-coated nanowire PCL scaffold has exhibited survival, differentiation, and migration of these cells into the retina. Neural retinal cells expressed mature retinal proteins in response to interactions with the nanowire scaffolds [24]. This may be a promising technology for human use if additional short- and long-term results from animal studies continue to support such outcomes.

3.2.5 Poly(D,L-Lactic-*co*-Glycolic Acid) (PLGA), Poly(L-Lactic Acid) (PLLA), and Poly(L-Lactide-*co*-ε-Caprolactone) (PLCL)

Biodegradable polymers, poly-L-lactic (PLLA), poly(L-lactide-*co*-ε-caprolactone) (PLCL), and poly-DL-lactic-*co*-glycolic acid (85:15) (PLGA), have been studied for RPE monolayer culture and implantation in animal models [26, 28, 45–48]. Human SC-RPE and photoreceptor cells grew effectively on thin biodegradable PCLC and PLGA films [26, 49]. RPE cells better maintained their differentiated cell phenotype and function on PLLA, PLCL, and PLGA sheets (Fig. 1) [26, 28]. Also, scaffolds made of PLCL 200 nm nanofibers enhanced RPE growth (more pigmented cells with higher density of hexagonal cells), improved subretinal implantation, and showed good subretinal biocompatibility [26]. In addition, PLGA/PLLA composite membranes promote neural retinal cell differentiation to semifunctional photoreceptors with elaborate cellular processes and good alignment with degenerating retinal explants [17, 27].

A potential advantage for these materials includes the ability to manipulate the PLLA and PLGA degradation rate by changing the molecular weight and the ratio of lactic to glycolic units. However, PLGA/PLLA membranes share the same limitations with other biodegradable membranes for subretinal cell delivery, including their mechanical properties and thickness. For example, the PLGA/PLLA blend has been shown to be hard and brittle in comparison to retinal tissue limiting the ability to manipulate these scaffolds, which in turn may increase the likelihood of tissue injury during and after surgical implantation [60].

3.2.6 Poly Glycerol Sebacate (PGS)

PGS is a biodegradable substance that is tested for the delivery of neural retinal cells to the subretinal space in rodent and porcine models. PGS supported enhanced cell differentiation and survival upon transplantation [60]. A big advantage for PGS membranes is their slower degradation compared to PLGA membranes. Slower rate of mechanical strength loss after cell seeding was associated with decreased fragility of the scaffold at the time of implantation [60].

3.3 Chemical Treatment and Surface Modification

Bioengineering techniques have been employed to enhance cell adhesion and proliferation within or over the surface of the scaffolds tested for RPE and neural retinal cell transplantation. Surface topology of the scaffold is vital for cell adhesion and permeability. Ultrathin mesh-like parylene membranes with areas 0.3 μm thick are a unique design to enhance permeability of the membrane. PMMA membranes are pored to allow water, oxygen, nutrients, and waste transport across the scaffold. However, as stated before, cells are shown to grow within the pores with possible compromised permeability of the membrane. Surface treatment of the scaffold with poly(hydroxyl-butyrate-*co*-hydroxy) valerate, air plasma treatment of polyurethane membranes, and ammonia gas plasma treatment of expanded polytetrafluoroethylene (ePTFE) change the substrate's surface to make it more hydrophilic and improve cell attachment [61–63]. Also, laminin coating is frequently used to optimize cell survival. Selected surface modifications are highlighted in Table 2.

Table 2 Selected surface modifications to scaffolds

Surface modification during the microfabrication process	Advantage
Ultrathin mesh-like parylene membrane [20]	Provides stability to the membrane while maintaining ultrathin areas for sufficient permeability to water and nutrients
Oxygen plasma treatment of the scaffold surface [61]	Renders a more hydrophilic surface
Loading PLGA polymer surface with preactivated, MMP2 molecules [64]	Stimulates removal of the inhibitory extracellular matrix and enhances cell integration and repopulation
Micropatterned surfaces [47]	Provides improved control of RPE cell morphology
PLGA scaffolds with fiber topography [26]	Enhances subretinal biocompatibility
Laminin treatment [23]	Enhances cell survival
Small intestinal submucosa (SIS)-PLGA hybrid films [65]	Offers higher cell survival rates and stronger RPE phenotype expression compared to pure PLGA films

4 Translational Strategies

Subretinal implantation of stem cell-derived RPE is being tested for clinical use in multiple trials. Multiple lines of evidence support improved survival and functional outcomes for SC-RPE monolayers delivered on an artificial basement membrane compared to SC-RPE cells delivered as a suspension and seeded on recipient BM. Subretinal cell suspension injection is technically less challenging and thus may provide an increased likelihood for delivery of a targeted number of stem cells to the subretinal space. In addition, given minimal surgical trauma during implantation, the method confers a lower risk of inducing local inflammation, which may jeopardize the survival of implanted cells. However, in contrast to RPE monolayers on an artificial scaffold, SC-RPE cells delivered as a cell suspension have not been proven to create a polarized monolayer in the subretinal space [2]. It is also possible that SC-RPE cells in suspension may form multilayered clumps of cells that may damage the overlying retina. Furthermore, reflux of the implanted cells from the injection site is theoretically possible and may compromise the ultimate number of implanted cells and increase the risk of epiretinal membrane formation and proliferative vitreoretinopathy. In addition, SC-RPE cells in suspension show lower tolerance to oxidative stress compared to those in monolayers [66]. Finally, a major concern with transplantation of SC-RPE cells in suspension is that aged BM of patients with AMD may not adequately support the adhesion, survival, and function of transplanted SC-RPE cells [16]. As a result, the survival and long-term function of SC-RPE transplants delivered to the subretinal space via suspension are not clearly known [67]. Preimplantation BM debridement and overexpression of integrin family molecules in SC-RPE have been suggested to improve the binding and proliferation of RPE cells on the recipient's BM [14].

As mentioned earlier, implanting SC-RPE monolayers, in contrast to suspension, necessitates a supporting scaffold. While multiple biodegradable and nonbiodegradable substances with or without surface modifications have been tested in animal studies, RPE sheets with no scaffold [39], the ultrathin polyester substrate [57], and the parylene C membranes [52] are the only substances being used in a clinical trial to test a stem cell-derived tissue graft in the eye.

4.1 Immune Rejection Concerns

Immune rejection of stem cell transplants is a major clinical concern. A biologically derived scaffold upon which stem cell-derived RPE monolayers are grown and delivered into the subretinal space is a potential source of inflammation. For example, increased macrophage activity in response to the presence of an artificial material in the subretinal space could increase the likelihood of rejection of the co-grafted foreign cells [23]. In contrast, parylene-C, the scaffold currently being used in human subjects, is an immunologically inert biocompatible platform that has been extensively used in medical biodevices with no associated immune reactivity [54, 55]. This makes parylene strategically advantageous over other materials for a monolayer scaffold implant. However, as other biodegradable and nonbiodegradable scaffolds are being developed, a major focus should be on their inability to stimulate the recipient's immune system.

4.2 Surgical Issues

Surgical implantation of a stem cell-derived RPE monolayer scaffold is much more challenging technically as compared to delivery via injection as a suspension. With a pars plana vitrectomy approach, one major surgical issue has been the insertion of the transplant scaffold into the subretinal space. The transplant needs to be implanted with the correct polarity such that the scaffold lies over the damaged BM and the SC-RPE cells face the outer retina. In addition, if larger patches of cells or populations need to be implanted in the future, the sclerotomy wound needs to be manageable in size so that appropriate intraocular pressure is maintained throughout the surgery. As most patients with AMD and geographic atrophy tend to be over the age of 65, a significant drop in intraocular pressure can predispose them to choroidal effusions or hemorrhage.

To address these issues, a stem-cell scaffold injector has been developed by Regenerative Patch Technologies and is manufactured by Synergetics. Similar to the concepts employed in intraocular lens injectors, the stem-cell scaffold injector allows the scaffold patch to be retracted and loaded into the injector. The scaffold transplant patch enters "like a taco" inside the injector without folding over on itself. The patch is loaded under the microscope so that correct orientation and polarity of the cells are maintained. Like intraocular lens injectors, the scaffold patch injector has a cylindrical bore and an easily distinguishable orientation. The instrument can be inserted through a small sclerotomy and allows the transplant to be injected through a retinotomy into the subretinal space. The transplant scaffold unfolds during the injection process, similar to the unfolding of the intraocular lens. A small tab allows for manipulation of the patch once beneath the retina. This surgical instrument has greatly facilitated the ability of these scaffold patches to be inserted beneath the retina, improving intraocular fluidics and pressure control during surgery. The outcomes of this SC-RPE cell scaffold transplant and the surgical technique are currently being assessed in a clinical trial described next.

5 Clinical Trial: The California Project to Cure Blindness

The Phase I/IIa clinical trial, funded by California Institute for Regenerative Medicine and sponsored by Regenerative Patch Technologies, LLC, is designed to assess the feasibility of delivery and safety of human embryonic stem cell-derived RPE cells on a parylene membrane (CPCB-RPE1) in patients with advanced, dry age-related macular degeneration [52]. The primary objective of the study is to test the safety and tolerability of CPCB-RPE1 during and after subretinal implantation in patients with geographic atrophy with evidence of involvement of the central fovea. The frequency and severity of treatment-related adverse events are being assessed, including comparing procedure and immunosuppression-related adverse events in the implanted eye to those experienced in the non-treated eye. The secondary objective is to assess visual acuity, visual field, and retinal function after CPCB-RPE1 implantation. Visual acuity, visual field, and multifocal ERG measurements will be compared in the treated eye versus baseline and versus the non-treated eye [52]. So far, one case has been attempted and successfully completed with verification of proper placement of the scaffold and implanted cells via OCT imaging on postoperative patient follow-up.

6 Summary and Future Directions

Cell replacement therapy using a scaffold to support cell growth, maturation, and implantation is in its infancy and showing early promise. Preclinical and animal studies suggest superior survival and function of SC-RPE cells in a monolayer grown on a scaffold. However, efforts to find the "ideal" scaffold material and design for cell therapies in the eye continue. Multiple natural and synthetic biodegradable and nonbiodegradable materials, some with chemical and mechanical surface modifications, have been bioengineered and tested to find the optimal material and design. Biodegradable materials although have the theoretical advantage of not leaving a biologic footprint after being resorbed, the process of degradation itself may incite an unacceptable inflammatory reaction as well as the fact that the timing of degradation of the material is difficult to control and hence maybe too early preventing the RPE to form a monolayer. Biodegradable scaffolds coupled with BM rejuvenation could theoretically regenerate the RPE-BM complex to its original state. Bruch's membrane restoration is an exciting field that needs to be further expanded. For example, in an attempt to improve RPE survival, Sugiano et al. treated aged BM with a biologically synthesized extracellular matrix (bovine corneal endothelial cell extracellular matrix) prior to transplantation and showed that transplanted RPE survival enhanced over 200% [68]. Chemical, laser, or mechanical modification of recipient BM could potentially be combined with SC-RPE cell suspension transplantation too. This may enhance RPE attachment, attainment of monolayer status, and integration to photoreceptor layer [16]. Further work will prove if this strategy could improve SC-RPE cell suspension transplantation outcomes. However, all these efforts have to be mindful of the regulatory limitations imposed by agencies such as the FDA in the USA.

In contrast, there are many examples of successful ocular implants that involve nonbiodegradable materials, such as the intraocular lens. These materials also have the additional advantage that in the unlikely event the implanted cells need to be explanted, the implant can be grasped with forceps and removed as one piece. Also another advantage of these bioinert implants is our ability to micromachine and surface treat them to improve oxygen and macromolecule diffusion. Mechanical surface modification may improve cell maturation and configuration while improving scaffold robustness and flexibility. Three-dimensional printing technology may also prove useful in managing the production cost of such scaffolds. Additionally, as more patients are recruited and implanted with these stem-cell scaffolds in clinical trials, optimization of the implantation instrumentation and surgical delivery techniques is anticipated. Thus, it is projected that successive patients implanted with these scaffolds will have better outcomes. Moreover, once safety and optimal delivery of these bioimplants are established, patients with decreased disease burden (e.g., minimal area of geographic atrophy) will likely be recruited for future trials, with even better outcome expectations.

This is an exciting era for stem-cell therapeutics of the retina. Strong preclinical studies have allowed multiple groups to progress to clinical trials using SC-RPE, and outcomes will soon be available. These results will guide future development of therapies. Transplantation of neural retinal cells is nearing clinical trials, and in the future it may be possible to combine RPE and photoreceptors to rebuild the outer retina. There are many possible scaffolds that have yet to be translated. For example, microfabrication of an optimal biodegradable scaffold with mechanical or biological surface pretreatment, such as addition of growth factors/extracellular matrix, is worth exploring. Surgical delivery methods and instrumentation are under active development as well.

References

1. Nazari H, Zhang L, Zhu D, Chader GJ, Falabella P, Stefanini F, et al. Stem cell based therapies for age-related macular degeneration: the promises and the challenges. Prog Retin Eye Res. 2015;48:1–39.
2. Schwartz SD, Regillo CD, Lam BL, Eliott D, Rosenfeld PJ, Gregori NZ, et al. Human embryonic stem cell-derived retinal pigment epithelium in patients with age-related macular degeneration and Stargardt's macular dystrophy: follow-up of two open-label phase 1/2 studies. Lancet. 2015;385(9967):509–16.
3. Thomas BB, Zhu D, Zhang L, Thomas PB, Hu Y, Nazari H, et al. Survival and functionality of hESC-derived retinal pigment epithelium cells cultured as a monolayer on polymer substrates transplanted in RCS rats. Invest Ophthalmol Vis Sci. 2016;57(6):2877–87.
4. Pennington BO, Clegg DO. Pluripotent stem cell-based therapies in combination with substrate for the treatment of age-related macular degeneration. J Ocul Pharmacol Ther. 2016;32(5):261–71.
5. Gamm DM, Wong R, and the AGIWP. Report on the National Eye Institute Audacious Goals Initiative: photoreceptor regeneration and integration workshop. Transl Vis Sci Technol. 2015;4(6):2.
6. Whiting P, Kerby J, Coffey P, da Cruz L, McKernan R. Progressing a human embryonic stem-cell-based regenerative medicine therapy towards the clinic. Philos Trans R Soc Lond Ser B Biol Sci. 2015;370(1680):20140375.
7. Di Foggia V, Makwana P, Ali RR, Sowden JC. Induced pluripotent stem cell therapies for degenerative disease of the outer retina: disease modeling and cell replacement. J Ocul Pharmacol Ther. 2016;32(5):240–52.
8. Hotaling NA, Khristov V, Wan Q, Sharma R, Jha BS, Lotfi M, et al. Nanofiber scaffold-based tissue-engineered retinal pigment epithelium to treat degenerative eye diseases. J Ocul Pharmacol Ther. 2016;32(5):272–85.
9. Strauss O. The retinal pigment epithelium in visual function. Physiol Rev. 2005;85(3):845–81.
10. Booij JC, Baas DC, Beisekeeva J, Gorgels TG, Bergen AA. The dynamic nature of Bruch's membrane. Prog Retin Eye Res. 2010;29(1):1–18.
11. Pauleikhoff D, Harper CA, Marshall J, Bird AC. Aging changes in Bruch's membrane. A histochemical and morphologic study. Ophthalmology. 1990;97(2):171–8.
12. Moore DJ, Clover GM. The effect of age on the macromolecular permeability of human Bruch's membrane. Invest Ophthalmol Vis Sci. 2001;42(12):2970–5.
13. Moore DJ, Hussain AA, Marshall J. Age-related variation in the hydraulic conductivity of Bruch's membrane. Invest Ophthalmol Vis Sci. 1995;36(7):1290–7.
14. Afshari FT, Kwok JC, Andrews MR, Blits B, Martin KR, Faissner A, et al. Integrin activation or alpha 9 expression allows retinal pigmented epithelial cell adhesion on Bruch's membrane in wet age-related macular degeneration. Brain. 2010;133(Pt 2):448–64.
15. Hussain AA, Starita C, Hodgetts A, Marshall J. Macromolecular diffusion characteristics of ageing human Bruch's membrane: implications for age-related macular degeneration (AMD). Exp Eye Res. 2010;90(6):703–10.
16. Gullapalli VK, Sugino IK, Van Patten Y, Shah S, Zarbin MA. Impaired RPE survival on aged submacular human Bruch's membrane. Exp Eye Res. 2005;80(2):235–48.
17. Tomita M, Lavik E, Klassen H, Zahir T, Langer R, Young MJ. Biodegradable polymer composite grafts promote the survival and differentiation of retinal progenitor cells. Stem Cells. 2005;23(10):1579–88.
18. Kador KE, Goldberg JL. Scaffolds and stem cells: delivery of cell transplants for retinal degenerations. Expert Rev Ophthalmol. 2012;7(5):459–70.
19. Kundu J, Michaelson A, Baranov P, Young MJ, Carrier RL. Approaches to cell delivery: substrates and scaffolds for cell therapy. Dev Ophthalmol. 2014;53:143–54.
20. Lu B, Zhu D, Hinton D, Humayun MS, Tai YC. Mesh-supported submicron parylene-C membranes for culturing retinal pigment epithelial cells. Biomed Microdevices. 2012;14(4):659–67.
21. Stanzel BV, Liu Z, Brinken R, Braun N, Holz FG, Eter N. Subretinal delivery of ultrathin rigid-elastic cell carriers using a metallic shooter instrument and biodegradable hydrogel encapsulation. Invest Ophthalmol Vis Sci. 2012;53(1):490–500.
22. Tao S, Young C, Redenti S, Zhang Y, Klassen H, Desai T, et al. Survival, migration and differentiation of retinal progenitor cells transplanted on micro-machined poly(methyl methacrylate) scaffolds to the subretinal space. Lab Chip. 2007;7(6):695–701.
23. Redenti S, Neeley WL, Rompani S, Saigal S, Yang J, Klassen H, et al. Engineering retinal progenitor cell and scrollable poly(glycerol-sebacate) composites for expansion and subretinal transplantation. Biomaterials. 2009;30(20):3405–14.
24. Redenti S, Tao S, Yang J, Gu P, Klassen H, Saigal S, et al. Retinal tissue engineering using mouse retinal progenitor cells and a novel biodegradable, thin-film poly(e-caprolactone) nanowire scaffold. J Ocul Biol Dis Infor. 2008;1(1):19–29.
25. Liu Y, Wang R, Zarembinski TI, Doty N, Jiang C, Regatieri C, et al. The application of hyaluronic acid hydrogels to retinal progenitor cell transplantation. Tissue Eng Part A. 2013;19(1–2):135–42.
26. Liu Z, Yu N, Holz FG, Yang F, Stanzel BV. Enhancement of retinal pigment epithelial culture characteristics and subretinal space tolerance of scaffolds with 200 nm fiber topography. Biomaterials. 2014;35(9):2837–50.
27. Lavik EB, Klassen H, Warfvinge K, Langer R, Young MJ. Fabrication of degradable polymer scaffolds to direct the integration and differentiation of retinal progenitors. Biomaterials. 2005;26(16):3187–96.
28. Hadlock T, Singh S, Vacanti JP, McLaughlin BJ. Ocular cell monolayers cultured on biodegradable substrates. Tissue Eng. 1999;5(3): 187–96.
29. Worthington KS, Wiley LA, Guymon CA, Salem AK, Tucker BA. Differentiation of induced pluripotent stem cells to neural retinal precursor cells on porous poly-lactic-co-glycolic acid scaffolds. J Ocul Pharmacol Ther. 2016;32(5):310–6.
30. Ulbrich S, Friedrichs J, Valtink M, Murovski S, Franz CM, Muller DJ, et al. Retinal pigment epithelium cell alignment on nanostructured collagen matrices. Cells Tissues Organs. 2011;194(6):443–56.
31. Bhatt NS, Newsome DA, Fenech T, Hessburg TP, Diamond JG, Miceli MV, et al. Experimental transplantation of human retinal pigment epithelial cells on collagen substrates. Am J Ophthalmol. 1994;117(2):214–21.
32. Del Priore LV, Tezel TH, Kaplan HJ. Survival of allogeneic porcine retinal pigment epithelial sheets after subretinal transplantation. Invest Ophthalmol Vis Sci. 2004;45(3):985–92.

33. Gabrielian K, Oganesian A, Farrokh-Siar L, Rezai KA, Verp MS, Patel SC, et al. Growth of human fetal retinal pigment epithelium as microspheres. Graefes Arch Clin Exp Ophthalmol. 1999;237(3):241–8.
34. Oganesian A, Gabrielian K, Ernest JT, Patel SC. A new model of retinal pigment epithelium transplantation with microspheres. Arch Ophthalmol. 1999;117(9):1192–200.
35. Hsiue GH, Lai JY, Lin PK. Absorbable sandwich-like membrane for retinal-sheet transplantation. J Biomed Mater Res. 2002;61(1):19–25.
36. Lai JY. Biocompatibility of chemically cross-linked gelatin hydrogels for ophthalmic use. J Mater Sci Mater Med. 2010;21(6):1899–911.
37. Xiang P, Wu KC, Zhu Y, Xiang L, Li C, Chen DL, et al. A novel Bruch's membrane-mimetic electrospun substrate scaffold for human retinal pigment epithelium cells. Biomaterials. 2014;35(37):9777–88.
38. Lu JT, Lee CJ, Bent SF, Fishman HA, Sabelman EE. Thin collagen film scaffolds for retinal epithelial cell culture. Biomaterials. 2007;28(8):1486–94.
39. Kamao H, Mandai M, Okamoto S, Sakai N, Suga A, Sugita S, et al. Characterization of human induced pluripotent stem cell-derived retinal pigment epithelium cell sheets aiming for clinical application. Stem Cell Reports. 2014;2(2):205–18.
40. Reardon S, Cyranoski D. Japan stem-cell trial stirs envy. Nature. 2014;513(7518):287–8.
41. Hartmann U, Sistani F, Steinhorst UH. Human and porcine anterior lens capsule as support for growing and grafting retinal pigment epithelium and iris pigment epithelium. Graefes Arch Clin Exp Ophthalmol. 1999;237(11):940–5.
42. Nicolini J, Kiilgaard JF, Wiencke AK, Heegaard S, Scherfig E, Prause JU, et al. The anterior lens capsule used as support material in RPE cell-transplantation. Acta Ophthalmol Scand. 2000;78(5):527–31.
43. Lee CJ, Fishman HA, Bent SF. Spatial cues for the enhancement of retinal pigment epithelial cell function in potential transplants. Biomaterials. 2007;28(13):2192–201.
44. Kiilgaard JF, Wiencke AK, Scherfig E, Prause JU, la Cour M. Transplantation of allogenic anterior lens capsule to the subretinal space in pigs. Acta Ophthalmol Scand. 2002;80(1):76–81.
45. Lu L, Yaszemski MJ, Mikos AG. Retinal pigment epithelium engineering using synthetic biodegradable polymers. Biomaterials. 2001;22(24):3345–55.
46. Thomson HA, Treharne AJ, Backholer LS, Cuda F, Grossel MC, Lotery AJ. Biodegradable poly(alpha-hydroxy ester) blended microspheres as suitable carriers for retinal pigment epithelium cell transplantation. J Biomed Mater Res A. 2010;95(4):1233–43.
47. Lu L, Nyalakonda K, Kam L, Bizios R, Gopferich A, Mikos AG. Retinal pigment epithelial cell adhesion on novel micropatterned surfaces fabricated from synthetic biodegradable polymers. Biomaterials. 2001;22(3):291–7.
48. Klassen H. Transplantation of cultured progenitor cells to the mammalian retina. Expert Opin Biol Ther. 2006;6(5):443–51.
49. Nadri S, Kazemi B, Eslaminejad MB, Yazdani S, Soleimani M. High yield of cells committed to the photoreceptor-like cells from conjunctiva mesenchymal stem cells on nanofibrous scaffolds. Mol Biol Rep. 2013;40(6):3883–90.
50. Kubota A, Nishida K, Yamato M, Yang J, Kikuchi A, Okano T, et al. Transplantable retinal pigment epithelial cell sheets for tissue engineering. Biomaterials. 2006;27(19):3639–44.
51. von Recum H, Kikuchi A, Yamato M, Sakurai Y, Okano T, Kim SW. Growth factor and matrix molecules preserve cell function on thermally responsive culture surfaces. Tissue Eng. 1999;5(3):251–65.
52. von Recum HA, Okano T, Kim SW, Bernstein PS. Maintenance of retinoid metabolism in human retinal pigment epithelium cell culture. Exp Eye Res. 1999;69(1):97–107.
53. Yaji N, Yamato M, Yang J, Okano T, Hori S. Transplantation of tissue-engineered retinal pigment epithelial cell sheets in a rabbit model. Biomaterials. 2009;30(5):797–803.
54. Lu B, Tai YC, Humayun MS. Microdevice-based cell therapy for age-related macular degeneration. Dev Ophthalmol. 2014;53:155–66.
55. Hu Y, Liu L, Lu B, Zhu D, Ribeiro R, Diniz B, et al. A novel approach for subretinal implantation of ultrathin substrates containing stem cell-derived retinal pigment epithelium monolayer. Ophthalmic Res. 2012;48(4):186–91.
56. Regenerative Patch Technologies L. Study of Subretinal Implantation of Human Embryonic Stem Cell-Derived RPE Cells in Advanced Dry AMD. 2015. https://clinicaltrials.gov/ct2/show/study/NCT02590692?term=retina+stem+cells&rank=112016. Accessed 25 Aug 2015.
57. Carr AJ, Smart MJ, Ramsden CM, Powner MB, da Cruz L, Coffey PJ. Development of human embryonic stem cell therapies for age-related macular degeneration. Trends Neurosci. 2013;36(7):385–95.
58. Pfizer. A study of implantation of retinal pigment epithelium in subjects with acute wet age related macular degeneration. 2016. http://www.clinicaltrials.gov/ct2/show/NCT01691261?term=rpe&rank=11. Accessed 16 Aug 2016.
59. Pete Coffey. 'Miracle' cells could cure blindness. 2016. http://www.cnn.com/2016/03/07/health/using-stem-cells-to-cure-blindness/2016.
60. Neeley WL, Redenti S, Klassen H, Tao S, Desai T, Young MJ, et al. A microfabricated scaffold for retinal progenitor cell grafting. Biomaterials. 2008;29(4):418–26.
61. Tezcaner A, Bugra K, Hasirci V. Retinal pigment epithelium cell culture on surface modified poly(hydroxybutyrate-co-hydroxyvalerate) thin films. Biomaterials. 2003;24(25):4573–83.
62. Williams RL, Krishna Y, Dixon S, Haridas A, Grierson I, Sheridan C. Polyurethanes as potential substrates for sub-retinal retinal pigment epithelial cell transplantation. J Mater Sci Mater Med. 2005;16(12):1087–92.
63. Krishna Y, Sheridan C, Kent D, Kearns V, Grierson I, Williams R. Expanded polytetrafluoroethylene as a substrate for retinal pigment epithelial cell growth and transplantation in age-related macular degeneration. Br J Ophthalmol. 2011;95(4):569–73.
64. Tucker BA, Redenti SM, Jiang C, Swift JS, Klassen HJ, Smith ME, et al. The use of progenitor cell/biodegradable MMP2-PLGA polymer constructs to enhance cellular integration and retinal repopulation. Biomaterials. 2010;31(1):9–19.
65. Lee GY, Kang SJ, Lee SJ, Song JE, Joo CK, Lee D, et al. Effects of small intestinal submucosa content on the adhesion and proliferation of retinal pigment epithelial cells on SIS-PLGA films. J Tissue Eng Regen Med. 2017;11(1):99–108.
66. Hsiung J, Zhu D, Hinton DR. Polarized human embryonic stem cell-derived retinal pigment epithelial cell monolayers have higher resistance to oxidative stress-induced cell death than nonpolarized cultures. Stem Cells Transl Med. 2015;4(1):10–20.
67. Diniz B, Thomas P, Thomas B, Ribeiro R, Hu Y, Brant R, et al. Subretinal implantation of retinal pigment epithelial cells derived from human embryonic stem cells: improved survival when implanted as a monolayer. Invest Ophthalmol Vis Sci. 2013;54(7):5087–96.
68. Sugino IK, Sun Q, Cheewatrakoolpong N, Malcuit C, Zarbin MA. Biochemical restoration of aged human Bruch's membrane: experimental studies to improve retinal pigment epithelium transplant survival and differentiation. Dev Ophthalmol. 2014;53:133–42.

Surgical Approaches for Cell Transplantation in Cell Replacement Therapy

Priya Sharma, Jayanth Sridhar, and Carl D. Regillo

1 Problem and Hypothesis

Modern medicine has limited treatment for many degenerative conditions, and in the majority of instances treatment goals are directed towards slowing progression of disease rather than curative therapy. For example, Alzheimer's disease remains the leading cause of adult dementia in the United States despite decades and millions of dollars spent on research. Management of type 1 diabetes mellitus has relied on supplementation with exogenous insulin rather than regenerating functional pancreatic islet cells. Similarly, most retinal dystrophies are both slowly progressive and irreversible, resulting in an inevitable loss of patient function and quality of life. Therefore, it is no surprise that there has been much interest towards utilizing stem cells to treat degenerative disease processes of the retina.

Embryonic stem cells are derived from blastocysts and have the ability to self-renew [1, 2]. They are pluripotent and can differentiate into any of the cell types of all three germ layers: ectoderm, mesoderm, or endoderm [1]. These cells can provide a theoretically inexhaustible supply of cells for transplantation. Embryonic stem cells were first isolated in murine models in 1981 [3, 4], and this knowledge was later applied in the late 1990s to create human embryonic stem cells [2, 5]. Since then there has been a surging interest in their use for tissue engineering and regenerative medicine.

With the correct culture conditions, stem cells can be maintained in vitro and stimulated to differentiate into various target cells. These cells can then be delivered to areas in a patient that are dysfunctional or missing, thereby reversing or slowing down an otherwise uncurable disease process. For example, stem cells can be used to create myocardium in a patient with a myocardial infarction, pancreatic islet cells in a patient with diabetes, or neurons in a patient with Alzheimer's disease. Injections or surgical transplantation can directly deliver these cells to the localized areas of dysfunction.

Retinal disorders in particular can be well suited towards embryonic stem cell therapy, as many retinal degenerations stem from single mutations or single-cell dysfunctions, such as macular degeneration, Stargardt's dystrophy, retinitis pigmentosa, cone dystrophy, and vitelliform dystrophy. These conditions harbor individual mutations or cellular dysfunctions which, if reversed, can theoretically result in restored visual function. While there are a multitude of other ocular conditions with individual cellular mutations, including lenticular, corneal, and neuro-ophthalmic disorders, retinal degenerative disorders often present with subtle findings early in the disease course, so that intervention is theoretically possible prior to permanent disability.

Although embryonic stem cell therapy is promising, there are many limitations that must be taken into consideration, including ethical and safety considerations. Ethically, there is much debate surrounding embryonic stem cells and the use of human embryos, including the ability to clone human beings, make human embryos, and the fear of creating unregulated tumors [6]. Furthermore, to date, there has been no approach in differentiation of embryonic stem cells that has yielded a 100% pure population of differentiated cells [7]. Therefore, there must be extreme caution to avoid implantation of undifferentiated cell lineages, as these could worsen the disease process or lead to proliferation or teratoma formation [7]. Finally, allogeneic implants can undergo rejection, which can be worse than the disease process that was being treated in terms of creating damage and loss of function, and can be life-threatening in some instances [7]. To help avoid these risks, early experimentation with embryonic stem cell transplantation has involved areas of the body that are relatively sequestered to immunogenicity, with one potentially ideal area being the eye.

The authors have no financial interest on the devices or medications in this document. The authors have no conflict of interest.

P. Sharma, M.D.
Wills Eye Hospital, 840 Walnut Street, Suite 800, Philadelphia, PA 19107, USA
e-mail: priyasharma141@gmail.com

J. Sridhar, M.D. • C.D. Regillo, M.D. (✉)
Midatlantic Retina of Wills Eye Hospital, 840 Walnut Street, Suite 1020, Philadelphia, PA 19107, USA
e-mail: jsridhar119@gmail.com; cregillo@aol.com

S.D. Schwartz et al. (eds.), *Cellular Therapies for Retinal Disease*, DOI 10.1007/978-3-319-49479-1_5

Abbreviations

ABCA4	ATP-binding cassette transporter
AREDS	Age-Related Eye Disease Study
hESC	Human embryonic stem cells
hESC-RPE	Human embryonic stem cell-derived retinal pigment epithelium
iPSC	Induced pluripotent stem cell
NIH	National Institute of Health
RCS	Royal College of Surgeons
RPE	Retinal pigment epithelium
VEGF	Vascular endothelial growth factor

2 Scientific Basis for Intervention

The subretinal space is an area of the body that is relatively sequestered from the immune system by the blood-ocular barrier. Not only are the ocular structures protected from systemic immunogenicity, but also systemic spread of medications delivered directly to the eye is limited [8]. Due to the unique nature of the blood-ocular barrier, the subretinal space and diseases that target the subretinal space serve as an ideal target for stem cell transplantation.

Degenerative ocular conditions that cause otherwise irreversible vision loss serve as an excellent target for stem cell trials in hopes that replacing dysfunctional cells or providing cells that can emit trophic factors will support cell survival and lead to noticeable improvements in anatomic and visual function. The retinal pigment epithelium (RPE) is a layer of cells that nourishes photoreceptor cells and metabolizes toxins. It serves to phagocytose shed photoreceptor segments, metabolize vitamin A derivatives, and recycle photopigments [9, 10]. Dysfunction of the RPE itself leads to photoreceptor loss with subsequent atrophy of surrounding retinal layers, which is irreversible once occurs. Two conditions that share this pathophysiology are age-related macular degeneration and Stargardt's macular dystrophy, the former being the leading cause of adult blindness in the United States and the latter representing the most common hereditary maculopathy. In age-related macular degeneration, genetic factors, environmental toxins, and oxidative stress are surmised to together result in retinal pigment epithelial damage, dysfunction of which leads to photoreceptor loss [8]. There are two main types of age-related macular degeneration—non-exudative and exudative, with the latter consisting of choroidal neovascular membrane formation resulting in macular exudation, hemorrhage, and scarring. Currently, the only known treatment to decrease the risk of progression of non-exudative, or dry, age-related macular degeneration is the Age-Related Eye Disease Study (AREDS) formula of vitamins and minerals, a formulation thought to decrease oxidative stress on the RPE [11]. There is no known treatment for Stargardt's macular dystrophy, which is thought of as the juvenile form of macular degeneration. In the most common form of Stargardt's macular dystrophy, mutations in the ATP-binding cassette transporter (ABCA4) gene in photoreceptors themselves leads to formation of toxic vitamin A dimers, which then build up into lipofuscin granules that get deposited in the retinal pigment epithelium [12]. This causes degeneration of the retinal pigment epithelium, leading to photoreceptor loss. The ABCA4 gene is too large for adeno-associated virus vectors, which have shown some success in treating other disorders caused by ocular genes [12]. Therefore, replacement and regeneration of the retinal pigment epithelium could serve as a treatment for both diseases, while concurrently taking advantage of the blood-ocular barrier. In order to achieve optimal outcomes, RPE cells must be delivered appropriately and must attach to the underlying Bruch's membrane and form polarized monolayers to have the most function and allow for phagocytosis of photoreceptor outer segments and formation of the blood ocular barrier [1].

3 Preclinical Data

Initial preclinical trials first experimented with autologous retinal pigment epithelial (RPE) cell transfer and injection of trophic factors in Royal College of Surgeon (RCS) rats [13, 14]. Royal College of Surgeon (RCS) rats are a widely studied model of recessively inherited retinal degeneration, caused by a small deletion of DNA that disrupts the gene encoding the receptor tyrosine kinase Mertk. This leads to the inability of cells to phagocytose photoreceptor outer segments, leading to photoreceptor cell atrophy [13, 14]. This model is a classic model of recessively inherited retinal degeneration, and thus serves as a unique framework to study retinal degenerations.

Initial studies with rats demonstrated that transplants of normal rat neonatal RPE cells into the retinas of Royal College of Surgeon (RCS) were successful [15, 16]. In these studies, freshly isolated retinal pigment epithelial cells from neonatal pigmented Long Evans rats were transplanted into the subretinal space of dystrophic RCS rats, and following transplantation, rescue of photoreceptor cells and regeneration of photoreceptor outer segments were noted [15, 16]. This was demonstrated by an increase in the histological thickness of the outer nuclear layer [15] along with photoreceptor outer segment growth in the region of RPE transfer [16].

Autologous RPE cells are limited in supply and harbor the same inherent genetic predisposition to disease. Therefore, attention has been directed towards in-vitro cultures of RPE from embryonic stem cells. RPE cells were first differentiated from primate embryonic stem cells in 2002 [17]. Similar principles have since been applied to create human embryonic stem cell-derived retinal pigment epithelium (hESC-RPE).

Early preclinical models in RCS rats showed that subretinal transplantation of hESC-RPE can rescue photoreceptors and prevent vision loss, resulting in marked improvement in visual function [18, 19]. In a study by Lund et al., hESC were cultured as per protocol and allowed to spontaneously differentiate into RPE clusters [18]. These RPE cells were subcultured multiple times, pooled together, and analyzed by polymerase chain reaction. This pure mixture of human embryonic cells was delivered to 14 dystrophic RCS rats. First, the rats were anesthesized and received topical ophthaine for corneal anesthesia and tropicamide for pupillary dilation. The eyes were proptosed slightly with a thread, and a small scleral incision was made in the upper temporal area, with the mixture of stem cells delivered to the subretinal space using a fine glass pipette attached to a 10 μL syringe. The cornea was simultaneously punctured to reduce intraocular pressure and limit efflux of cells [20]. Eight dystrophic RCS rats acted as control subjects and received subretinal injections of the medium alone in a similar fashion as described. All twenty rats received injection at postnatal day 21–23, at an age when photoreceptor development had yet to occur. After transplantation, all animals received dexamethasone injections daily and were maintained on cyclosporine-A until euthanization. Electroretinogram responses were recorded at both P60 and P90 to measure the electrical activity of the outer (a-wave) and inner (b-wave) retina [18]. These signs, which were severely depleted in dystrophic RCS rats, were significantly better in the study rats that had received transplantation of hESC-RPE [18, 21]. Anatomically, examination of the retinas showed that donor hESC-RPE cells were distributed in the subretinal space adjacent to the host RPE, with extensive photoreceptor rescue [18]. Therefore, this study showed that hESC-RPE delivered to the subretinal space in RCS rats was capable of significant improvement of visual function in an animal model of retinal disease and provided a viable method of delivery.

Lu et al. further elucidated on this study to show a dose-specific response to visual rehabilitation [19]. In their model of RCS rats, they found that increased hESC-RPE injections from 5000 to 50,000 cells/eye also rescued visual function in a dose-dependent fashion, as measured by visual acuity and luminance testing. However, the same response did not also pertain at 100,000 cells/eye, suggesting that there is no decreased efficacy of higher doses, but that there is a level at which the effectiveness of therapy peaks [19]. This demonstration of the safety of increased dosages of hESC-RPE injections helped validate similar dosage-dependent models in human studies [8].

Further surgical transplantation of human RPE was performed in the owl monkey, in which open-sky technique was used. In this method, the anterior segment was opened, a small retinal flap was created, and the underlying RPE cells were slowly removed using trypsinization and gentle manual traction [22, 23]. Cultured radioactive labeled human RPE cells were then placed onto the denuded Bruch's membrane, and showed signs of polarity and attachment [22, 23]. These promising results led to closed-eye methods of transplantation of cultured rabbit RPE cells. In these studies, a conjunctival flap was formed in an anesthesized and dilated rabbit eye, and a pars plana incision was made. A cannula measuring approximately 100 μm was inserted through the pars plana incision and was used to create a small bleb detachment 2–3 disc diameters from the optic nerve. Care was taken to keep the retinal hole as small as possible to avoid vitreous seeding. The conjunctival flap and sclerotomy site were closed [24]. In the closed-sky technique, the neural retina spontaneously reattached within 24–28 h, bringing photoreceptor segments in close contact with transplanted cells, allowing for partial RPE-function as early as 24 h after the surgery [24].

In summary, preclinical studies with both autologous RPE transplantation and hESC-RPE transplantation demonstrate rescue of vision in varying degrees. It is unclear whether it is the RPE cells itself or the paracrine effects of growth factors secreted by RPE cells that results in photoreceptor rescue, but the preclinical studies do demonstrate promising results in animal models of retinal disease [1].

4 Translation Strategies (Patient Selection, Safety Considerations, and Outcomes)

Translation of animal research to humans encompasses many additional concerns. Primarily, safety remains a major concern with stem cell transplantation. Concerns around teratoma formation and rejection exist, and cells must still adhere to current Good Manufacturing Practices to ensure lack of infectious agents. Therefore, in a preclinical model in RCS rats, Lu et al. augmented previously published RPE-derivation protocols [18] and implemented appropriate controls around cell manufacturing to achieve full compliance with current Good Manufacturing Practices [19]. This group also studied the long-term risk of teratoma formation in the National Institute of Health (NIH) III immune-deficient mouse model. In NIH III mice, injection of human embryonic stem cell-derived RPE did not result in any teratoma or tumor formation over the 9-month life span of the animal [19].

When applying similar study protocols to humans, there must be many stringent variables when it comes to patient selection, with the goal of proving safety before proving efficacy. First, the vision in the study eye must be poor enough to warrant initial trials with stem cell therapy, as there are risks in the transplantation process that may result in worsening of vision. Furthermore, there must not be any concurrent unrelated retinal disease, including diabetic retinopathy or retinal vein occlusion that might alter imaging results or visual acuity measurements. There must not be any other confounding factors in visual determination, including presence of optic nerve disease or occipital lobe disease. Presence of uveitis or other inflammatory conditions could also alter results and potentiate possible immune rejection, and therefore these patients would also have to be excluded. Once these inclusion and exclusion criteria have been met, patients must be willing to enroll in the study for many months, for the purpose of not only monitoring for efficacy, but also monitoring to ensure early intervention for any complications that may occur. Furthermore, as with any procedure, the risks of the procedure itself must be considered in light of the patient's condition. For example, the risk of retinal detachment or endophthalmitis in the setting of vitrectomy should be considered and weighed against the benefit of stem cell transplantation. Resultant data should also not be skewed by the visual loss from cataract that inevitably develops after vitrectomy.

Transplantation of stem cell-derived RPE cells also must be delivered to the subretinal space safely. There are two main approaches to delivery—injection of a stem cell-derived suspension or engrafting a monolayer of stem cell-derived RPE cells that are seeded on a supporting membrane [1]. To deliver cell suspensions, primate and rabbit models initially delivered RPE cells via open-sky vitrectomy techniques, which later evolved into the more common closed-eye models observed in humans today [22–24]. Currently, two methods are employed to deliver cell suspensions. The first delivery approach utilizes a microcatheter delivery system to access the subretinal space externally by creating a localized peritomy followed by dissection of the dorsal sclera and choroid and passage of a blunt-tipped needle through this dissected space (Fig. 1). Indirect ophthalmoscopic examination is utilized to visualize the beacon tip as saline is first injected to create a local retinal detachment, followed by injection of the cellular suspension into the subretinal space [25]. Given most vitreoretinal surgeons are not comfortable performing procedures with indirect visualization, the microcatheter external delivery option is not ideal. The second approach consists of pars plana vitrectomy with a small, self-sealing retinotomy and injection of the cellular suspension to create a small subretinal bleb [25]. Twenty-three or twenty-five gauge cannula is typically used with use of a 38 gauge cannula to access the subretinal space via a small retinotomy. The surgeon may first create a "prebleb" with injection of balanced salt solution into the subretinal space to ensure that subsequent injected cellular suspension is in the proper plane. The advantages of this approach are it is easier technically for most vitreoretinal surgeons and visualization is superior. The disadvantages of this approach include the standard risks of pars plana vitrectomy, namely cataract formation, endophthalmitis, retinal detachment, and intraocular bleeding. Additionally, care must be taken to create a small retinotomy to avoid seeding the vitreous with RPE suspension, which can lead to subsequent proliferative vitreoretinopathy. However, given the greater technical ease, the latter approach is currently primarily employed in translation to human research.

The site of cell suspension delivery must be selected carefully, as it typically should allow for access to compromised retina but also harbor good vascular supply. Therefore, cellular suspensions are commonly being injected at the junction between atrophic photoreceptor changes and compromised retina. Presurgical planning of site selection can be optimized by assessing areas of atrophy using both fundus autofluorescence and en face optical coherence tomography imaging. Due to the nascent nature of stem cell transplantation, most patients currently studied are those with extensive geographic atrophy in the macula with extensive concurrent visual dysfunction. Although this is not the ideal patient for visual rehabilitation due to irreversibility of retinal atrophy, patients with extensive geographic atrophy still serve to provide proof of concept of stem cell transplantation and help to monitor reactions or adverse events to immunosuppressive therapy.

An alternative to cellular suspensions is using premade sheets with engrafted RPE stem cells. In this method, 23 or 25 gauge cannula is typically used with use of a 33 or 38 gauge cannula to access the subretinal space via a small retinotomy. Similarly to described prior, the surgeon may first create a "prebleb" with injection of balanced salt solution into the subretinal space. After this, a tapered pipette may be used to deliver the cellular sheet through the retinotomy into the subretinal space under direct visualization, taking care to visualize the RPE sheet as it unfolds. The advantage of this approach is the enhanced function of RPE cells when delivered in a monolayer. The disadvantages of this approach include the standard risks of pars plana vitrectomy as mentioned before, along with the enhanced technical difficulty in delivering premade cellular sheets of RPE via pars plana vitrectomy due to the size and bulk of the cellular sheets. Additionally, there can be issues with unfolding of the sheet in the correct orientation and biocompatibility of the sheet with human ocular tissue.

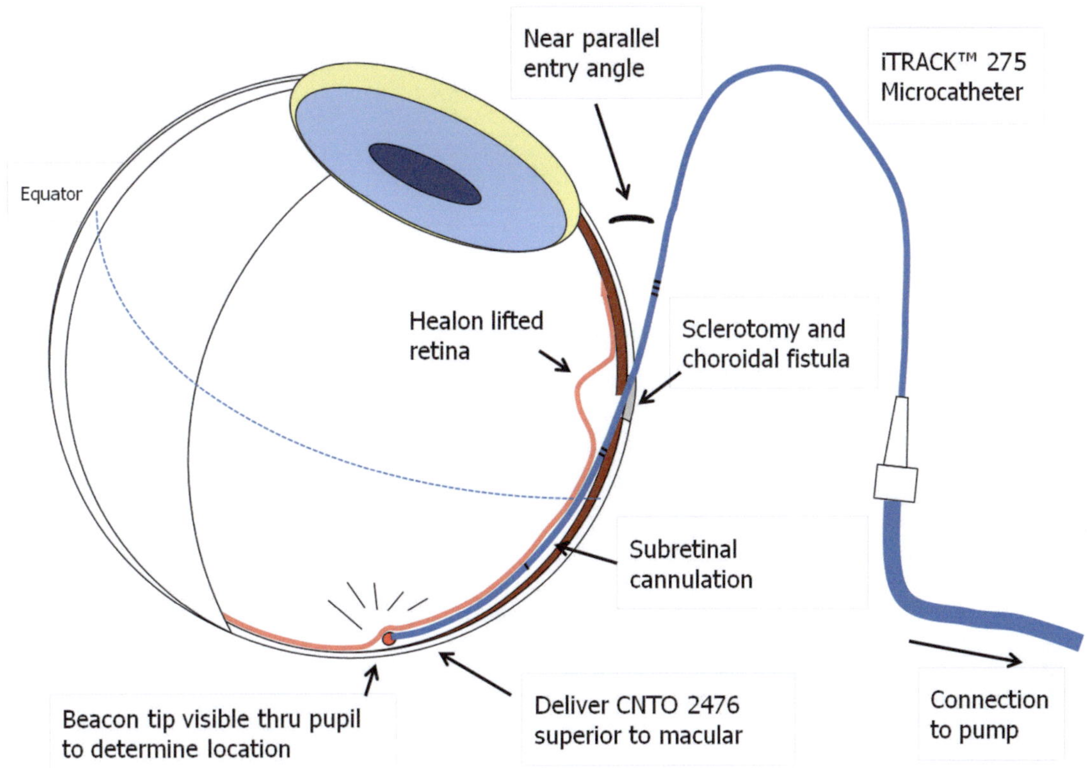

Fig. 1 Microcatheter delivery system illustration demonstrating ability to deliver human embryonic cells to the subretinal space via external approach

5 Human Data

Human trials with autologous retinal pigment epithelial (RPE) cell transfer have been modestly successful. Autologous RPE transplants retain the unique advantage of decreased rejection due to the lack of foreign immunogenicity. Binder et al. reported on a series of 53 eyes with foveal choroidal neovascularization that necessitated membrane excision [26]. Thirty-nine of these eyes had concurrent RPE transplantation with autologous RPE cells at the time of membrane excision, and the remaining 14 had membrane excision alone. In the 39 study patients, vitrectomy and posterior hyaloid removal was performed, and then nasal retinotomy and bleb detachment was achieved with balanced salt solution to free the retina from the pigment epithelium. A bent 20 gauge cannula was used to aspirate RPE cells nasal to the optic disc, and this suspension was handed to a pathologist available in the operating room, who centrifuged the material and confirmed the presence of at least 1500 RPE cells. This suspension was then diluted with sterile saline. A second retinotomy was performed surrounding the foveal choroidal neovascular membrane, and the membrane was removed. After achieving hemostasis and performing endolaser to the retinotomy sites, the RPE suspension was delivered to the subretinal space via injection. The remaining fluid-air exchange was conducted as per protocol. The 14 eyes without RPE transplantation had choroidal neovascular membrane removal in the same method, without harvesting of autologous RPE from the nasal retina [26]. This study showed good short-term success over 8–12 months, with an overall trend towards improved visual acuity in those patients with RPE transplantation and membrane peel as compared to membrane peel alone (54% vs. 21% with an improvement of two or more lines in vision; $p = 0.97$). There was also a statistically significant improvement in reading acuity (1.85 ± 0.42 vs. 0.43 ± 0.47; $p = 0.001$) [26].

Similarly, freshly isolated autologous iris pigment epithelial cells have been translocated to the subretinal space during surgical removal of choroidal neovascularization [27, 28]. In these studies, a large iridectomy was performed on the eye undergoing vitrectomy, and the iris pigment epithelial cells were isolated. After choroidal neovascular membrane removal, saline was injected to create a small retinal detachment in the area of foveal atrophy and choroidal neovascularization. The iris pigment epithelial cells were then injected into the subretinal space, and fluid-air or fluid-gas exchange was performed by protocol. These cells were shown to be well tolerated for 3 years and resulted in stabilization of vision [28].

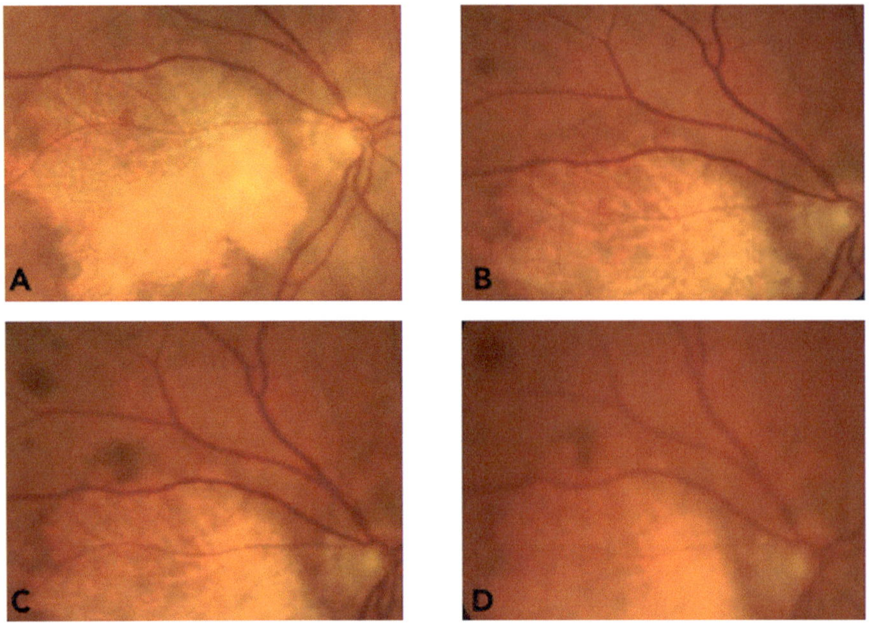

Fig. 2 Patient A with age-related macular degeneration, part of a multicenter phase 1/2 study with human embryonic stem cell-derived retinal pigment epithelium (hESC-RPE) transplantation. (**a**) Preoperative photograph of the surgical site prior to injection of hESC-RPE suspension. Note the absence of pigment around the area of atrophic retina. (**b**) Photograph from postoperative month 1, which shows an area of pigmentation, corresponding to the area of hESC-RPE transplantation, that becomes larger at postoperative month 3 (**c**) and postoperative month 6 (**d**, view slightly obscured by cataract) [8]

However, trials with long-term outcomes have not been as promising, showing only transient rescue of visual function. MacLaren et al. reported long-term follow-up data on four patients who had initially successful macular retinal pigment epithelium translocation surgery [29]. In these patients, the standard vitrectomy with posterior hyaloid removal was performed, along with retinotomy and injection of balanced salt solution to create a localized retinal detachment. However, these patients had full-thickness RPE-choroid autologous grafts harvested from an area surrounding foveal RPE atrophy, and these grafts were translocated to the site of foveal RPE atrophy and choroidal neovascularization. Despite initial improvement in visual acuity, these patients had declining visual acuity and lack of function of the grafts at 4–6 years [29].

These three studies show that autologous RPE grafts can provide some form of visual rehabilitation, which is striking considering these RPE cells harbor the same inherent genetic risks. It is unclear whether it is the RPE cells themselves or the secretion of growth factors that provides the most benefit, but regardless, these studies provide proof of concept that healthy RPE cells may have the potential to sustain and revitalize dying photoreceptor cells.

Recently, two multicenter phase 1/2 studies were performed in the United States with hESC-RPE, in which 18 patients (nine with age-related macular degeneration (Fig. 2a–d) and nine with Stargardt's macular dystrophy (Fig. 3a–d)) were selected to receive hESC-RPE transplantation [8]. The hESC-RPE cells were developed in accordance with current Good Manufacturing Practices and were assessed for safety and purity. Vials of hESC-RPE were delivered to the operating room and 150 μL of the formulation was delivered directly into the subretinal space via pars plana vitrectomy, a small self-sealing retinotomy (via a 39 gauge translocation needle), and localized injection of the cell suspension (Fig. 4) [8]. The cell suspension was injected at the junction between atrophic photoreceptor changes and compromised retina, preselected at this location since it was thought to still harbor good vascular supply and outer retinal anatomy, and, therefore, could optimize the chances of transplant integration. All patients received systemic immunosuppression with tacrolimus and mycophenolate mofetil 1 week before the procedure and continued this immunosuppression for 12 weeks after the procedure [8].

After surgery, optical coherence tomography imaging of transplanted areas showed a layer of cells lining aspects of Bruch's membrane in 13 of 18 patients (72%). In terms of safety and complications, 3 of 18 eyes also had a layer of preretinal, pigmented cell growth that did not seem to contract or cause any macular pucker. Four of eighteen eyes developed visually significant progression of cataract, with all four patients undergoing eventual cataract surgery. One eye developed acute postoperative endophthalmitis and one eye developed an inferior transvitreal band at posttransplant week 3 that resolved spontaneously by month 6. No signs of acute transplant rejection were noted in any of the 18 eyes. There was no retinal or subretinal whitening in the transplanted area, and

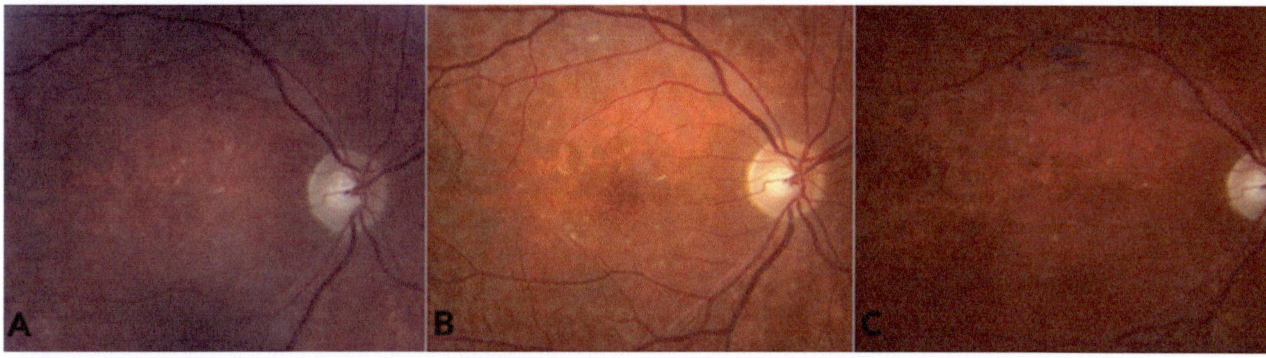

Fig. 3 Patient B with Stargardt's degeneration. (**a**) Preoperative photograph prior to injection of human embryonic stem cell-derived retinal pigment epithelium (hESC-RPE) suspension. Note the absence of pigment around the large central area of atrophic retina. (**b**) Photograph from postoperative week 1, which shows a very light area of pigmentation, corresponding to the area of hESC-RPE transplantation, that becomes larger and more pronounced at postoperative week 6 (**c**) [8]

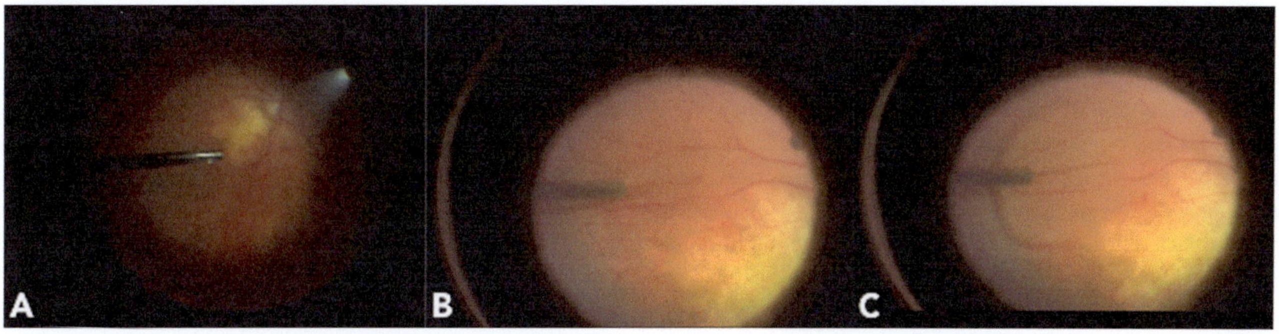

Fig. 4 Intraoperative pictures during surgical transplantation of hESC-RPE transplantation. (**a**–**c**) Vials of human embryonic stem cell-derived retinal pigment epithelium (hESC-RPE) were delivered to the operating room. After a standard three-port pars plana vitrectomy (**a**), a small self-sealing retinotomy via a 39 gauge translocation needle was achieved (**b**). One-hundred-fifty microliter of the stem cell formulation was delivered directly into the subretinal space with localized injection of the cell suspension at the junction between atrophic retina and healthy retina (**c**) [8]

there were no occurrences of growth of transplanted tissue suggestive of a teratoma, ectopic tissue, or other adverse hESC-related growth [8].

In terms of visual acuity, despite transplantation at transition zone of atrophic and compromised retina instead of directly at the fovea, eight of eighteen patients had an improvement in visual acuity of at least 15 letters during the first year after surgery. Participants in this study also reported their perceived visual changes with the National Eye Institute Visual Function Questionnaire 25, which has vision subscales for general vision, peripheral vision, near activities, and distance activities [30, 31]. Participants in this study generally reported improvement in all parameters.

Overall, this study was the first of its kind to use hESC-RPE in human subjects. It showed that hESC-derived cells were well tolerated for up to 37 months after transplantation and there were no serious adverse safety events attributable to hESC use [8].

Despite good overall outcomes with hESC-RPE suspensions delivered into the subretinal space, it should be stressed that RPE cells must attach to the underlying Bruch's membrane and form polarized monolayers to have the most function and allow for phagocytosis of photoreceptor outer segments and formation of the blood ocular barrier [1]. However, it is unclear whether injected RPE suspensions into patients with aged Bruch's membrane provide the most optimal support for RPE survival [32–34]. An aged Bruch's membrane may also hinder the initial formation of functional polarized monolayers [35] that is crucial to RPE physiology. Several studies are underway to see if RPE cells can be induced to overexpress integrin family molecules to help with adherence to Brunch's membrane [36, 37], but until there is a good solution for RPE adherence to Bruch's membrane, subretinal injected suspensions of hESC-RPE cells may suffer from gradual loss and dysfunction of implanted cells [38]. Recent animal work has shown improved survival of hESC-RPE polarized monolayers as compared to cell suspensions, although delivery of monolayers is technically more difficult [38]. Furthermore, studies have also shown injection of human umbilical tissue-derived cells, which secrete specific bridge molecules, can repair RPE function in RCS RPE cells in vitro, demonstrating the potential for using non-RPE-based cells for the treatment of retinal degenerative diseases [39].

6 Future Directions

There is much potential for future therapies utilizing human embryonic stem cell transplants. Major challenges in bringing therapies to the patient exist, as stem cell-based therapies are costly and labor-intensive. Although allogeneic or autologous cell sources can be used, the advantage of human embryonic stem cells is that they allow for mass production for wide scale clinical use [18]. Furthermore, it is difficult to assess for safety in allogeneic transfers, whereas embryonic stem cells can be derived and stimulated to differentiate under controlled conditions [18], with multiple steps in place to ensure purity of culture, as described previously.

Although human clinical trials with embryonic stem cell-derived RPE suspensions have been studied, there yet remains a clinical trial examining transplantation of polarized RPE monolayers. A clinical trial is currently being planned in which hESC-RPE monolayers will be seeded on an ultrathin membrane at the University of Southern California Eye Institute. This membrane will, in theory, support the hESC-RPE cells as an artificial Bruch's membrane and will provide the RPE cells a framework on which to attain a uniform, differentiated monolayer [1]. Additionally, the University College London is conducting a trial using hESC-RPE cell lines as monolayer scaffolds for wet age-related macular degeneration, to assess the utility of RPE transplantation in the setting of neovascular disease (London Project to Cure Blindness, Pfizer) [1].

As techniques refine, the isolation of pure derivatives of human embryonic stem cells will increase, allowing for greater safety and reproducibility. To date, there have been no reported cases of tumor formation, but follow-up at this time still remains limited. As patients with hESC-RPE transplantation continue to be monitored, more will be known about the long-term safety of these transplantations over 10–20 years, with larger trials also helping to accurately assess safety concerns on a larger scale. The continued ability of RPE cells to adhere to the underlying Bruch's membrane will also be monitored over time. Additionally, more will be known about supporting materials that can help with adherence of these cells, including synthetic monolayers and matrices that can act as an artificial Bruch's membrane on which RPE monolayers can be seeded. Furthermore, functional outcomes may differ depending on the stage of macular degeneration, and, therefore, more trials may start to enroll patients with earlier macular degeneration, prior to complete atrophy of the retinal pigment epithelium. Finally, stem cell therapy can extend beyond treatment of retinopathy to treatment of neuropathy, providing a possible treatment for patients afflicted with Leber's hereditary optic neuropathy, traumatic optic neuropathy, ischemic optic neuropathy, glaucomatous optic neuropathy, or other forms of irreversible visual loss.

As hESC-RPE transplantation is performed, better antirejection medications may be developed to help sustain these cell populations. Despite the subretinal space being a relatively immunoprivileged site, surgical trauma from hESC-RPE transplantation can break down the blood-ocular barrier and lead to immune recognition. Perhaps local antirejection medications can be delivered to the eye, to decrease rejection but also avoid systemic immunosuppression. In addition, more work is being done with induced pluripotent stem cells (iPSCs), which are derived from somatic cells but are stimulated to express factors that induce pluripotency [40]. The derivation of pluripotent stem cells from a patient's own somatic cells is promising as these cells possess autologous immune markers and thereby can evade immune detection. However, much research is still needed regarding iPSCs and their safety, and little is known if similar genetic defects persist in autologous iPSCs that would predispose them to degenerative disease [1, 40].

There are additional considerations to optimize the surgical efficacy and side effects of stem cell transplantation. Reflux of injected cells into the vitreous is a potential problem, given that these cells may trigger fibrous proliferation of proliferative vitreoretinopathy although this did not occur to any significant degree in the phase 1/2 hESC-RPE studies [1]. Significant care must be taken to reduce postoperative cystoid macular edema or inflammation, as the presence of these etiologies could further potentiate rejection. Furthermore, as further studies enroll, the efficacy of direct injection of hESC-RPE via a subretinal bleb versus implantation via retinotomy and then localized injection will become clear, with the former perhaps being more advantageous due to decreased surgical trauma. Smaller-gauge vitrectomy instrumentation may also limit the drawbacks of the retinotomy approach. Little is known as to how further surgeries such as cataract extraction may affect RPE transplants. There are also many new surgical techniques being developed that can help deliver RPE suspensions to the subretinal space in a more successful and less traumatic fashion. Hydrogel-encapsulated rigid-elastic ultrathin carriers have been shown to have easier injection and better cellular protection [41]. Platform devices to deliver RPE cells in a monolayer have proven successful for delivery of RPE cells into the subretinal space in a polarized fashion [42]. Additionally, technological advances allow for better in-vivo imaging [43], with intraoperative microscope-mounted OCT [44, 45] being a crucial advance that can confirm successful subretinal RPE transplantation in real time.

The eventual goal is to create a reproducible and marketable method to help patients with visual loss from degenerative ocular conditions. Much like anti-vascular endothelial growth factor (VEGF) injections revolutionized the treatment of exudative age-related macular degeneration, retinal vein occlusion, and diabetic macular edema, stem-cell therapy has the potential to bring new momentum to retinal diseases that are currently inexorably progressive.

References

1. Nazari H, Zhang L, Zhu D, Chader GJ, Falabella P, Stefanini F, et al. Stem cell based therapies for age-related macular degeneration: the promises and the challenges. Prog Retin Eye Res. 2015;48:1–39.
2. Thomson JA, Itskovitz-Eldor J, Shapiro SS, Waknitz MA, Swiergiel JJ, Marshall VS, et al. Embryonic stem cells derived from human blastocysts. Science. 1998;282(5391):1145–7.
3. Evans MJ, Kaufman MH. Establishment in culture of pluripotential cells from mouse embryos. Nature. 1981;292(5819):154–6.
4. Martin G. Isolation of a pluripotent cell line from early mouse embryos cultured in medium conditioned by teratocarcinoma cells. Proc Natl Acad Sci U S A. 1981;78(12):7634–8.
5. Reubinoff BE, Pera MF, Fong CY, Trounson A, Bongso A. Embryonic stem cell lines from human blastocysts; somatic differentiation in vitro. Nat Biotechnol. 2000;18(4):399–404.
6. Zheng YL. Some ethical concerns about human induced pluripotent stem cells. Sci Eng Ethics. 2016;22(5):1277–84.
7. Rippon HJ, Bishop AE. Embryonic stem cells. Cell Prolif. 2004;37:23–34.
8. Schwartz SD, Regillo CD, Lam BL, Eliott D, Rosenfeld PJ, Gregori NZ, et al. Human embryonic stem cell-derived retinal pigment epithelium in patients with age-related macular degeneration and Stargardt's macular dystrophy: follow-up of two open-label phase 1/2 studies. Lancet. 2015;385(9967):509–16.
9. Sparrow JR, Hicks D, Hamel CP. The retinal pigment epithelium in health and disease. Curr Mol Med. 2010;10(9):802–23.
10. Strauss O. The retinal pigment epithelium in visual function. Physiol Rev. 2005;85(3):845–81.
11. Chew EY, Clemons TE, Agrón E, Sperduto RD, Sangiovanni JP, Davis MD, et al. Ten-year follow-up of age-related macular degeneration in the age-related eye disease study: AREDS report no. 36. JAMA Ophthalmol. 2014;132(3):272–7.
12. Han Z, Conley SM, Naash MI. Gene therapy for Stargardt disease associated with ABCA4 gene. Adv Exp Med Biol. 2014;801:719–24.
13. Strauss O, Stumpff F, Mergler S, Wienrich M, Widerholt M. The Royal College of Surgeons rat: an animal model for inherited retinal degeneration with a still unknown genetic deficit. Acta Anat. 1998;162(2–3):101–11.
14. D'Cruz PM, Yasumura D, Weir J, Matthes MT, Abderrahim H, LaVail MM, et al. Mutation of the receptor tyrosine kinase gene Mertk in the retinal dystrophic RCS rat. Hum Mol Genet. 2000;9(4):645–51.
15. Sheedlo HJ, Li L, Turner JE. Photoreceptor cell rescue at early and late RPE-cell transplantation periods during retinal disease in RCS dystrophic rats. J Neural Transplant Plast. 1991;2(1):55–63.
16. Lin N, Fan W, Sheedlo HJ, Aschenbrenner JE, Turner JE. Photoreceptor repair in response to RPE transplants in RCS rats: outer segment regeneration. Curr Eye Res. 1996;15(10):1069–77.
17. Kawasaki H, Suemori H, Mizuseki K, Watanabe K, Urano F, Ichinose H, et al. Generation of dopaminergic neurons and pigmented epithelia from primate ES cells by stromal cell-derived inducing activity. Proc Natl Acad Sci U S A. 2002;99(3):1580–5.
18. Lund RD, Wang S, Klimanskaya I, Holmes T, Ramos-Kelsey R, Lu B, et al. Human embryonic stem cell-derived cells rescue visual function in dystrophic RCS rats. Cloning Stem Cells. 2006;8(3):189–99.
19. Lu B, Malcuit C, Wang S, Girman S, Francis P, Lemieux L, et al. Long-term safety and function of RPE from human embryonic stem cells in preclinical models of macular degeneration. Stem Cells. 2009;27(9):2126–35.
20. Wang S, Lu B, Wood P, Lund RD. Grafting of ARPE-19 and Schwann cells to the subretinal space in RCS rats. Invest Ophthalmol Vis Sci. 2005;46(7):2552–60.
21. Pinilla I, Lund RD, Lu B, Suavé Y. Measuring the cone contribution to the ERG b-wave to assess function and predict anatomical rescue in RCS rats. Vis Res. 2005;45(5):635–41.
22. Gouras P, Flood MT, Kjeldbye H. Transplantation of cultured human retinal cells to monkey retina. An Acad Bras Cienc. 1984;56(4):431–43.
23. Gouras P, Flood MT, Kjedbye H, Bilek MK, Eggers H. Transplantation of cultured human retinal epithelium to Bruch's membrane of the owl monkey's eye. Curr Eye Res. 1985;4(3):253–65.
24. Lopez R, Gouras P, Brittis M, Kjeldbye H. Transplantation of cultured rabbit retinal epithelium to rabbit retina using a closed-eye method. Invest Ophthalmol Vis Sci. 1987;28(7):1131–7.
25. Da Cruz L, Chen FK, Ahmado A, Greenwood J, Coffey P. RPE transplantation and its role in retinal disease. Prog Retin Eye Res. 2007;26(6):598–635.
26. Binder S, Krebs I, Hilgers RD, Abri A, Stolba U, Assadoulina A, et al. Outcome of transplantation of autologous retinal pigment epithelium in age-related macular degeneration: a prospective trial. Invest Ophthalmol Vis Sci. 2004;45(11):4151–60.
27. Thumann G, Aisenbrey S, Schraermeyer U, Lafaut B, Esser P, Walter P, et al. Transplantation of autologous iris pigment epithelium after removal of choroidal neovascular membranes. Arch Ophthalmol. 2000;118(10):1350–5.
28. Aisenbrey S, Lafaut BA, Szurman P, Hilgers RD, Esser P, Walter P, et al. Iris pigment epithelial translocation in the treatment of exudative macular degeneration: a 3-year follow up. Arch Ophthalmol. 2006;124(2):183–8.
29. Maclaren RE, Bird AC, Sathia PJ, Aylward GW. Long-term results of submacular surgery combined with macular translocation of the retinal pigment epithelium in neovascular age-related macular degeneration. Ophthalmology. 2005;112(12):2081–7.
30. Mangione CM, Berry S, Spritzer K, Janz NK, Klein R, Owsley C, et al. Identifying the content area for the 51-item National Eye Institute Visual Function Questionnaire: results from focus groups with visually impaired persons. Arch Ophthalmol. 1998;116(2):227–33.
31. Mangione CM, Lee PP, Gutierrez PR, Spritzer K, Berry S, Hays RD, et al. Development of the 25-item National Eye Institute Visual Field Questionnaire. Arch Ophthalmol. 2001;119(7):1050–8.
32. Gullapalli VK, Sugino IK, Van Patten Y, Shah S, Zarbin MA. Retinal pigment epithelium resurfacing of aged submacular human Bruch's membrane. Trans Am Ophthalmol Soc. 2004;102:123–37.
33. Gullapalli VK, Sugino IK, Van Patten Y, Shah S, Zarbin MA. Impaired RPE survival on aged submacular human Bruch's membrane. Exp Eye Res. 2005;80:235–48.
34. Pan CK, Heilweil G, Lanza R, Schwartz SD. Embryonic stem cells as a treatment for macular degeneration. Expert Opin Biol Ther. 2013;13(8):1125–33.
35. Sugino IK, Sun Q, Wang J, Nunes CF, Cheewatrakoolpong N, Rapista A, et al. Comparison of FRPE and human embryonic stem cell-derived RPE behavior on aged human Bruch's membrane. Invest Ophthalmol Vis Sci. 2011;52(8):4979–97.
36. Fang IM, Yang CH, Yang CM, Chen MS. Overexpression of integrin alpha 6 and beta 4 enhances adhesion and proliferation of human retinal pigment epithelial cells on layers of porcine Bruch's membrane. Exp Eye Res. 2009;88(1):12–21.
37. Afshari FT, Kwok JC, Andrews MR, Blits B, Martin KR, Faissner A, et al. Integrin activation or alpha 9 expression allows retinal

pigmented epithelial cell adhesion on Bruch's membrane in wet age-related macular degeneration. Brain. 2010;133:448–64.
38. Diniz B, Thomas P, Thomas B, Ribeiro R, Hu Y, Brant R, et al. Subretinal implantation of retinal pigment epithelial cells derived from human embryonic stem cells: improved survival when implanted as a monolayer. Invest Ophthalmol Vis Sci. 2013;54(7): 5087–96.
39. Cao J, Murat C, An W, Yao X, Lee J, Santulli-Marotto S, et al. Human umbilical tissue-derived cells rescue retinal pigment epithelium dysfunction in retinal degeneration. Stem Cells. 2016;34(2):367–79.
40. Takahashi K, Yamanaka S. Induction of pluripotent stem cells from mouse embryonic and adult fibroblast cultures by defined factors. Cell. 2006;126(4):663–76.
41. Stanzel V, Liu Z, Brinken R, Braun N, Holz FG, Eter N. Subretinal delivery of ultrathin rigid-elastic cell carriers using a metallic shooter instrument and biodegradable hydrogel encapsulation. Invest Ophthalmol Vis Sci. 2012;53(1):490–500.
42. Hu Y, Liu L, Lu B, Zhu D, Ribeiro R, Diniz B, et al. A novel approach for subretinal implantation of ultrathin substrates containing stem cell-derived retinal pigment epithelium monolayer. Ophthalmic Res. 2012;48(4):186–91.
43. Butler MC, Sullivan JM. A novel, real-time, in vivo mouse retinal imaging system. Invest Ophthalmol Vis Sci. 2015;56(12):7159–68.
44. Ehlers JP, Tao YK, Farsiu S, Maldonado R, Izatt JA, Toth CA. Integration of a spectral domain optical coherence tomography system into a surgical microscope for intraoperative imaging. Invest Ophthalmol Vis Sci. 2011;52(6):3153–9.
45. Ehlers JP, Srivastava SK, Feiler D, Noonan AI, Rollins AM, Tao YK. Integrative advances for OCT-guided ophthalmic surgery and intraoperative OCT: microscope integration, surgical instrumentation, and heads-up display surgeon feedback. PLoS One. 2014;9(8):e105224.

Cell Transplantation Therapy for Glaucoma

Xiong Zhang, Praseeda Venugopalan, and Jeffrey L. Goldberg

1 Problem and Hypothesis

Glaucoma, a leading cause of irreversible blindness worldwide, is a group of chronic, multifactorial neurodegenerative diseases characterized by progressive optic nerve degeneration and retinal ganglion cell (RGC) death [1]. Clinically, RGC loss is evident in patients with optic nerve head cupping, disc excavation, and retinal nerve fiber layer thinning [2]. Since the progressive vision loss is mostly painless and unnoticeable until after significant damage occurs, glaucoma is sometimes called "silent disease." Clinical research has shown that elevated intraocular pressure (IOP) is the most relevant, treatable factor for the development of glaucoma [3]. The current treatment for slowing RGC death in glaucoma is lowering IOP surgically or pharmacologically [4]. However, in some patients, ocular hypotensive therapy fails to stop progressive glaucomatous vision loss [5], and in other patients, glaucoma presents at a late stage and extensive numbers of RGCs have been lost already. Unfortunately, RGC loss and vision impairment are normally irreversible in adult mammals including humans, and can lead to bilateral blindness in as many as 14% of all diagnosed patients [6]. Therefore, novel treatments to prevent RGC loss, halt glaucoma progression, or even restore vision in advanced disease after RGC death are urgently needed.

Cell transplantation therapies provide new hope for treating glaucoma and other optic neuropathies. Cell therapyin glaucoma has two major goals: (1) to provide neuroprotective support to surviving neurons and slow the course of disease and (2) to replace neurons that have already been lost and restore visual function by transplanting RGCs or RGC-like cells derived from stem cells. In prior studies using transplanted cells in the CNS, neural precursors and differentiated neurons survive transplantation and form appropriate synaptic connections, integrating into the host circuit to a certain degree [7–9]. In the retina, attempts at cell transplantation to repopulate the photoreceptor layer have had modest success [10–13], leading to some level of restored vision in disease models. But little progress has been made for RGC transplantation and regeneration. RGC replacement is complex, as it requires both local, retinal integration, and axon regeneration down the optic nerve to appropriate targets in the brain. Recently, our group has discovered that transplanted RGCs can integrate into mature retina, respond to light, and reach synaptic targets in the brain in some cases [14]. These studies represent a promising approach to develop cell transplantation therapies for glaucoma and other neurodegenerative diseases.

X. Zhang, Ph.D. • P. Venugopalan, Ph.D.
J.L. Goldberg, M.D., Ph.D. (✉)
Byers Eye Institute at Stanford University,
2452 Watson Court, Palo Alto, CA 94303, USA
e-mail: jlgoldbe@stanford.edu

2 Scientific Basis for Intervention

The mammalian retina has a highly organized structure with different cell types occupying distinct layers in the tissue (Fig. 1). The retinal pigmented epithelium (RPE) is a monolayer of pigmented cells that form part of the blood/retina barrier and provide nutrition and other critical functions to the adjacent photoreceptors. The outer nuclear layer (ONL) is the outermost layer of cells for neurosensory retina comprising cell bodies of the photoreceptors. The rod and cone photoreceptors are capable of sensing light due to the photopigments in their outer segments. These cells hyperpolarize in response to light causing a reduction of glutamate release from their synaptic terminals. This reduction of release produces responses in the postsynaptic cells, namely the bipolar cells and horizontal cells expressing glutamate receptors. The postsynaptic bipolar and horizontal cells have their dendrites within the outer plexiform layer (OPL) and their cell bodies in the inner nuclear layer (INL). The inner nuclear layer also houses a large proportion of amacrine cells in the retina. Amacrine cells are interneurons that process visual

S.D. Schwartz et al. (eds.), *Cellular Therapies for Retinal Disease*, DOI 10.1007/978-3-319-49479-1_6

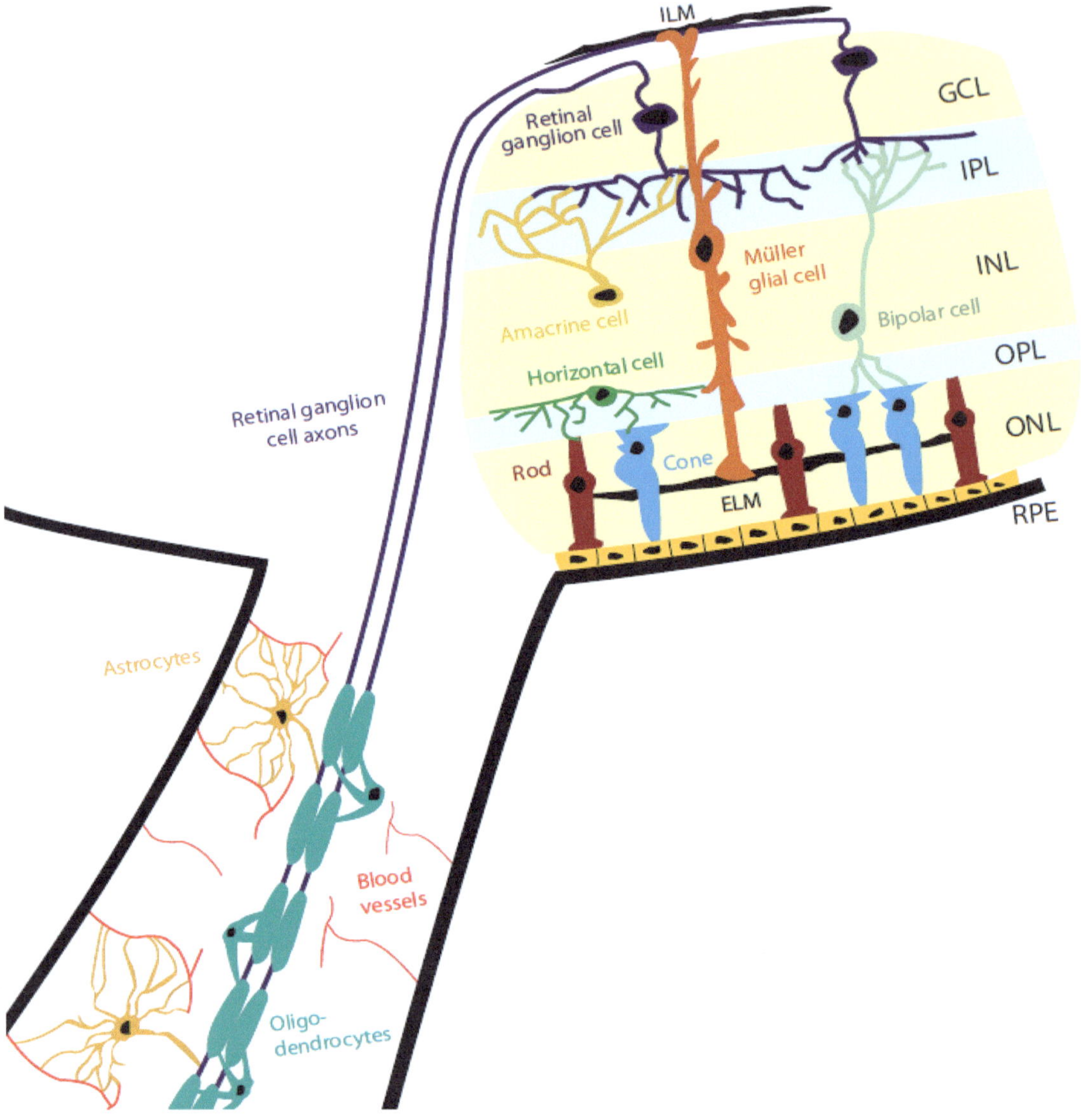

Fig. 1 Schematic representation of a cross section of the eye at the optic nerve head. Retinal neurons occupy precise layers in the retina. Axons from RGCs exit the eye at the optic nerve head and form the optic nerve

information on the way to the retinal ganglion cells (RGCs), with whom they interact in the inner plexiform layer (IPL). Within the IPL, RGCs, the output neurons of the retina, receive input from their presynaptic partners and are responsible for transmitting visual information to distinct targets in the brain via optic nerve.

The retina has a highly ordered structure and well-studied physiology, making it an attractive target for cell therapy. The retina is amenable to surgical intervention and both intravitreal and subretinal delivery methods have been established. In addition, the retina is an immune-privileged site in normal conditions that may permit allogeneic cell therapy [15]. In the disease state or after surgery, however, this immune-privilege may be compromised [16]. Noninvasive technologies such as optical coherence tomography (OCT) and electroretinogram (ERG) allow analysis of structural and functional recovery after cell transplantation. In degenerative retinal diseases such as Stargardt macular dystrophy and age-related macular degeneration (AMD), in which photoreceptors and their intimately associated functional partners, RPE cells, are lost leading to severe vision impairment, significant progress has been made using RPE transplantation to restore function in human clinical trials [17–19]. In addition, preclinical studies of photoreceptor transplantation have yielded promising results, where transplanted cells survive and integrate with host retina, with some level of functional recovery [10–13, 20]. In AMD and other diseases affecting the outer retina and RPE, however, transplanted RPE cells mainly need to make contact with adjacent photoreceptors, and transplanted photoreceptors similarly just need to form local synaptic connections with bipolar and horizontal cells.

Could the same approach to cell replacement be applied to RGCs? Cell connectivity and physiology of photoreceptors or retinal pigment epithelium is different than that of RGCs, which are CNS projection neurons with complex phenotypes. RGCs have to form synaptic connections with local retinal interneurons (primarily amacrine cells), thereby collecting visual information. Then, to relay visual information, they have to send lengthy axons towards the optic nerve head, along the optic nerve, to targets in the brain such as the lateral geniculate nucleus (LGN) and the superior colliculus (SC). In addition, there are many subtypes of RGCs with distinct functions and cell type-specific targets in the brain [21]. For example, intrinsically photosensitive RGCs (ipRGCs) express the photopigment melanopsin, project to the suprachiasmatic nucleus (SCN) and are critical for circadian rhythm entrainment [22]. Whether RGCs cross the optic chiasm and project ipsilaterally or contralaterally, target selection and more specific synapse formation in the brain, combined with their selective input and sensitivity derived from intra-retinal circuitry, all play critical roles in RGC function and ultimately in vision [23–25]. These may all prove important considerations for RGC transplantation as cell replacement therapy.

There has been some evidence for transplanted cells providing neuroprotection of RGCs after optic nerve insult [26]; for example, transplantation of stem cell-derived retinal progenitors and embryonic retinal progenitors can improve optic nerve regeneration [27, 28] and visual function [29] in RGC-depleted animals. While cell transplantation for neuroprotection has been progressing towards human clinical trials, cell therapy approaches for RGC replacement have not yet progressed as far.

3 Preclinical Data and Early Translational Efforts

3.1 Cell Transplantation for Neuroprotection in Glaucoma

Multiple factors contribute to RGC death, including failure of axon transport, loss of local and target-derived neurotrophic factors, dysfunction of mitochondria, and oxidative stress [1]. In the progression of glaucoma, deprivation of neurotrophic factors and the subsequent activation of RGC apoptosis may play key roles in stress-induced RGC loss [1]. Of particular importance is brain-derived neurotrophic factor (BDNF). In rodent brain, BDNF is strongly expressed in the lateral geniculate nucleus and the superior colliculus—both major brain target regions for RGC projections—and is retrogradely transported to the retina within RGC axons [1]. BDNF can potently promote RGC survival via activation of the tropomyosin related kinase (Trk) family of receptor tyrosine kinases during development and after injuries such as optic nerve crush or chronic increase in intraocular pressure (IOP) [30]. Other neurotrophic factors, such as ciliary neurotrophic factor (CNTF) and glial cell line-derived neurotrophic factor (GDNF), are also promising neurotrophic factors for exogenously promoting RGC survival and growth after injury, both in vitro and in animal models [1, 30]. New therapies directed at supplementing these neurotrophic factors in glaucoma are under development, but it is likely that sustained delivery will be needed to achieve significant long-term effects [31]. Therefore, cell-based therapy that allows constant secretion of neurotropic factors is an enticing approach (Fig. 2). Many types of stem cells can secrete growth factors including neurotrophic factors that are proposed to activate multiple pathways to protect against neuronal stress and injury [32].

Mesenchymal stem cells (MSCs) or bone marrow stromal cells are a diverse population of adult stem cells of mesodermal origin [33, 34]. Originally identified in the bone marrow, MSCs represents 0.001–0.01% of the cell population in the bone marrow [33]. MSCs can be found in other tissue or organs in the body, such as adipose tissue, dental pulp, and umbilical cord [34]. MSCs can be identified with a few defining features such as expression of cell surface markers CD105, CD73, and CD90; and differentiation potential to osteoblasts, adipocytes, and chondroblasts in vitro [35]. MSCs also have remarkable paracrine effects in secreting multiple growth factors, promoting cell survival, and recruiting repair mechanisms. In addition, they have strong immunosuppressive properties by inhibiting pro-inflammatory cytokine release [33]. These features make MSC transplantation an attractive potential therapy to treat degenerative CNS disorders such as glaucoma.

A number of preclinical studies in rodent models of glaucoma have demonstrated proof of concept feasibility and efficacy of MSC transplantation, and shed light on molecular mechanisms of action. Potential MSC-based glaucoma therapy includes MSCs derived from bone marrow (BMSCs) [36–39], umbilical cord blood (UCB-MSC) [40, 41], adipose tissues (AD-MSC) [28, 42], and dental pulp (DP-MSC) [26, 28]. In rodent models of glaucoma, MSCs support RGC survival and promote axon maintenance and even regeneration through the secretion of growth factors, such as BDNF, CNTF, and GDNF [26, 39, 43], as well as other key pro-survival factors including platelet-derived growth factor (PDGF) [43] and vascular endothelial growth factor (VEGF) [26]. This paracrine effect of MSCs is the main reason for neuroprotection of RGCs, as MSC-conditioned media have similar protective effects. Notably, the activity of Trk receptors are critical for RGC protection, with significantly diminished effects after treatment with Trk inhibitors [26]. In addition, MSC-derived neuroprotection can involve other

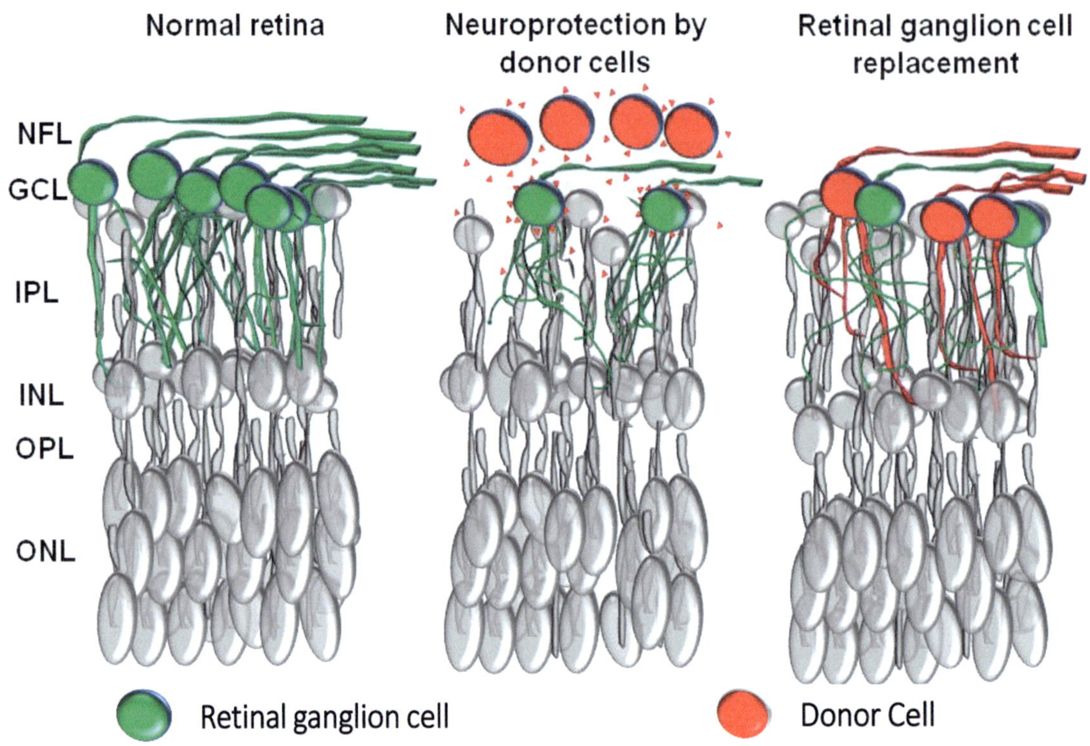

Fig. 2 Strategies for treating RGC degeneration: neuroprotection and cell replacement

indirect mechanisms, including secretion of extracellular matrix proteins involved in cell adhesion and growth, and alternation of the microenvironment to recruit and support internal repair mechanisms such as from adult neural progenitor cells [26, 43, 44]. After intravitreal injection, MSCs survive at least 3 weeks [26, 37, 39], providing sustained release of neurotrophic factors to support RGC survival after injury. In addition, MSCs are "immune privileged," regulate immune response with secreted factors, suppress inflammation, and can promote endogenous cellular repair mechanisms [45]. Together these preclinical rodent studies suggest that cell-based therapy using MSCs is a potential therapy to treat glaucoma for neuroprotection and repair.

These important properties have led to clinical trials using allogeneic MSC transplantation to treat steroid-resistant, severe, acute graft-versus-host diseases [46] and autologous MSC transplantation to treat chronic ischemic cardiomyopathy [47], among many other trials. These studies showed some evidence of safety and efficacy of MSC treatment [46, 47]. Therefore, systemic or local delivery of allogeneic or autologous MSCs is expected to be safe for testing in degenerative diseases. The vitreous cavity is an immune-privileged environment and amiable for cell transplantation [48], and there are over ten ongoing clinical trials on intravitreal injection of MSCs to treat degenerative eye diseases [48]. Two of these studies have been published: in one study, the safety of intravitreal injection of autologous bone marrow-derived mononuclear cells in patients with retinitis pigmentosa or cone–rod dystrophy was studied. Patients were injected with ten million cells intravitreally in one study and followed up for 10 months. No adverse events were reported but of course the small sample size precludes any definitive conclusions on efficacy [49]. In another study, intravitreal transplantation of autologous bone marrow-harvested $CD34^+$ cells was studied for treatment of ischemic/degenerative retinal diseases in six patients. Autologous $CD34^+$ cells were isolated, purified by MACS sorting, and one to seven million of cells were injected per eye. Again, no adverse events were observed and the therapy appeared to be feasible and tolerated. On the other hand, patients being treated with purported MSCs outside of clinical trials under adequate ethics board and regulatory oversight are at significant risk of negative outcomes, as recently observed in three patients blinded bilaterally [50].

In addition to MSCs, neural progenitor cell (NPC) transplantation has also been studied in neuroprotection [19], and human NPCs engineered to secrete IGF-1 were shown to improve RGC survival and protect against RGC loss in a rodent model of glaucoma [51]. Some non-stem cells, including glial and RPE cells that have also been demonstrated to confer some measure of neuroprotection, can likewise be engineered to secrete extra neurotrophic factors and be transplanted intravitreally for neuro-protection in glaucoma [32]. For example, olfactory ensheathing cells (OECs), which normally wrap olfactory nerve axons and naturally produce factors such as BDNF, CNTF, and GDNF [52], were shown in vitro [53] and in vivo [54] to protect against RGC loss and promote axonal regeneration. Genetically modified CNTF-secreting RPE cells has been tested as a therapy for retinitis pigmentosa [55] and a phase-I clinical trial for glaucoma (NCT01408472; http://clinicaltrials.gov/) has been completed with no serious adverse events and encouraging results (our unpublished data).

While the eye is an accessible, relatively immune-privileged site ideal for cell therapy, and as aforementioned, there are very encouraging preclinical studies transplanting different cell types for neuroprotection in glaucoma, several factors need to be addressed and reliably defined to advance a cell therapy. These factors include the cell source, cell culture conditions, and rigorous adherence to current Good Manufacturing Practice (cGMP) with minimized batch-to-batch variation and vigorous quality control. Cells can be encapsulated in a removable device to protect against immune response and other unwanted interactions or complications. Immunogenicity should be minimized, and tumorigenicity and genetic alteration needs to be avoided. For example, transplanted cells genetically engineered using lentiviral infection to increase secretion of neurotrophic factors may acquire mutations after random gene insertion into grafted cells. Moreover, if long-term survival of graft is expected, transplant-host interaction mechanisms need to be studied to assure safety and efficacy. For example, transplanted neural progenitor cells can activate glial cells and induce matrix metalloprotease-2 secretion. These in turn promote an environment supportive of neurite outgrowth [56]. In contrast, another study found that intravitreal MSC injection can induce reactive gliosis, which is associated with gliosis-mediated retinal folding, recruitment of macrophages, and upregulation of intermediate filaments. These responses created a barrier for the engraftment of transplanted cells, and could be attenuated by STAT3 inhibition [43]. Thus, cell source and subtle forms of long-term toxicity need to be carefully considered, even when transplanting cells purely for their secreted delivery of neuroprotective factors.

3.2 Cell Transplantation for RGC Replacement and Function Restoration

While there is accumulating evidence that neuroprotective cell therapy for glaucoma and other diseases is promising, such interventions are of potential human use only if the disease is diagnosed in early stages. If by the time the disease is diagnosed a patient already has considerable loss of RGCs, neuroprotection may be of limited use. Instead, cell replacement becomes a more attractive option, wherein dead or dying cells may be replenished through transplantation (Fig. 2).

In the retina, attempts at cell transplantation to repopulate the RPE layer have led the field with RPE transplantation under phase II clinical trial to treat AMD and RP [17–19]. Significant progress has been made in photoreceptor layer transplantation as well [10–13]. Similar strategies may be applied to replenish the ganglion cell layer, bearing in mind that RGC structure and physiology are more varied and complex than that of the RPE or photoreceptors. Transplanted RGCs need to not only form local connections within the retina, they also have to grow long axons towards the optic nerve head, extend down the optic nerve, and connect with corresponding targets in the brain. Thus far there have been only a few studies investigating cell replacement for RGCs.

One of the primary hurdles for implementing cell replacement strategies is identifying optimal donor cell types. Compared with fully differentiated cell types, stem cells and progenitor cells from various sources are more readily available cell types that could be promising as donor cells. In many cases, sub-retinal and intravitreal injections of neural progenitor cells (such as brain progenitor cells and retinal progenitor cells) resulted in survival of the donor cells within host tissue [57, 58]. One such study indicated that donor adult hippocampal precursor cells (AHPCs) survived in 100% of the neonatal hosts and about 65% of adult hosts [59], but it was not determined if the transplanted cells could contribute to functional recovery. Müller glial cells with stem cell characteristics can differentiate into RGC precursors in vitro and partially restore visual function in vivo [29]. When transplanted into the retinas of RGC-depleted rats using *N*-methyl-D-aspartate treatment, the cells improved their visual function as assessed by electroretinograms (ERGs). Whether this effect is the result of a neuroprotective effect or true integration is unclear but offers a path forward for studying cell replacement for RGC loss in glaucoma.

However, transplanting dividing cells comes with potential tumorigenic risk in addition to differentiation into unwanted cell types that may negatively affect function. Differentiated neurons derived from stem cells or progenitor cells may be much more conducive to targeted therapy for neurodegenerative diseases where one specific cell population is depleted. Such experiments in the retina have likely been limited by our inability to differentiate stem cells into RGCs with high efficiency. In the studies attempted so far, survival of donor cells in the host retina did not always correlate with appropriate retinal marker expression [60, 61] and synaptic integration has not been reported. Another important consideration in studying stem cell-derived donor neurons is whether differentiated RGCs are capable of integrating into a preexisting neural network.

The age of the host animals may greatly influence the ability of donor neurons to functionally integrate following transplantation. While some studies suggest that adult hosts may not be an ideal environment for transplantations, others indicate that the age of the donor cells themselves is of greater importance. However, these data are in some cases contradictory, suggesting in some cases that adult hosts were associated with lower success rates [62, 63] and in others

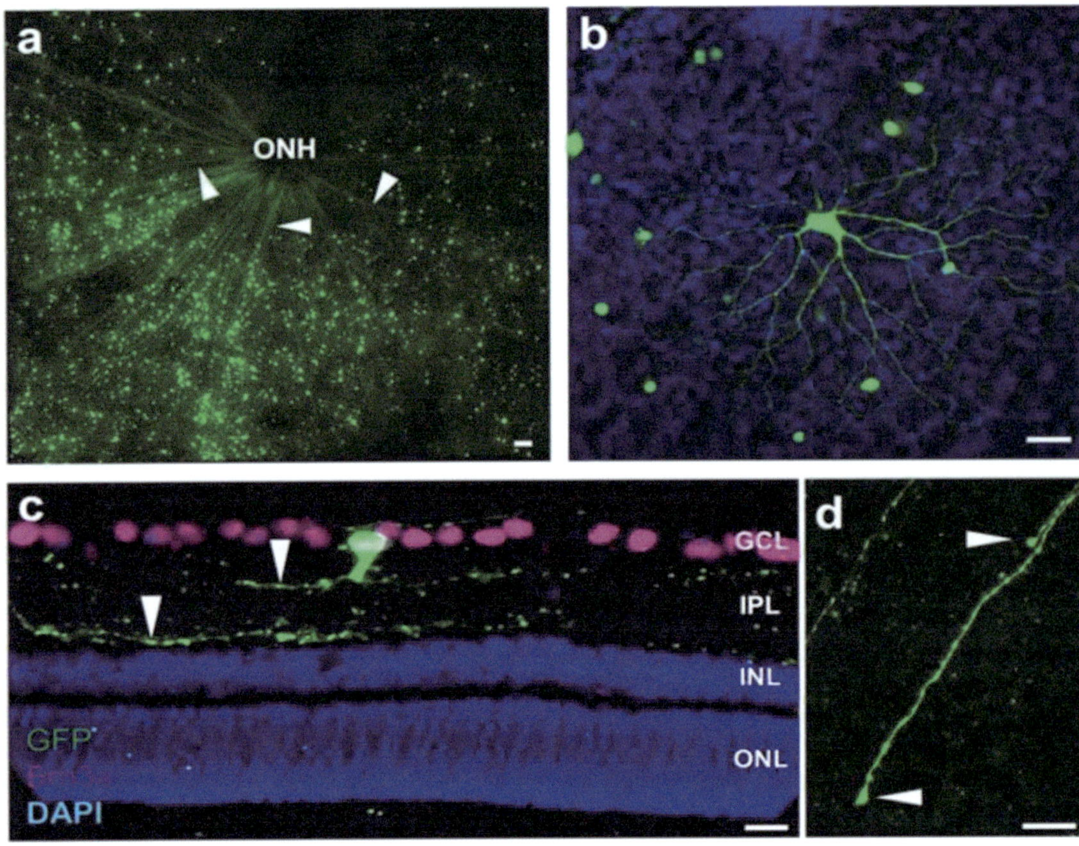

Fig. 3 Morphology of transplanted GFP⁺ RGCs in the host retina. Data from Venugopalan et al. [14] showing dendrites and axons from GFP⁺ donor RGCs growing within the host retina. (**a**) 3 weeks after transplantation, GFP⁺ mouse RGCs were observed in host rat retinal explants with putative axons (*arrow ahead*) extending towards the optic nerve head (ONH). (**b**) Dendrite architecture of donor RGCs with morphologies similar to those of endogenous subtypes of RGCs. (**c**) GFP⁺ RGCs were seen in the host ganglion cell layer (GCL) with neurites (*arrow head*) extending to the inner plexiform layer (IPL) in retinal sections from host eyes. (**d**) One week after transplantation, growth cone-like terminal endings (*arrow heads*) were observed in some GFP⁺ axons in the nerve fiber layer of host retinal explants. Scale bar 25 μm

that graft survival can be similar in both neonatal and adult host retinas [10, 64]. These differences highlight a need for careful study of the contribution of host environment and donor cell properties towards the success of transplants. Since the elderly population represents the majority of patients who could benefit from cell therapy, researchers need to produce high quality donor cells that can survive and integrate in the host environment of elderly patients, or change the host environment to mimic the neonatal environment that is amiable for cell transplantation.

Such studies have mainly focused on survival and differentiation of transplanted cells within host tissue. It is important to also understand to what capacity surviving cells can function in this environment. For example, purified RGCs transplanted onto retinal explants in culture could not only survive but also grow neurites [65]. This suggested that it was possible for donor RGCs to acquire the morphological characteristics of neurons needed for function, namely dendrite and axon growth. But can this lead to functional integration? Recent studies conducted in vivo show the promise of using post-mitotic RGCs as potential donor cells in cell replacement therapies [14]. In a subset of experiments, mouse RGCs transplanted into uninjured rat eyes intravitreally migrated to the correct layer of the host retina, the ganglion cell layer (GCL), and acquired the general morphology of RGCs (Fig. 3). In addition, different morphological subtypes appeared in transplanted RGCs in vivo. Importantly, donor RGCs were electrically active and capable of responding to light stimulation of the host retinas. These results suggest that adult retinas are receptive to differentiated donor neurons and that post-mitotic RGCs have the potential to form functional synapses within adult retinas. The data describing light responses in transplanted cells is encouraging but represents only one aspect of the connectivity required by RGCs for their function [14]. Other important areas include the function of multiple RGC subtypes and the connectivity and projection of different RGC subtypes in vivo.

It may be useful, in addition, to investigate ways to enhance the number of cells retained in the host retina. Simultaneous injection of growth factors and donor cells

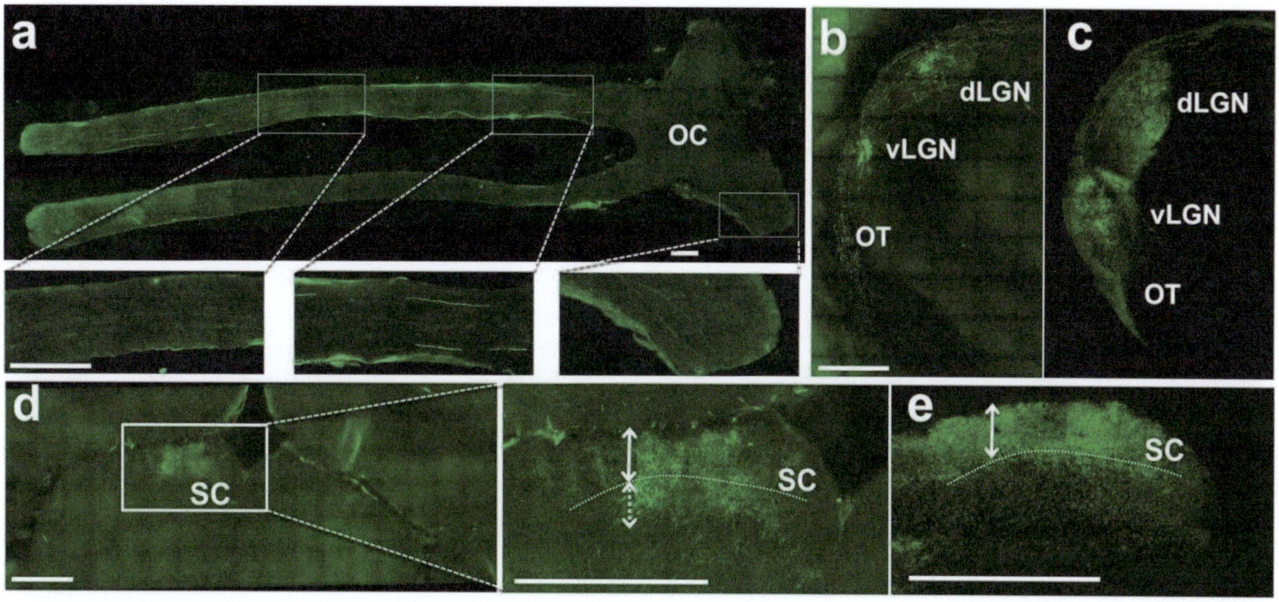

Fig. 4 Transplanted RGCs extend axons through host optic nerves and into known visual targets in the brain. RGCs purified from GFP^{+} mice were transplanted into uninjured rat eyes intravitreally. One month after the unilateral transplantation, optic nerves and brains of host animals were fixed and sectioned. (**a**) GFP^{+} axons extend to the optic chiasm (OC) and cross over in fixed optic nerve sections. (**b**) These axons are visible within the optic tract (OT) and nuclei in the ventral or dorsal lateral geniculate nucleus (LGN), known RGC targets in the brain. (**c**) RGC axon tracts in age-matched control brains labeled by intravitreally injected AAV-GFP for comparison. (**d**) In deeper layers of the superior colliculus (SC), known RGC target in the brain, GFP^{+} axon terminals exhibit an immature innervation pattern. (**e**) For comparison, AAV-GFP labeled RGC axon innervation of superficial layers of the SC in age-matched control animals is more restricted to those layers. Medial-lateral axis of the SC from right to left. Scale bars: 500 μm. Data from Venugopalan et al. [14]

may improve immediate survival and neurite growth of the cells post-transplantation. Enzymatic pretreatment to loosen the vitreous fluid or outer limiting membrane could make it easier for the donor cells to reach the retina and migrate into the ganglion cell layer. Further experiments could study specific donor cell types that can more efficiently form functional synapses with the host retina and connect with their postsynaptic partners in the brain. Moreover, in order to restore vision, transplanted RGCs need to send long axons towards the optic nerve head, along the optic nerve, and reach targets in the brain. Understanding the molecular mechanism of target guidance, and the ability to engineer or modulate a host environment amiable for target reaching is critical for the success of cell replacement therapy.

As projection neurons, however, dendritic growth alone is not sufficient for functional recovery following RGC loss. Axon regeneration to appropriate regions in the brain is equally critical. In a few surveyed tissues, regenerating axons were found to extend from the donor RGCs into the host optic nerve and brain (Fig. 4). It remains to be seen if the growing axons make functional connections with their postsynaptic partners in the brain. These studies establish a promising first step in cell replacement therapy for glaucoma: primary RGCs can survive, integrate local connection in the retina and reach targets in the brain. A critical next step will be to study cell transplantation in animal models of RGC degeneration.

Other cells are also potentially sources for RGC replacement therapy. Retinal progenitor cells from fetal rodents possess the potential to differentiate into all cell types in the retina, can partially protect against photoreceptor loss in animal model of AMD or RP, and appear to be immune-privileged to survive allografting without immune suppression [19, 66, 67]. Müller glia, the major glial component of the retina [68], when cultured in neural stem cell conditions, demonstrate stem cell potential, and can differentiate into RGC-like progeny and partially restore RGC function after transplantation in vivo [29, 69]. In addition, ciliary margin stem cells, multipotent adult stem cells located at the ciliary margin, can grow in stem cell characteristic neurospheres and differentiate into many neuron types in the retina [70, 71]. More recently, a subset of amacrine cells expressing stem cell marker Lgr5 was demonstrated to possess regenerative potential including a capacity to differentiate into other retinal neurons [72].

Retinal ganglion cells derived from human embryonic stem cells (ESCs) or human-induced pluripotent stem cells (iPSCs) may have even greater potential, since these cell sources could ostensibly generate an unlimited supply of RGCs. hESCs are pluripotent cells derived from preimplantation embryos.

hESCs are able to proliferate indefinitely and can differentiate to any cell types of the adult body [73]. iPSCs are functionally similar to hESCs, but are derived from somatic cells such as fibroblasts, which are induced to a stem cell phenotype by expressing specific combinations of transcription factors such as Oct4, Sox2, Klf4, and c-Myc [74]. Additional experiments in the last 10 years established reprogramming methods without viral integration of transcription factors and built a foundation for clinical application of iPSCs [75]. Thus, iPSCs can be derived from individual patients and provide an unlimited cell source for personalized therapy [74, 76, 77].

ESC and iPSCs probably cannot be directly transplanted into patients for cell therapy, as they can create teratomas [74, 78]. Instead, they should be differentiated into specific cell types first, ideally minimizing transfer of pluripotent cells. The development pathway of the major types of cells in the retina is well studied and defined [79, 80]. The neurons and most glia within the retina are generated from a pool of multipotent retinal progenitor cells (RPCs) [79]. The cell fate of sequentially differentiated cell types is restricted by their time of birth from the RPCs. This occurs through a combination of extrinsic and intrinsic cues that activate cohorts of different transcription factors that drive differentiation towards specific cell fates [79, 81]. Following these important advances in our understanding of retinal development, significant progress has been made in differentiating ESCs or iPSCs towards retinal lineage progeny including RPCs [82], photoreceptors [13, 20], RPE [17], and RGCs [83–85]. Differentiation methods have evolved from an attached cell culture monolayer method [13, 82] to a 3-D method [86, 87] more closely mimicking the interactions of normal retinal development. RGCs are one of the first cell types to differentiate from RPCs, and the master regulators of RGC differentiation, including Math5, Brn3, Isl1, and Sox4/11, are defined [88–92]. Significant progress has been made to enhance the efficiency of generating RGCs from ESCs or iPSCs: in early studies, RGCs were generated at a relatively low efficiency in conditions based on growth factor induction [85]. More recently, chemicals such as the Notch inhibitor *N*-[*N*-(3,5-difluorophenacetyl)-L-alanyl]-*S*-phenylglycine *t*-butyl ester (DAPT) were shown to promote RGC-like differentiation from stem cells [93]. In addition, higher efficiency of RGC differentiation was achieved by mono-layer culture as well as 3D culture in different groups [94–96]. Tremendous progress has been made in differentiating RGCs from human iPSCs that express the mature RGC markers and show some functional potency for disease modeling [94–96], yet the purity and maturation stage of the cells, which will likely influence their ability to survive, integrate, and rescue function in disease, is still an area in desperate need of investigation.

Finally, the route and method of delivery may prove critical to the success of cell replacement therapy for glaucoma. As RGCs are on the vitreal face of the retina, intravitreal injection may be the best option, although disadvantages may include the larger vitreous volume that might dilute donor cells away from the retina, and the internal limiting membrane (ILM), which may reduce the number of donor cells able to migrate into the retina. Delivering donor cells on a scaffold could address the first potential problem [97], as such a scaffold could help retain the donor cells in a proximate location next to the retina while they have time to migrate and integrate locally. Peeling the ILM, as is performed by vitreoretinal surgeons today for a variety of indications [98], could address this should it prove limiting. Subretinal delivery, as is done for RPE and photoreceptor transplants [13, 17], could provide another option, but this would then require donor cells to migrate through the retina to take up residence in the ganglion cell layer.

In summary, primary RGC transplantation provide promising evidence that RGC cell replacement therapy is possible. Yet, due to limited tissue supply and complicated procedures, primary cells have limited potential for translation to human therapy. Human ESCs or iPSCs provide an unlimited supply for cell replacement therapy, with significant progress being made in the differentiation of stem cells into RGCs. Therefore, this approach holds great therapeutic promise. However, further study is needed of the quality, differentiation, maturation, functional profile, and transplantation of hESC/hiPSC derived RGCs, before moving to human testing.

3.3 Harnessing Endogenous Regenerative Potential

The discovery of stem cells in the adult retina may suggest a new potential avenue for regenerative medicine. The promising discoveries of an adult stem cell niche in the mammalian ciliary margin [70, 71], Müller glia cells possessing stem cell like potential [29, 69], and subset of amacrine cells with multipotency to generate retinal neurons [72] are all promising sources for RGC replacement therapy. For example, ATOH7 was shown to convert Müller glia cells-derived retinal progenitor cells into RGC-like fate [99]. An intricate network of transcription factors, including ATOH7, Brn3b, Isl1, Sox4, and Sox11, can define RGC fate [88–92]. With the progress of transdifferentiation, pioneered by the generation of neurons from fibroblasts with transduction of defined factors [100], we propose that transduction of those RGC factors might convert the aforementioned endogenous progenitor cells into RGCs. However, a number of questions remain to be answered: How to activate the proliferation potential of the adult stem cells in the retina? How to promote cell migration to the injury site? How to change the cell fate from one type of neuron or glia into RGC and particularly if RGC subtypes need to be specified?

How to encourage newly converted RGCs to restore function, form local synaptic connections, and send lengthy axons to specific targets in the brain?

4 Conclusions

The mammalian retina has a highly organized structure with different types of cells occupying distinct layers. Visual information are captured in photoreceptors in the outmost layer of the neurosensory retina, relayed through and processed by interneurons including horizontal cells, bipolar cells, and amacrine cells, and collected by RGCs for transmitting to distinct target in the brain. Muller glia, a type of glia cells in the retina, serves as support cells for retinal neurons. Outside of the neurosensory retina, a RPE layer provides nutrients for photoreceptors and form part of the blood-retina barrier critical for normal function of the retina. In developed country, the top causes of visual impairment are glaucoma and age-related macular degeneration (AMD): Glaucoma is characterized by progressive optic nerve degeneration and loss of RGCs. While AMD is represented by loss of RPE function and death of photoreceptors. For this chapter, we focus on cell therapy for glaucoma treatment.

The eye is a highly accessible organ for cell therapy with well-studied physiology, established surgical intervention methods, and noninvasive technologies for functional surveillance. In addition, it's an immune-privileged site for allogeneic transplantation. There are two major goals of cell transplantation in the eye: neuro-protection and cell replacement. Transplanted neuro-progenitor cells or mesenchymal stem cells may secrete growth factors that are critical to support surviving neurons and slow disease progression. While transplanted cells may replace RPEs or retinal neurons that have already been lost and restore visual function.

Neurotropic factors, such as BDNF, CNTF, and GDNF, are important for RGC survival. Deprivation of these factors may lead to RGC apoptosis and contribute to disease progression and vision loss in glaucoma. Transplanted cells such as MSCs or NPCs may provide those critical neurotrophic factors, support RGC survival, and maintain axon structure and function. This protective effect has been validated in preclinical MSC transplantation studies in rodent model of glaucoma and is being tested in human clinical trials for degenerative retinal diseases. While accumulating evidence in these studies show that cell transplantation for neuroprotection is promising in glaucoma therapy, a few questions need to be addressed: How to consistently produce high quality, functional donor cells under cGMP? How to thoroughly study the transplant-host interaction in the long term, to minimize immune response, gliosis and tumorigenicity? What are the molecular mechanisms and detailed pathways of neuroprotection associated with cell transplantation?

While neuroprotective cell therapy can protect RGC and delay disease progression, RGC loss is irreversible in most of the cases. Therefore, cell replacement therapy is a more attractive option. In the eye, RPE replacement therapy is leading the field, with cell therapy using hESC-derived RPEs under clinical trial for AMD and RP [17, 18]. In addition, photoreceptor replacement has made significant progress in preclinical rodent models [10, 13, 101]. However, for glaucoma therapy, RGC replacement is more complex: Transplanted RGCs need to survive in the RGC layer, form local connections with retinal interneurons to receive visual information, and extend long axons to the optic nerve head, along the optic nerve, cross the optic chiasm and reach specific targets in the brain. Recent studies from our group showed promising results that post-mitotic RGCs can survive, integrate in host, normal retina and send axons to brain target [14]. Some big questions remain for further investigation: Is the retina in disease state receptive to RGC replacement therapy? How to promote RGC survival and form local, functional connections? Does the transplant RGCs include or adopt different RGC subtypes for function? How to guide axons of transplanted RGCs to their specific brain targets and restore vision?

Cell therapy for glaucoma and other retinal disease require a reliable, high quality source of donor cells. While adult tissue is limited, fetal progenitor tissues are expandable and a potential source. The retina may have some endogenous regenerative potential and cell fate conversion from other types of neuron to RGCs might be possible. Human ESC and iPSC provide an unlimited source of donor cells. In the past 10 years, tremendous progress has been made in generating iPSCs without viral integration, and efficiently producing retinal specific cell types from hESCs or iPSCs [75, 77]. These studies paved the way for human clinical trials for degenerative retinal diseases. Yet, a few factors need to be considered and thoroughly studied: Generating hiPSCs/hESCs and differentiating those cells to RGCs in large quantity with high purity under cGMP conditions; the maturity and functional potency of hESC/hiPSC derived RGCs; the subtypes of these RGCs and the ability to integrate, form functional connections, and reach brain targets; long-term safety, immune-reactions, and tumorigenicity.

In summary, glaucoma is a major cause of blindness with a tremendous unmet medical need for therapies that can slow degeneration or restore vision loss. Cell therapy has great promise for treating glaucoma and the eye is an ideal target organ for cell therapy. As summarized above, cell transplantation for neuroprotection is more realistic for the short-term therapy and is currently translating towards clinical testing. Yet cell replacement and restoration of cell function is an important long-term goal and potential solution for glaucoma. We are on the cusp today, as the coming research progress will continue to focus on achieving the goal of curing glaucoma using cell therapy.

References

1. Almasieh M, Wilson AM, Morquette B, Cueva Vargas JL, Di Polo A. The molecular basis of retinal ganglion cell death in glaucoma. Prog Retin Eye Res. 2012;31:152–81.
2. Weinreb RN, Tee Khaw P. Primary open-angle glaucoma. Lancet. 2004;363:1711–20.
3. Friedman DS, Wilson MR, Liebmann JM, Fechtner RD, Weinreb RN. An evidence-based assessment of risk factors for the progression of ocular hypertension and glaucoma. Am J Ophthalmol. 2004;138:S19–31.
4. Much JW, Liu C, Piltz-Seymour JR. Long-term survival of central visual field in end-stage glaucoma. Ophthalmology. 2008;115:1162–6.
5. Walland MJ, et al. Failure of medical therapy despite normal intraocular pressure. Clin Exp Ophthalmol. 2006;34:827–36.
6. Shields MB. Normal-tension glaucoma: is it different from primary open-angle glaucoma? Curr Opin Ophthalmol. 2008;19:85–8.
7. Shin JJ, et al. Transplanted neuroblasts differentiate appropriately into projection neurons with correct neurotransmitter and receptor phenotype in neocortex undergoing targeted projection neuron degeneration. J Neurosci. 2000;20:7404–16.
8. Ideguchi M, Palmer TD, Recht LD, Weimann JM. Murine embryonic stem cell-derived pyramidal neurons integrate into the cerebral cortex and appropriately project axons to subcortical targets. J Neurosci. 2010;30:894–904.
9. Englund U, Bjorklund A, Wictorin K, Lindvall O, Kokaia M. Grafted neural stem cells develop into functional pyramidal neurons and integrate into host cortical circuitry. Proc Natl Acad Sci U S A. 2002;99:17089–94.
10. MacLaren RE, et al. Retinal repair by transplantation of photoreceptor precursors. Nature. 2006;444:203–7.
11. Luo J, et al. Human retinal progenitor cell transplantation preserves vision. J Biol Chem. 2014;289(10):6362–71.
12. Pearson RA. Advances in repairing the degenerate retina by rod photoreceptor transplantation. Biotechnol Adv. 2014;32:485–91.
13. Lamba DA, Gust J, Reh TA. Transplantation of human embryonic stem cell-derived photoreceptors restores some visual function in Crx-deficient mice. Cell Stem Cell. 2009;4:73–9.
14. Venugopalan P, et al. Transplanted neurons integrate into adult retinas and respond to light. Nat Commun. 2016;7:10472.
15. Streilein JW. Ocular immune privilege: therapeutic opportunities from an experiment of nature. Nat Rev Immunol. 2003;3:879–89.
16. Ambati J, Atkinson JP, Gelfand BD. Immunology of age-related macular degeneration. Nat Rev Immunol. 2013;13:438–51.
17. Schwartz SD, et al. Embryonic stem cell trials for macular degeneration: a preliminary report. Lancet. 2012;379:713–20.
18. Schwartz SD, et al. Human embryonic stem cell-derived retinal pigment epithelium in patients with age-related macular degeneration and Stargardt's macular dystrophy: follow-up of two open-label phase 1/2 studies. Lancet. 2015;385:509–16.
19. Klassen H. Stem cells in clinical trials for treatment of retinal degeneration. Expert Opin Biol Ther. 2015;2598:1–8.
20. Gonzalez-Cordero A, et al. Photoreceptor precursors derived from three-dimensional embryonic stem cell cultures integrate and mature within adult degenerate retina. Nat Biotechnol. 2013;31:741–7.
21. Sanes JR, Masland RH. The types of retinal ganglion cells: current status and implications for neuronal classification. Annu Rev Neurosci. 2014;38:150421150146009.
22. Hattar S, Liao HW, Takao M, Berson DM, Yau KW. Melanopsin-containing retinal ganglion cells: architecture, projections, and intrinsic photosensitivity. Science. 2002;295:1065–70.
23. Erskine L, Herrera E. The retinal ganglion cell axon's journey: insights into molecular mechanisms of axon guidance. Dev Biol. 2007;308:1–14.
24. Petros TJ, Rebsam A, Mason CA. Retinal axon growth at the optic chiasm: to cross or not to cross. Annu Rev Neurosci. 2008;31:295–315.
25. Dhande OS, Huberman AD. Retinal ganglion cell maps in the brain: implications for visual processing. Curr Opin Neurobiol. 2014;24:133–42.
26. Mead B, Logan A, Berry M, Leadbeater W, Scheven BA. Intravitreally transplanted dental pulp stem cells promote Neuroprotection and axon regeneration of retinal ganglion cells after optic nerve injury. Invest Ophthalmol Vis Sci. 2013;54:7544–56.
27. Cho J-H, Mao C-A, Klein WH. Adult mice transplanted with embryonic retinal progenitor cells: new approach for repairing damaged optic nerves. Mol Vis. 2012;18:2658–72.
28. Mead B, Logan A, Berry M, Leadbeater W, Scheven BA. Paracrine-mediated neuroprotection and neuritogenesis of axotomised retinal ganglion cells by human dental pulp stem cells: comparison with human bone marrow and adipose-derived mesenchymal stem cells. PLoS One. 2014;9:e109305.
29. Singhal S, et al. Human Muller glia with stem cell characteristics differentiate into retinal ganglion cell (RGC) precursors in vitro and partially restore RGC function in vivo following transplantation. Stem Cells Transl Med. 2012;1:188–99.
30. Johnson TV, Bull ND, Martin KR. Neurotrophic factor delivery as a protective treatment for glaucoma. Exp Eye Res. 2011;93:196–203.
31. Ko ML, Hu DN, Ritch R, Sharma SC, Chen CF. Patterns of retinal ganglion cell survival after brain-derived neurotrophic factor administration in hypertensive eyes of rats. Neurosci Lett. 2001;305:139–42.
32. Johnson TV, Martin KR. Cell transplantation approaches to retinal ganglion cell neuroprotection in glaucoma. Curr Opin Pharmacol. 2013;13:78–82.
33. Ng TK, Fortino VR, Pelaez D, Cheung HS. Progress of mesenchymal stem cell therapy for neural and retinal diseases. World J Stem Cells. 2014;6:111–9.
34. Mendicino M, Bailey AM, Wonnacott K, Puri RK, Bauer SR. MSC-based product characterization for clinical trials: an FDA perspective. Cell Stem Cell. 2014;14:141–5.
35. Dominici M, et al. Minimal criteria for defining multipotent mesenchymal stromal cells. The International society for cellular therapy position statement. Cytotherapy. 2006;8:315–7.
36. Yu S, Tanabe T, Dezawa M, Ishikawa H, Yoshimura N. Effects of bone marrow stromal cell injection in an experimental glaucoma model. Biochem Biophys Res Commun. 2006;344:1071–9.
37. Johnson TV, et al. Neuroprotective effects of intravitreal mesenchymal stem cell transplantation in experimental glaucoma. Invest Ophthalmol Vis Sci. 2010;51:2051–9.
38. Harper MM, et al. Transplantation of BDNF-secreting mesenchymal stem cells provides neuroprotection in chronically hypertensive rat eyes. Investig Ophthalmol Vis Sci. 2011;52:4506–15.
39. Levkovitch-Verbin H, et al. Intravitreal injections of neurotrophic factors secreting mesenchymal stem cells are neuroprotective in rat eyes following optic nerve transaction. Investig Ophthalmol Vis Sci. 2010;51:6394–400.
40. Zhao T, et al. Protective effects of human umbilical cord blood stem cell intravitreal transplantation against optic nerve injury in rats. Graefes Arch Clin Exp Ophthalmol. 2011;249:1021–8.
41. Zwart I, et al. Umbilical cord blood mesenchymal stromal cells are neuroprotective and promote regeneration in a rat optic tract model. Exp Neurol. 2009;216:439–48.
42. Sugitani S, et al. The potential neuroprotective effect of human adipose stem cells conditioned medium against light-induced retinal damage. Exp Eye Res. 2013;116:254–64.
43. Johnson TV, et al. Identification of retinal ganglion cell neuroprotection conferred by platelet-derived growth factor through analysis of the mesenchymal stem cell secretome. Brain. 2014;137:503–19.
44. Dreixler JC, et al. Delayed Administration of Bone Marrow Mesenchymal Stem Cell Conditioned Media Significantly

Improves Outcome after retinal ischemia in rats. Invest Ophthalmol Vis Sci. 2014; doi:10.1167/iovs.13-11683.
45. Singer NG, Caplan AI. Mesenchymal stem cells: mechanisms of inflammation. Annu Rev Pathol. 2011;6:457–78.
46. Le Blanc K, et al. Mesenchymal stem cells for treatment of steroid-resistant, severe, acute graft-versus-host disease: a phase II study. Lancet. 2008;371:1579–86.
47. Heldman AW, et al. Transendocardial mesenchymal stem cells and mononuclear bone marrow cells for ischemic cardiomyopathy: the TAC-HFT randomized trial. JAMA. 2014;311:62–73.
48. Mead B, et al. Stem cell treatment of degenerative eye disease. Stem Cell Res. 2015;14:243–57.
49. Siqueira RC, Messias A, Voltarelli JC, Scott IU, Jorge R. Intravitreal injection of autologous bone marrow–derived mononuclear cells for hereditary retinal dystrophy. Retina. 2011;31:1207–14.
50. Kuriyan AE,et al. Vision Loss after intravitreal injection of autologous "Stem Cells" for AMD. N Engl J Med. 2017; 376(11):1047-1053.
51. Ma J, et al. Transplantation of human neural progenitor cells expressing IGF-1 enhances retinal ganglion cell survival. PLoS One. 2015;10(4):e0125695.
52. Woodhall E, West AK, Chuah MI. Cultured olfactory ensheathing cells express nerve growth factor, brain-derived neurotrophic factor, glia cell line-derived neurotrophic factor and their receptors. Brain Res Mol Brain Res. 2001;88:203–13.
53. Dai C, Qin Yin Z, Li Y, Raisman G, Li D. Survival of retinal ganglion cells in slice culture provides a rapid screen for olfactory ensheathing cell preparations. Brain Res. 2010;1354:40–6.
54. Wu MM, et al. Death of axotomized retinal ganglion cells delayed after intraoptic nerve transplantation of olfactory ensheathing cells in adult rats. Cell Transplant. 2010;19:159–66.
55. Sieving PA, et al. Ciliary neurotrophic factor (CNTF) for human retinal degeneration: phase I trial of CNTF delivered by encapsulated cell intraocular implants. Proc Natl Acad Sci U S A. 2006;103:3896–901.
56. Zhang Y, Klassen HJ, Tucker BA, Perez M-TR, Young MJ. CNS progenitor cells promote a permissive environment for Neurite outgrowth via a matrix metalloproteinase-2-dependent mechanism. J Neurosci. 2007;27:4499–506.
57. Sakaguchi DS, et al. Transplantation of neural progenitor cells into the developing retina of the Brazilian opossum: an in vivo system for studying stem/progenitor cell plasticity. Dev Neurosci. 2004;26:336–45.
58. Klassen H, et al. Neural precursors isolated from the developing cat brain show retinal integration following transplantation to the retina of the dystrophic cat. Vet Ophthalmol. 2007;10:245–53.
59. Mellough CB, Cui Q, Harvey AR. Treatment of adult neural progenitor cells prior to transplantation affects graft survival and integration in a neonatal and adult rat model of selective retinal ganglion cell depletion. Restor Neurol Neurosci. 2007;25:177–90.
60. Young MJ, Ray J, Whiteley SJ, Klassen H, Gage FH. Neuronal differentiation and morphological integration of hippocampal progenitor cells transplanted to the retina of immature and mature dystrophic rats. Mol Cell Neurosci. 2000;16:197–205.
61. Qiu G, et al. Photoreceptor differentiation and integration of retinal progenitor cells transplanted into transgenic rats. Exp Eye Res. 2005;80:515–25.
62. Sakaguchi DS, Van Hoffelen SJ, Young MJ. Differentiation and morphological integration of neural progenitor cells transplanted into the developing mammalian eye. Ann N Y Acad Sci. 2003;995:127–39.
63. Van Hoffelen SJ, Young MJ, Shatos MA, Sakaguchi DS. Incorporation of murine brain progenitor cells into the developing mammalian retina. Invest Ophthalmol Vis Sci. 2003;44:426–34.
64. Bartsch U, et al. Retinal cells integrate into the outer nuclear layer and differentiate into mature photoreceptors after subretinal transplantation into adult mice. Exp Eye Res. 2008;86:691–700.
65. Hertz J, et al. Survival and integration of developing and progenitor-derived retinal ganglion cells following transplantation. Cell Transplant. 2013; doi:10.3727/096368913X667024.
66. Klassen H, Sakaguchi DS, Young MJ. Stem cells and retinal repair. Prog Retin Eye Res. 2004;23:149–81.
67. Hori J, et al. Neural progenitor cells lack immunogenicity and resist destruccion as allografts. Stem Cells. 2003;21:405–16.
68. Goldman D. Müller glial cell reprogramming and retina regeneration. Nat Rev Neurosci. 2014;15:431–42.
69. Becker S, et al. Allogeneic transplantation of Müller-derived retinal ganglion cells improves retinal function in a feline model of ganglion cell depletion. Stem Cells Transl Med. 2015; doi:10.5966/sctm.2015-0125.
70. Ballios BG, van der Kooy D. Biology and therapeutic potential of adult retinal stem cells. Can J Ophthalmol. 2010;45:342–51.
71. Balenci L, van der Kooy D. Notch signaling induces retinal stem-like properties in perinatal neural retina progenitors and promotes symmetric divisions in adult retinal stem cells. Stem Cells Dev. 2014;23:230–44.
72. Chen M, et al. $Lgr5^+$ amacrine cells possess regenerative potential in the retina of adult mice. Aging Cell. 2015; doi:10.1111/acel.12346.
73. Thomson JA, et al. Embryonic stem cell lines derived from human blastocysts. Science. 1998;282:1145–7.
74. Takahashi K, et al. Induction of pluripotent stem cells from adult human fibroblasts by defined factors. Cell. 2007;131:861–72.
75. Hockemeyer D, Jaenisch R. Induced pluripotent stem cells meet genome editing. Cell Stem Cell. 2016;18:573–86.
76. Yu J, et al. Induced pluripotent stem cell lines derived from human somatic cells. Science. 2007;318:1917–20.
77. Al-Shamekh S, Goldberg JL. Retinal repair with induced pluripotent stem cells. Transl Res. 2013; doi:10.1016/j.trsl.2013.11.002.
78. Ben-David U, Benvenisty N. The tumorigenicity of human embryonic and induced pluripotent stem cells. Nat Rev Cancer. 2011;11:268–77.
79. Livesey FJ, Cepko CL. Vertebrate neural cell-fate determination: lessons from the retina. Nat Rev Neurosci. 2001;2:109–18.
80. Bassett EA, Wallace VA. Cell fate determination in the vertebrate retina. Trends Neurosci. 2012;35:565–73.
81. Edlund T, Jessell TM. Progression from extrinsic to intrinsic signaling in cell fate specification: a view from the nervous system. Cell. 1999;96:211–24.
82. Lamba DA, Karl MO, Ware CB, Reh TA. Efficient generation of retinal progenitor cells from human embryonic stem cells. Proc Natl Acad Sci U S A. 2006;103:12769–74.
83. Chen M, et al. Generation of retinal ganglion-like cells from reprogrammed mouse fibroblasts. Invest Ophthalmol Vis Sci. 2010;51:5970–8.
84. Jagatha B, et al. In vitro differentiation of retinal ganglion-like cells from embryonic stem cell derived neural progenitors. Biochem Biophys Res Commun. 2009;380:230–5.
85. Parameswaran S, et al. Induced pluripotent stem cells generate both retinal ganglion cells and photoreceptors: therapeutic implications in degenerative changes in glaucoma and age-related macular degeneration. Stem Cells. 2010;28:695–703.
86. Eiraku M, et al. Self-organizing optic-cup morphogenesis in three-dimensional culture. Nature. 2011;472:51–6.
87. Nakano T, et al. Self-formation of optic cups and storable stratified neural retina from human ESCs. Cell Stem Cell. 2012;10:771–85.
88. Wang SW, et al. Requirement for math5 in the development of retinal ganglion cells. Genes Dev. 2001;15:24–9.

89. Mu X, et al. A gene network downstream of transcription factor Math5 regulates retinal progenitor cell competence and ganglion cell fate. Dev Biol. 2005;280:467–81.
90. Xiang M, Jiang H, Jin K, Qiu F. *Molecular control of retinal ganglion cell specification and differentiation. cdn.intechweb.org*. 2011. Available at http://cdn.intechweb.org/pdfs/23817.pdf.
91. Jiang Y, et al. Transcription factors SOX4 and SOX11 function redundantly to regulate the development of mouse retinal ganglion cells. J Biol Chem. 2013;288:18429–38.
92. Wu F, et al. Two transcription factors, Pou4f2 and Isl1, are sufficient to specify the retinal ganglion cell fate. Proc Natl Acad Sci U S A. 2015;112(13):E1559–68.
93. Riazifar H, Jia Y, Chen J, Lynch G, Huang T. Chemically induced specification of retinal ganglion cells from human embryonic and induced pluripotent stem cells. Stem Cells Transl Med. 2014; doi:10.5966/sctm.2013-0147.
94. Sluch VM, et al. Differentiation of human ESCs to retinal ganglion cells using a CRISPR engineered reporter cell line. Sci Rep. 2015;5:16595.
95. Tanaka T, et al. Generation of retinal ganglion cells with functional axons from human induced pluripotent stem cells. Sci Rep. 2015;5:1–11.
96. Ohlemacher SK, et al. Stepwise differentiation of retinal ganglion cells from human pluripotent stem cells enables analysis of glaucomatous neurodegeneration. Stem Cells. 2016;34(6):1553–62.
97. Kador KE, Goldberg JL. Scaffolds and stem cells: delivery of cell transplants for retinal degenerations. Expert Rev Ophthalmol. 2012;7:459–70.
98. Almony A, et al. Techniques, rationale, and outcomes of internal limiting membrane peeling. Retina. 2012;32:877–91.
99. Song W-T, Zhang X-Y, Xia X-B. Atoh7 promotes the differentiation of retinal stem cells derived from Müller cells into retinal ganglion cells by inhibiting Notch signaling. Stem Cell Res Ther. 2013;4:94.
100. Vierbuchen T, et al. Supplementary data – direct conversion of fibroblasts to functional neurons by defined factors. Nature. 2010;463:1035–41.
101. Lamba DA, et al. Generation, purification and transplantation of photoreceptors derived from human induced pluripotent stem cells. PLoS One. 2010;5:e8763.

Part II

Cell-based Neuroprotection

Autologous Bone Marrow-Derived Cell Therapies for Retinal Disease

Elad Moisseiev and Susanna S. Park

1 Problem and Hypothesis

1.1 Retinal Disease Is a Significant Burden on Global Health, and There Is a Need for Improved Alternative Treatments

Retinal diseases are a major cause of blindness and visual impairment worldwide. According to The Global Burden of Disease Study, macular diseases were the third most common cause of vision loss worldwide between 1990 and 2010, surpassed only by cataract and under-corrected refractive error [1]. As these latter causes are reversible, it can be inferred that macular diseases are the most common cause of irreversible vision loss. A recent meta-analysis by The Vision Loss Expert Group of The Global Burden of Disease Study reported that macular diseases, excluding diabetic retinopathy, were responsible for 6.6% of blindness (defined as visual acuity worse than 20/400) in the world. It was also noted that the number of people blinded due to macular diseases had increased by 36% from 1990 to 2010, and that the proportion of macular diseases in causing blindness was larger in more developed countries and high-income regions [2].

Age-related macular degeneration (AMD) has been recognized as the leading cause of irreversible severe vision loss in people over 50 years of age in the developed world [3]. Several population studies have shown that it is responsible for up to 8.7% of blindness worldwide, and that its incidence rises considerably with age [4–9]. As the world's population continues to enlarge and age, the prevalence of AMD and its role as a major cause of blindness are expected to increase significantly in the future, with 288 million people projected to have AMD by 2040 [10].

Diabetic retinopathy, which was excluded in the abovementioned Global Burden of Disease Study reports, is the leading cause of blindness in the younger working-age population in the developed world [11, 12]. Globally, the prevalence of diabetes mellitus is on the rise, in both developed and developing countries, and the prevalence of diabetic retinopathy and associated vision loss is expected to grow with it [13–16].

Besides AMD and diabetic retinopathy, other common retinal diseases that often cause significant visual loss include retinal vascular occlusions, which are considered to be the second most common vascular retinopathy after diabetic retinopathy [17], and inherited retinal degenerations such as retinitis pigmentosa and Stargardt's disease, which have an estimated prevalence of 1:4000 and 10,000, respectively [18, 19]. Additionally, myopic degeneration is increasing and becoming a common cause for vision loss, especially in Asian populations [20–23].

As the world population increases in number, age, and rate of diabetes mellitus and myopia, it is likely that retinal diseases will only become a more prevalent cause of irreversible visual loss. Research is currently focused on elucidating the complex mechanisms of pathogenesis of these complex diseases, as well as designing safe and effective treatment methods for them. Currently, the majority of available treatments are based on interfering with molecular pathways that have been identified as having a role in disease progression and development. This chapter will focus on the clinical applications of autologous bone marrow-derived stem cell therapy in retinal diseases, which is based on the hypothesis that these cells are capable of repairing or replacing cells in the retina that have been damaged by various disease processes.

2 Scientific Basis for Intervention

2.1 Pathogenesis and Current Treatment of Retinal Diseases

A common pathway for vision loss associated with a majority of retinal disease is loss of retinal neurons and/or retinal pigment epithelial cells, regardless of their different mechanisms of pathogenesis. However, conventional treatment

E. Moisseiev, M.D. • S.S. Park, M.D., Ph.D. (✉)
Department of Ophthalmology & Vision Science, University of California Davis Eye Center, 4860 Y St., Suite 2400, Sacramento, CA 95817, USA
e-mail: sscpark@ucdavis.edu

S.D. Schwartz et al. (eds.), *Cellular Therapies for Retinal Disease*, DOI 10.1007/978-3-319-49479-1_7

strategies are not directed at restoring these cells that are critical for functional vision. Rather, the current therapeutic approaches influence the retinal environment in an attempt to indirectly delay progression of the degenerative or ischemic process.

For example, AMD is primarily a degenerative disease with multiple risk factors for progression, such as age, smoking, and genetics [24–28]. Approximately 20% of patients with AMD advance to the exudative form and develop choroidal neovascularization, which is responsible for almost 90% of severe vision loss associated with this disease [4, 29]. The current mainstay of treatment for exudative AMD is intravitreal injections of anti-vascular endothelial growth-factor (anti-VEGF) agents, such as Ranibizumab (Lucentis) or Aflibercept (Eylea). The visual benefit of intravitreal anti-VEGF for exudative AMD has been established in several large, prospective, randomized, controlled multicenter studies [30–34]. Bevacizumab (Avastin) has also been shown to be effective for the treatment of neovascular AMD [32, 35], and its use is very common worldwide for economic reasons [36].

The majority of AMD patients have non-exudative disease, which can also lead to significant visual loss, especially if the disease progresses to geographic atrophy, which is characterized by loss of photoreceptors and retinal pigment epithelium in the macula (Fig. 1). It has been shown that the prevalence of geographic atrophy increases with age [9, 37, 38], and it constitutes a major cause of vision loss for which there is currently no available treatment. In fact, the only available treatment for patients with non-exudative AMD at the present is antioxidant nutritional supplementation, which has been shown to reduce the risk of severe vision loss among patients with intermediate AMD [39–41].

On the other hand, the pathogenesis of diabetic retinopathy is more complex, with the primary mechanism of pathogenesis being retinal ischemia from retinal vascular damage. Numerous pathways have been implicated in the pathogenesis of diabetic retinopathy, including VEGF formation, inflammation, protein kinase C activation, hemodynamic changes, and oxidative stress [42]. The current mainstay of treatment for diabetic retinopathy is improved blood glucose

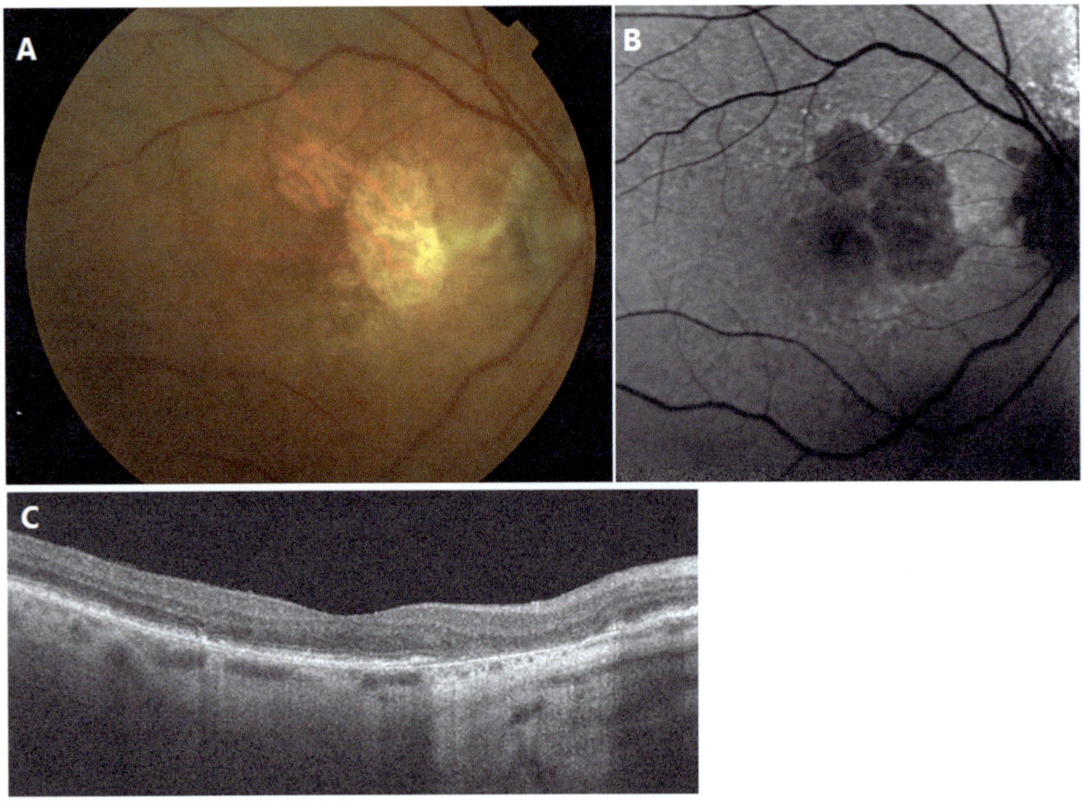

Fig. 1 Photoreceptor and retinal pigment epithelium loss in the macula of the right eye of a 79-year-old man with geographic atrophy from non-exudative age-related macular degeneration (AMD). (**a**) Fundus photograph demonstrates the circumscribed central area of macular photoreceptor and retinal pigment epithelial atrophy, allowing visualization of the underlying choroidal vessels. (**b**) Autofluorescence imaging of the same eye demonstrates hypoautofluorescence in the area of geographic atrophy from loss of retinal pigment epithelium. At the rim of geographic atrophy, some hyperautofluorescence is seen indicating that the retinal pigment epithelium is dysfunctional at the edges of the geographic atrophy. (**c**) Macular OCT line scan demonstrates the central focal loss of photoreceptor, retinal pigment epithelium, and choriocapillaris in the area of geographic atrophy. Note the increased light penetration within the area of atrophy

control and control of concurrent cardiovascular risk factors. Intravitreal injections of anti-VEGF agents or corticosteroids have been established as an effective treatment for diabetic macular edema [43–52]. Laser treatment also is available to minimize risk of long-term vision loss associated with diabetic macular edema and proliferative diabetic retinopathy [45, 53]. However, vision loss is generally progressive in eyes with diabetic retinopathy despite these treatments. This is because there is no treatment available to reverse the retinal ischemia associated with diabetic retinopathy and other retinal vasculopathy. In addition, some of the vision loss associated with diabetic retinopathy is believed to result from retinal degeneration associated with the diabetic environment or secondary to the retinal ischemia (Fig. 2). Electroretinography abnormalities have been associated with diabetes even before retinopathy is clinically diagnosed [54].

Thus, the majority of retinal diseases that result in irreversible vision loss or blindness currently are degenerative and/or ischemic conditions, for which AMD and diabetic retinopathy are the respective paradigms. Intravitreal anti-VEGF agents or corticosteroids that are used currently for associated complications such as macular edema or ocular neovascularization may minimize the degree of acute vision loss associated with these retinal conditions. However, these therapies do not influence the underlying degenerative or ischemic processes. In fact, progressive geographic atrophy leading to progressive vision loss was observed in many eyes with exudative AMD after long-term anti-VEGF therapy [55].

2.2 Stem Cell Therapy as a Treatment Strategy for Retinal Disease

Intravitreal stem cell therapy is an attractive potential therapy to explore as a treatment for ischemic or degenerative retinal diseases, as it may offer several advantages over treatment options currently available. First, treating retinal disease at the cellular level may prove to be more effective as it is a more comprehensive approach to prevention of disease progression. Introduction of stem cells to the diseased retina may have a paracrine trophic affect in the damaged degenerative or ischemic tissue. Second, the stem cells are viable

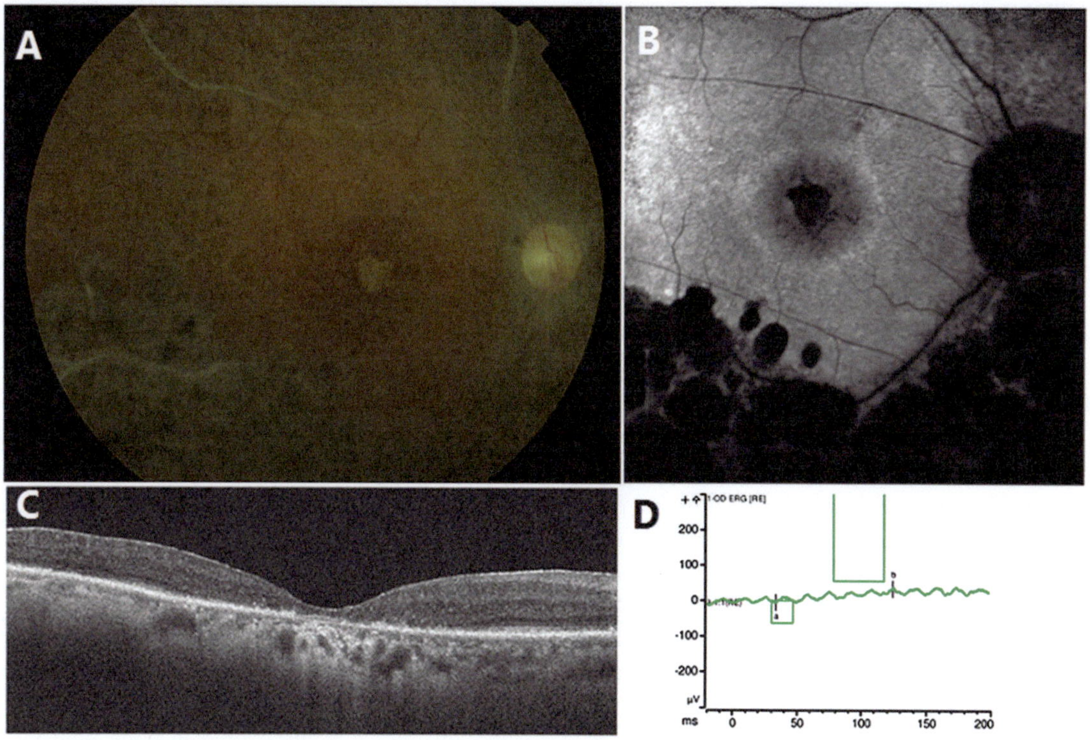

Fig. 2 Retinal degeneration and outer retinal atrophy associated with diabetic retinopathy. (**a**) Fundus photograph of the right eye of a 78-year-old woman with inactive proliferative diabetic retinopathy and clinically significant macular edema after panretinal photocoagulation and macular laser. Diffuse retinal ischemia is evident in the diffusely sclerosed retinal vessels. Pigmented laser scars are also seen. (**b**) Autofluorescence imaging of the macula of the same eye demonstrates areas of hypo-autofluorescence from loss of the retinal pigment epithelium in the central macula from atrophy. In additions hypo-autofluorescence is also noted in areas of old laser scars and around the disc from peripapillary atrophy. (**c**) Macular OCT line scan image demonstrates the focal central loss of photoreceptor and retinal pigment epithelium in the macula. (**d**) Electroretinography of the same shows diffuse loss of inner and outer retinal function from retinal degeneration and ischemia, demonstrated by the virtually flat ERG signal (*dark* adapted)

cells which may incorporate into the retina long-term and function in the eye longer than currently available intravitreally administered drugs which have relatively short half-lives [56–61] and require repeated injections. Third, it is possible that stem cell therapy will not only be effective in preventing disease progression but also may be effective in repairing or regenerating damaged cells or tissue within the retina or retinal pigment epithelium. If the stem cells are able to integrate or promote regeneration of the damaged cells, they may be capable of restoring vision for patients for whom there are currently no treatment options. Fourth, stem cell therapy may be the platform leading to development of other potentially novel effective therapies, e.g., gene therapy. At present, gene therapy requires integration of vector DNA into the retinal cells, and its delivery into the correct cells without causing DNA alterations is difficult and not yet perfected [62–64]. It has been shown that despite initial improvement in VA, patients treated by RPE65 gene therapy eventually undergo photoreceptor degeneration [65]. Introduction of desired genes via stem cells into the eye may prove to be a superior method for gene therapy, as it can overcome potential obstacles such as undesired DNA alterations that may contribute to subsequent degeneration.

This chapter is devoted to autologous bone marrow stem cell therapy. Autologous stem cell therapy offers additional advantages. Since the stem cells are derived from the patient, rejection is not an issue. Ocular or systemic immunosuppression is not needed, as may be the case when allogeneic cells are used. Also, no ethical and regulatory concerns are associated with autologous cell therapy, unlike embryonic or fetal stem cell therapy.

3 Preclinical Data

3.1 Classification of Stem Cells

Stem cells are defined as cells that possess both a capacity for unlimited self-renewal and for differentiation into multiple cell types. Stem cells are generally classified according to their origin. Embryonic stem cells are harvested from the inner cell mass of the blastocyst of 5-day-old preimplantation embryos, and are pluripotent, meaning they can differentiate into cells of ectodermal, mesodermal, and endodermal origin [66]. These cells may serve as a potential treatment for numerous diseases, and theoretically can be used to replace several types of cells in the retina or retinal pigment epithelium. However, the use of embryonic stem cells raises some ethical concerns. In addition, because of their unlimited capability to differentiate into multiple cell types and regenerate, these cells can form teratomas or other tumors or incite inflammatory reactions [67–69]. Fetal stem cells are harvested from fetal tissue later in its development, and are multi-potent, meaning they are capable of differentiating into a smaller number of cell types [66]. These cells are harder to obtain and use. Autologous therapy is not possible using embryonic or fetal stem cells.

Adult stem cells are present in small number and may be found in most tissues and organs. They are also multi-potent cells but more limited in their differentiation capacity than fetal stem cells.

The bone marrow was the first tissue where adult stem cells were first recognized, as cells derived from bone marrow were shown to be able to differentiate into several types of cells. Hematopoietic stem cells from bone marrow can be stimulated to differentiate into various blood cell lineages and have been used successfully for bone marrow transplantation to treat patients with various hematopoietic diseases [70, 71]. These cells are identified commonly in humans by their cell surface markers (e.g., CD34) but may consist of a mixture of cells including endothelial progenitor cells (EPS). The bone marrow also contains stem cells that can be cultured and stimulated to differentiate in cells of mesenchymal origin [72]. These adult stem cells from bone marrow are called mesenchymal stem cells and their multi-potent differentiation capability has been well established [73–75]. More recently, mesenchymal stem cells have been found in other tissues including adipose tissue, fetal cord blood, liver, and periosteum [76–78]. These cells can be stimulated to differentiate into different cell types by varying the culture conditions [78, 79].

Stem cell research has been revolutionized by the discovery that mature cells can be reprogrammed to become pluripotent [80, 81], a discovery that awarded its discoverers the Nobel Prize (2012). These induced pluripotent stem cells (iPSCs) behave similarly to embryonic stem cells [81, 82] and have significant therapeutic potential for many diseases [83]. The main advantage of using iPSCs over embryonic or fetal stem cells is that autologous treatment is possible. However, as their reprogramming requires exposure to c-Myc, they carry the potential risk of teratoma formation and other tumors, similar to embryonic stem cells [82, 84, 85].

3.2 Mesenchymal Stem Cell and Their Potential in Treating Retinal Disease

Although originally discovered in bone marrow, it is currently thought that mesenchymal stem cells are present in many organs on the body [86]. The two most common sources for isolation of these cells are bone marrow and adipose tissue [87]. Although bone marrow-derived mesenchymal stem cells (BM-MSCs) and adipose tissue-derived mesenchymal stem cells (AD-MSCs) share many similarities, there are several important differences between them.

First, BM-MSCs are much rarer, and constitute less than 0.01% of the nucleated cells in the bone marrow [78, 88]. AD-MSCs are much more common, and their number is about 500 times greater in an equivalent amount of adipose tissue [87, 89, 90]. Second, due to their larger quantity, AD-MSCs can potentially be used fresh after their isolation. BM-MSCs have to be cultured and expanded prior to their use [88]. Third, BM-MSCs are harvested by bone marrow aspiration, while AD-BMCs can be harvested more easily by a lipoaspiration procedure which is simpler to perform and more tolerable. Fourth, some differences in immunophenotype and gene expression profiles between BM-MSCs and AD-MSCs have been reported. For example, 13.2% of genes and 18% of total proteins expressed were found to be significantly different between these two types of cell populations [91]. The clinical significance of these differences in immunophenotype and gene expression profiles is unclear. However, these observed differences may explain the different responses of BM-MSCs and AD-MSCs in vitro to various growth factors and conditions [87]. Finally, some controversy exists regarding the differentiation potential of these mesenchymal stem cells. Several studies have shown that AD-MSCs are more prone to adipogenic differentiation while others have shown that BM-MSCs are more prone to osteogenic differentiation, and still others have reported no such differences [87, 91–95].

AD-MSCs have been shown to be capable of differentiating into retinal progenitor cells and retinal pigment epithelial cells in vitro [96–98]. Additionally, they have been shown to have a neuro-protective effect in vitro on retinal pigment epithelial cells via a paracrine effect [99]. It has been reported that BM-MSCs can be stimulated to differentiate towards retinal neural cells, photoreceptors, retinal progenitor cells, and retinal pigment epithelial cells in vitro [100–105]. Additionally, these cells have been shown to have a paracrine neuro-protective effect on retinal ganglion cells via secretion of neuro-protective cytokines, including platelet-derived growth factor [106]. Since both degenerative and ischemic retinal diseases can eventually result in photoreceptor and retinal pigment epithelial cell damage and loss, AD-MSCs and BM-MSCs are appealing potential candidates for cell therapy to explore.

3.3 Studies Using Mesenchymal Stem Cells for Retinal Disease

Animal studies have demonstrated that intravitreal injection of condition media from human BM-MSCs had a protective effect on light-induced retinal degeneration [107, 108]. In addition, various groups have shown that intravitreal or subretinal transplantation of BM-MSCs in animal models of retinal degeneration slowed down retinal degeneration [109, 110]. In addition, intravitreal injection of BM-MSCs was also reported to be successful in preventing retinal neuronal cells loss in a rat model of ischemia-reperfusion retinal injury, suggesting a potential use of these cells in treating retinal vascular disorders [111]. However, the intravitreally injected BM-MSC were detected mostly on top of the retina (on the internal limiting membrane), with only a few cells integrating into the inner retina following intravitreal injection. Most transplanted mesenchymal stem cells do not appear to readily incorporate into the degenerating or ischemic retina following intravitreal or subretinal transplantation. Thus, a paracrine effect is implicated. In fact, the mesenchymal cells often form a cellular clump in the vitreous in most animal models following intravitreal transplantation. This raises some safety concerns regarding clinical application.

Intravenous administration of mesenchymal stem cell could be an alternative route of administration of these cells that can be explored as treatment for eyes with retinal degeneration or ischemia. In a study using a rat model of inherited retinal degeneration, intravenous administration of BM-MSCs achieved preservation of rods and cone, as well as visual function [111]. In a study using a murine model of laser-induced choroidal neovascularization, intravenous administration of BM-MSCs resulted in the stem cells migrating into the choroidal neovascularization site and differentiation into several cell types [112]. There were no systemic adverse effects associated with intravenous mesenchymal stem cell administration. Utilizing this recruitment potential, this study further reported that BM-MSCs engineered to produce the anti-angiogenic pigment epithelial-derived factor (PEDF) had localized to the choroidal neovascular sites and inhibited their growth [112]. This is an important concept of stem cell therapy in general. Since these cells have the potential to home into sites of tissue damage, they can be pre-manipulated to express certain proteins to increase their desired local effect once they have reached their target. This approach has also being explored in fields outside retinal research, and several studies have reported on the use of MSCs as a vector for gene therapy. These studies include studies using mesenchymal stem cells transfected to express VEGF for the treatment of cardiac ischemia, interferon-β to treat tumors, and bone morphogenic proteins to promote bone production [113–118]. For retinal research, BM-MSCs transfected with a retroviral vector to produce brain-derived neurotrophic factor (BDNF) were injected into the subretinal space to increase the level of BDNF [119] in a rat model.

Human BM-MSCs have also been studied in animal models of retinal disease. Using cross-species cell transplantation has an innate limitation as the host animal is bound to reject the human cells and thereby require systemic immunosuppression. However, despite this limitation, such experiments are a necessary step in moving from using animal BM-MSCs to human BM-MSCs for clinical trial. Human BM-MSCs were shown to survive in mouse eyes following subretinal injection, and it was shown that their survival may vary according to the age of the animal at the time of transplantation [120]. In a study using a rat model of retinal degeneration, subretinal injection of human BM-MSCs achieved preservation of photoreceptors and visual function (assessed by luminance threshold recording), with no systemic adverse effects [121]. A recent study that compared intravitreal and subretinal transplantation of human BM-MSCs in a rat model of retinal degeneration reported that both routes of administration achieved preservation of photoreceptors and retinal function by ERG, with a longer lasting effect with subretinal administration [122]. One study in a rat model demonstrated that following intravitreal injection of human BM-MSCs, the human cells survived in the vitreous for up to 90 days. However, some human cells were integrated into other ocular structures, and had also bypassed the blood-retina barrier and migrated to nontarget tissues, a finding that raises significant safety concerns [123]. Therefore, despite their relative ease of acquisition and potential neuroprotective effects, detailed long-term safety studies are important before injecting these cultured mesenchymal stem cells into human eyes.

3.4 Animal Bone Marrow-Derived Stem Cells in Models of Retinal Disease

Promising results have been published using intravitreal administration of stem cells isolated from bone marrow but not cultured and expanded like mesenchymal stem cells. Rat and murine models of retinal injury, degeneration, and ischemia have been used.

In an early study using a rat model of mechanically induced retinal injury, intravitreal injection of a stem cell-enriched cellular fraction from bone marrow achieved accumulation of the cells primarily around the site of retinal injury. The injected cells were incorporated into the retina, and expressed calbindin and rhodopsin, which are retinal cell-specific proteins. The findings appear to suggest that they have differentiated into retinal neurons [124]. It should be noted that the injected cells were not pure stem cells as the purification techniques were not as developed then as they are today.

More recently, in a study using a murine model of hereditary retinal degeneration, intravitreal injection of autologous bone marrow-derived lineage negative hematopoietic stem cells delayed the retinal degenerative process. Treated mice had preservation of retinal function by electroretinography, preservation of the photoreceptor layer on histology, and down-regulation of apoptotic genes when compared to untreated control eyes [125]. A paracrine trophic effect of the cells was implicated since the injected cells were found localized in the retinal vasculature following intravitreal administration.

3.5 Endothelial Progenitor Cells and CD34+ Cells

Endothelial progenitor cells (EPCs) are a unique subtype of adult stem cells that deserve further elaboration on their potential as cell therapy for retinal diseases. This is a relatively rare cell population in the peripheral circulation, accounting for about 0.01% of all circulating cells (estimated numbers vary between 0.5 and 400 cells/ml) [126–129]. EPCs are precursors for angioblasts and believed to originate from the bone marrow where they are present in highest concentration [130, 131].

The exact definition of EPCs is somewhat controversial since the method of isolation is not uniform and the isolated population may consist of a mixture of cells. One method of isolating EPCs is by cell sorting based on the expression of cell surface markers. This is challenging as the EPCs are relatively rare in the circulation unless they are mobilized from bone marrow. Also, there is no consensus on the surface markers that should be used for their identification. There is a potential overlap between the cell surface markers used for EPCs and more mature cells in the circulation. Generally, sorting for CD34+ and CD45− cells is considered a reasonable approach to isolate EPCs, although the resulting cell population is not homogenous for EPCs [132]. It also should be noted that bone marrow CD34+ cells share some similar traits as CD34+ cells in the peripheral circulation [133]. Thus, bone marrow CD34+ cells are believed to contain precursors of EPCs found in the peripheral circulation.

An alternative method for EPC isolation from peripheral blood is by selective in vitro cultures of the mononuclear cell fraction from blood. Using differential adhesion characteristics and subsequent growth of derivative clones, investigators have isolated more homogenous cell clones in culture [132]. Using this technique, two distinct subtypes of EPCs have been identified from peripheral blood: early EPCs appear within 1 week in culture, and late EPCs (also termed outgrowth endothelial cells (OECs)) take longer to appear in culture [126, 132, 134, 135]. Further research demonstrated that these two subtypes of cultured EPCs have different functions, as their gene expression profiles are very different. Early EPCs were found to be primarily hematopoietic cells and involved in immunity and inflammation, while OECs appear to be primarily involved in angiogenesis and vascular development. OECs have a higher proliferation rate and are also capable of forming endothelial cells [127, 135]. Because OECs are capable of forming endothelial cells while early EPCs cannot, some researchers believe that OECs are more representative of true EPCs in peripheral blood [127, 129].

3.6 Endothelial Precursor Cells and CD34+ Cells and Retinal Disease

There are several advantages of using EPCs as potential therapy for retinal vascular disease. The EPCs can be isolated from peripheral blood and autologous therapy is possible. Obtaining EPCs from blood would be easier than obtaining cells from bone marrow or adipose tissue. These cells are capable of homing to damaged areas and secreting factors that promote tissue repair and regeneration [127, 128]. However, as mentioned earlier, they are present in very low concentration, and mobilization from bone marrow is required to obtain a high enough concentration of these cells for clinical therapy. In addition, some debate persists whether mobilized EPCs behave similar to native CD34+ cells in bone marrow or EPCs in peripheral blood in the absence of mobilization in terms of their homing and regenerative potential.

Early studies have demonstrated that EPCs injected into ischemic tissue in animal models of myocardial infarction or critical limb ischemia and animal models of tumors are capable of incorporating into the damaged vessels [129, 136–138]. The migration and integration of these cells into the damaged vasculature appear to be in response to both ischemia and endothelial cell apoptosis [129, 139, 140]. It has been suggested that early EPCs and OECs work in synergy, with OECs directly involved in forming and repairing the endothelium of damaged vessels, and early EPCs promoting the process via a local paracrine effect [127, 129, 132]. This concept is supported by a study that showed transplantation of both EPC subtypes achieved synergistic neovascularization in a murine model of critical limb ischemia [141].

CD34+ EPCs have been shown to originate from the bone marrow [142, 143]. Thus, EPCs in peripheral circulation can be considered a subtype of bone marrow-derived stem cells. EPCs are excellent candidates for stem cell therapy as they respond to ischemic and apoptotic stimuli, are capable of homing to damaged tissue sites, and have a reparative effect on the damaged tissue. However, it has been shown that with intravenous administration, the majority of EPCs accumulate in the liver and spleen [144]. Therefore, local administration close to the damaged tissue is required in order to achieve their therapeutic potential. This makes retinal disease a suitable target for EPC therapy, as intraocular administration is possible by simple intravitreal injection or subretinal surgical transplantation.

Our group has shown in a NOD-SCID mice model of acute retinal ischemia-reperfusion injury that intravitreal injection of human CD34+ bone marrow-derived stem cells resulted in human cells being incorporated into the mouse retinal vasculature up to 6 months following cell administration with no abnormal ocular or systemic cellular proliferation [145] (Fig. 3). The retinal vasculature appeared to have normalized and no systemic or ocular adverse effects were noted associated with the cell therapy.

It has been shown that EPCs isolated from peripheral blood of diabetic patients are impaired in their ability to home into the damaged tissue, have reduced vaso-reparative potential, and have premature senescence [134, 146, 147]. Several studies have found CD34+ EPC numbers to be increased in peripheral blood in diabetics [146–151], but several others reported reduced numbers [152, 153]. As mentioned previously, diabetic retinopathy is primarily ischemic in nature but concurrent retinal degeneration from ischemia or hyperglycemic effects can eventually lead to degeneration of photoreceptor and retinal pigment epithelium. It has been proposed that the basis for the ischemic process associated with diabetic retinopathy is a defect in vascular endothelial function [154]. It is

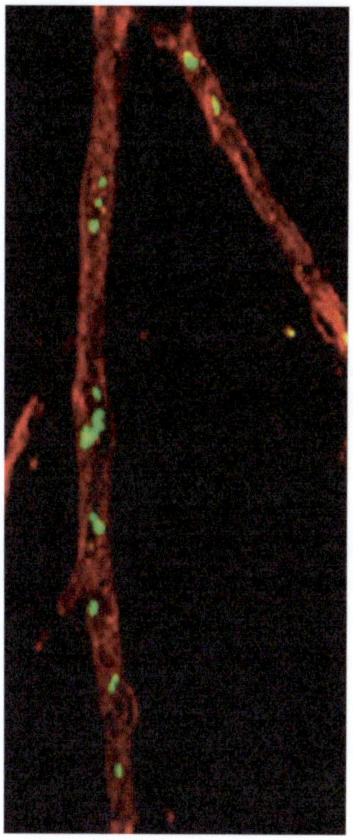

Fig. 3 Immunohistochemical staining of retinal whole mount perfused with rhodamine-conjugated dextran to examine retinal vessel patency showing incorporation of human cells (*green*) into a normal appearing retinal vasculature in a mouse eye 6 months after acute retinal ischemia-reperfusion injury and intravitreal injection of CD34+ cells from human bone marrow. [Adapted from Park SS et al., Long-term effects of intravitreal injection of GMP-grade bone marrow-derived CD34+ cells in NOD-SCID mice with acute ischemia-reperfusion injury. Invest Ophthalmol Vis Sci 2012;53:986–94. Copyright by The Association for Research in Vision and Ophthalmology and reproduced with permission]

known that both endothelial cells and pericytes are lost in diabetic retinopathy, but only the endothelial cells have the capacity for repair and regeneration [132, 155]. Thus, EPCs are logical candidates for cell therapy for diabetic retinopathy. However, there may be limitations of using autologous EPCs from peripheral blood given their impaired mobility and function. Bone marrow CD34+ cells may not be exposed to the diabetic environment and may provide an alternative source of autologous cell therapy for diabetic retinopathy.

The therapeutic potential of EPCs and CD34+ cells may not be limited to diseases associated with retinal ischemia. An increase in circulating CD34+ cells have been found in AMD patients with active choroidal neovascularization, and it has been suggested that they are recruited from the bone marrow in response to disease activity [156]. Additionally, damaged retinal pigment epithelial cells have been shown to secrete cytokines that serve as chemo-attractants for bone marrow stem cells to migrate towards the retina [157]. The number of circulating CD34+ cells was reported to be elevated also in patients with myopic and idiopathic choroidal neovascularization [158, 159], further supporting the notion that these cells are recruited to participate in tissue repair. Although it has been speculated that these CD34+ cells might play a role in the neovascular process, the potential trophic effect of these cells on the degenerative retinal conditions also needs to be explored since supported by preclinical studies.

4 Translational Strategies

Establishing a treatment program using stem cells of any kind in human patients can be very challenging. Data collected in experimental in vitro and animal studies and even observational studies in humans can provide convincing evidence that this strategy has the potential to be an effective treatment for patients with retinal disease. However, research in human subjects is still in its infancy using cell therapy, including autologous bone marrow stem cell therapy. In order to proceed from the laboratory bench to the clinic, two important aspects must be considered for human trials—patient selection and patient safety. In this section we will focus on these aspects for the use of autologous bone marrow adult stem cells for the treatment of patients with retinal disease. However, the guidelines can be extrapolated to other clinical trials for cell therapy.

4.1 Patient Selection

The following are general guidelines for patient selection for a clinical trial:

Baseline vision—For an early phase 1 clinical study designed to evaluate the safety of an experimental treatment, the enrolled patients should have baseline visual acuity that is low enough to warrant an intervention with an experimental therapy with unknown safety profile. In addition, the vision loss should be irreversible and not amendable to any available treatment. As a general rule, patients with good visual function are excluded from a phase 1 study as they may have more to lose potentially than to gain. Patients with visual acuity of light perception or worse may not be the ideal candidate for a phase 1 clinical trial as their damage may be too severe to obtain any significant benefit from the therapy. In addition, advanced vision loss at baseline might limit the ability to obtain reliable safety information regarding the cell therapy since it may be difficult to evaluate visual function as accurately with diagnostic tests.

Laterality—Only one eye should be treated with any experimental therapy, including stem cell therapy, since the safety profile is unknown. The eye selected for the study should be the patient's eye with the worse vision. This is to ensure that the experimental treatment does not have significant negative effect on the subject's quality of life or ability to function should an adverse effect of the experimental therapy be observed during the clinical trial.

Absence of other concurrent ocular diseases—It is preferable for a human trial to be focused on a single retinal disease such that the study enrolls only eyes with vision loss resulting from the retinal condition of interest. Enrolling eyes with concurrent pathology that affect vision would add compounding variables that may affect the study outcome or data analysis. The study eye also should not require any concomitant treatment or surgery during the study period. This is again to minimize compounding variables that may affect the study outcome or data analysis.

Systemic conditions—Systemic health should be considered prior to inclusion of a patient for any clinical trial. It is important to enroll subjects in good general health to ensure that the subject can maintain study follow-up as scheduled. In addition, in the case of stem cell therapy, it is important to exclude patients with history of malignancy since teratoma and other tumor formation can been associated with the use of stem cells. Although embryonic stem cells and iPSCs are more teratogenic than adult stem cells, teratogenicity and other cellular proliferative effects are potential serious adverse effects that may be associated with any stem cell therapy.

In the case of autologous bone marrow adult stem cell therapy, any patient with a systemic condition that may alter the composition of the bone marrow also should be excluded. This includes patients on chronic systemic immunosuppressive therapy or history of chemotherapy or radiotherapy. In addition, patients on chronic anti-coagulation may need to stop the anti-coagulation temporarily before study treatment since excess bleeding may be associated with the bone marrow aspiration or stem cell injection in the eye.

Needless to say, female patients of childbearing age must not be pregnant at study enrollment and during the study follow-up period. Thus, pregnancy test would need to be conducted at the time of study enrollment. In addition, contraception should be used following the study treatment since there may be unknown adverse effects of the stem cells on the fetus.

Because of these potential adverse effects that may be serious and associated with cell therapy, all study subjects should be of legal age and have no cognitive impairment such that they can give informed consent following a detailed explanation of the study's design, risks, and benefits.

4.2 Patient Safety

Because of potential serious adverse effects that may be associated with any stem cell therapy, including autologous bone marrow stem cell therapy, it is important to perform adequate long-term preclinical safety studies before embarking on a clinical trial that treats patients with an experimental therapy. Such a study was performed for intravitreal administration of bone marrow CD34+ cells using a NOD-SCID mouse model before the IND for the clinical trial was cleared by the Food and Drug Administration (FDA) [145]. Although animal studies do not always replicate the observations in human patients, preclinical long-term studies provide important additional safety screening regarding an experimental therapy.

Once a therapy has been approved for clinical trial by the FDA and local regulatory agencies, there are certain adverse effects that can be associated with stem cell therapy that should be included in the informed consent and discussed with the patients who are considering enrolling in the study and receiving treatment. They include the following:

1. Tumor formation and cellular proliferation in the eye or systemically are potential adverse effects of any stem cell therapy in the eye. Such adverse effects are more of a concern when using undifferentiated embryonic stem cells or iPSCs [67–69] and of less of a concern when using adult stem cells or differentiated cells derived from embryonic stem cells or iPSCs. Nonetheless, the concern remains with all cell therapies. Currently, there are limited clinical data regarding the use of cell therapies, including autologous bone marrow cell therapy. Thus, larger studies with longer follow-up are needed to evaluate for these potential adverse effects. Endothelial progenitor cells and CD34+ cells may be safer cells to use in clinical trial in that they have limited capacity to proliferate and expand in vivo and in vitro and already have been used safely clinically for non-ocular indications. Nonetheless, each human study subject should be followed for signs of cellular proliferation in the eye and systemically following cell therapy. This may manifest as a new tumor, epiretinal membrane, or ocular neovascularization, for example.
2. There is a risk of immune activation in response to the introduction of foreign cells into any tissue. Even with autologous bone marrow adult stem cell therapy, these cells are introduced into a foreign environment, the eye, and this might stimulate an inflammatory response in the eye or even systemically. Such a response could cause detrimental complications and irreversible loss of vision. Fibrovascular proliferation has been described following intravitreal injection of mesenchymal stem cells in animals although these cells are typically believed to be immunosuppressive. This observation raises concerns about suitability of these cells for human studies (personal communication, Jan Nolta, Ed Pugh, Robert Zawadzki, manuscript in preparation). No ocular adverse effects have been noted short-term or long-term following intravitreal injection of CD34+ bone marrow cells in animal models [145].
3. There may be adverse effects associated with the procedure for stem cell isolation and delivery. For intravitreal or subretinal administration of cells, there is always a small risk of complications such as endophthalmitis, retinal detachment, elevated intraocular pressure, intraocular bleeding, and traumatic cataract [160]. Subretinal transplantation requires surgery, and this carries the additional risks of anesthesia [161].
4. For autologous bone marrow stem cell therapy, there is the additional risk associated with the bone marrow aspiration procedure. Although the risk is small when the procedure is performed by an experienced hematologist, there is always a small risk of bleeding, pain, and infection. In rare cases, the procedure may be unsuccessful in obtaining adequate bone marrow aspirate for the cell therapy.

5 Human Data

5.1 Autologous Bone Marrow-Derived Stem Cells for the Treatment of Retinal Diseases

Bone marrow adult stem cells have several characteristics that make them very attractive candidates as cell therapy for treating retinal diseases. First, they are much easier to obtain than embryonic stem cells and require no in vitro manipulation to direct their differentiation prior to use. Second, these cells may be used autologously, thereby reducing the risk of rejection and obviate the need for systemic immunosuppression. Third, they can be administered by intravitreal injection, which is much simpler and safer than subretinal transplantation which would usually require vitrectomy surgery. These cells have been shown to home towards the damaged retina following intravitreal injection in a variety of disease models studied [125, 145]. They are believed to exert their reparative function via a paracrine trophic effect although some of the cells do incorporate into the retina.

As discussed earlier, EPCs in peripheral blood share some characteristics of bone marrow CD34+ cells and can be explored as alternative adult cell therapy. However, EPCs are present in very low concentration in peripheral blood, and pharmacologic mobilization of these cells from bone marrow to peripheral blood is required. This approach has been used for bone marrow transplantation and cell therapy for ischemic cardiomyopathy [126]. However, these mobilized EPCs in peripheral blood may not have the same characteristics as the native cells in bone marrow and unmobilized peripheral blood. Thus, the current clinical trials for treatment of retinal disease concentrate on the use of bone marrow-derived stem cells [162–166].

Based on the long-term studies conducted in NOD-SCID mice showing no safety issues associated with intravitreal administration of CD34+ cells from bone marrow [145], the FDA has cleared one human trial investigating intravitreal injection of autologous bone marrow CD34+ cells in patients with ischemic and degenerative retinal disorders (registered with clinicaltrials.gov, NCT01736059). These cells are isolated from the mononuclear cellular fraction of the bone marrow aspirate by positive selection using a magnetic cell sorter under Good Manufacturing Practice conditions. Since it is autologous therapy, the patients are not immunosuppressed. A recent publication reported on the 6 months results of the first 6 patients that enrolled and received the cell therapy [162]. This report included 2 patients with Stargardt's disease, 2 patients with AMD, 1 patient with retinitis pigmentosa, and 1 patient with a combined central retinal artery and vein occlusion (CRAO/CRVO). In 4 of 6 eyes, visual acuity improved 2 or more lines during the course of the study. The most dramatic improvement in vision was achieved in the patient with CRAO/CRVO, where that pathogenesis of vision loss is more ischemic than degenerative. No ocular or systemic complications were recorded in any of the study subjects. In addition, adaptive optics optical coherence tomography imaging of the macula of a treated eye with hereditary macular degeneration demonstrated intraretinal hyperreflective changes that were suggestive of intraretinal incorporation of the injected cells [162] (Fig. 4). These optical coherence tomography findings have been reproduced in a murine model of retinal degeneration, and identified by immunohistochemistry to be CD34+ cells that entered the retinal layers (Elad Moisseiev, Susanna Park, Robert Zawadzki, Jan Nolta, et al., Invest Ophthalmol Vis Sci., manuscript in press).

The outcome of autologous intravitreal bone marrow-derived mononuclear cell therapy has been reported in a few patients in Brazil and Germany [163–166]. The mononuclear cell fraction of bone marrow is a crude cellular fraction that contains about 2% CD34+ cells. This cell fraction has been used in clinical trial as potential therapy for ischemic cardiomyopathy [167]. The intravitreal administration of mononuclear cells from bone marrow was well tolerated in both studies. Larger studies are planned and ongoing for the treatment of retinitis pigmentosa, non-exudative AMD, and ischemic retinopathies (registered with clinicaltrials.

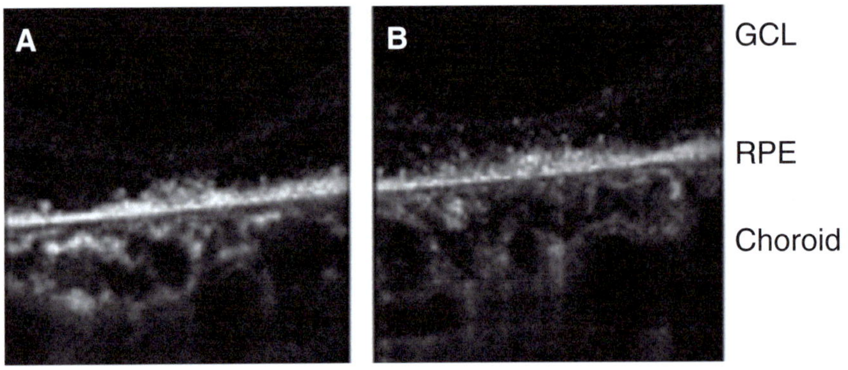

Fig. 4 Adaptive optics optical coherence tomography (AO-OCT) imaging of the central macula of the study eye a subject with Stargardt's disease, that was treated with an intravitreal injection of autologous CD34+ stem cells. (**a**) Prior to this treatment, an irregular hyperreflective RPE layer was demonstrated, with the overlying photoreceptors not well visualized due to atrophy. (**b**) One month after intravitreal CD34+ cell injection new multiple hyperreflective (*white*) deposits were demonstrated within the retinal layers, suggestive of intraretinal migration of the cells. [Adapted from Park SS et al., Intravitreal autologous bone marrow CD34+ cell therapy for ischemic and degenerative retinal disorders: preliminary phase 1 clinical trial findings. Invest Ophthalmol Vis Sci 2015;56:81–9. Copyright by The Association for Research in Vision and Ophthalmology and reproduced with permission]

gov, NCT01068561, NCT01518127 and NCT01518842, respectively). The initial published report with this therapy included 5 patients (3 with retinitis pigmentosa and 2 with cone-rod dystrophy) that were treated with mononuclear cell therapy and followed for 10 months. A maximum improvement in visual acuity of 1 line was noted in four of the five study patients [163]. There were no associated safety issues locally or systemically.

More recently, intravitreal injection of these bone marrow mononuclear cells was reported to be associated with resolution of refractory cystoid macular edema in a patient with retinitis pigmentosa [164]. It also has been reported to be associated with visual improvement and reduction in cystoid macular edema in two patients with CRVO [165]. It should be noted that these are preliminary pioneering studies that have reported on small numbers of patients and had no control arms. However, based on these observations, autologous bone marrow stem cell therapy should be further explored in larger studies as potential treatment for a variety of retinal conditions. Early phase clinical studies have been initiated and are ongoing in a few centers, and future studies are likely to expand as our understanding of the bone marrow stem cell biology continues to evolve.

6 Future Directions

As research on stem cell therapy for retinal diseases advance, the therapeutic potentials and limitations of the various cell therapies will become more evident. It is possible that stem cell therapy may become a clinical reality that can be routinely offered to patients with different degenerative or ischemic retinal conditions that are currently untreatable. In the case of autologous adult bone marrow stem cell therapy, the use of these cells for treatment of retinal diseases in human patients is in early stage clinical trial. Thus, considerable future research is required to determine the full potential of this cell therapy. The following are directions for future research that may maximize the therapeutic potential of bone marrow adult stem cells as therapy for retinal disorders.

Stem cell type selection—The best cell type to be used for treating various retinal diseases is still not established. Further research will help determine which cell therapy offers the best safety and therapeutic profile for patients. Intravitreal autologous bone marrow mononuclear cells and CD34+ cells have been used in an early clinical trial without any safety issues. Future studies characterizing the various cell populations in bone marrow would help identify the ideal cell therapy candidate. Whether further selective isolation of the cells would be desired for an enhanced treatment effect is not known. Isolation and selection of the desired subpopulation of cells from bone marrow would allow a higher number of the effective cells to be delivered into the eye for therapy. However, such a selective cell delivery may not necessarily produce a better treatment outcome since interplay among cells in bone marrow and resulting synergism cannot be neglected [127, 129, 132, 141].

Source of cells—Another factor to be investigated is the ideal source of adult stem cells for treating various retinal conditions. This chapter concentrates on bone marrow cell therapy. However, other sources of adult stem cells also are being explored including adipose, peripheral blood, and placenta/cord blood.

Dosage—Published reports using intravitreal bone marrow mononuclear cells used ten million cells per eye since higher cell numbers often resulted in cell clumping (RC Siqueira, personal communication). The current phase 1 clinical study using intravitreal autologous CD34+ bone marrow stem cell therapy explores a range of cell doses since the ideal dose has not been determined. In the ongoing phase 1 study, patients were treated with all the CD34+ cells that were isolated from the bone marrow aspiration (ranging from one to seven million cells). This approach allows indirect dose-response information to be obtained [162]. Once the optimum dose range of cell therapy is established, future research can explore the relative efficacy of set doses of cell therapy in a larger clinical trial.

Method of administration—Although the current clinical trials using autologous bone marrow cells have concentrated on intravitreal administration, the relative efficacy of intravitreal administration should be compared to intravenous and subretinal routes of administration to determine the most effective route of therapy for the various retinal conditions. It is possible a different route of cell administration might be preferred for different retinal conditions. Intravitreal injection is the preferred method, if possible, since it is simpler and easier to administer. In addition, it eliminates the possible complications of vitreoretinal or trans-scleral surgery. However, whether this mode of delivery is as effective as subretinal implantation or intravenous administration has yet to be determined.

Timing of intervention—It is possible that the therapeutic effect of any stem cell therapy may be limited by the severity of the preexisting condition. Earlier intervention may show more therapeutic effect. Unfortunately, since only early phase clinical trials have been initiated, only eyes with advanced vision loss have been treated so far with autologous intravitreal bone marrow cell therapy and other experimental cell therapies. Thus, the full therapeutic potential of cell therapy may not have been appreciated thus far. The therapeutic effect of stem cell therapy may manifest not only as improvement in vision from regenerative effects but also may manifest as stabilization of a progressive retinal condition. The optimal timing of intervention to observe the maximum therapeutic effects on the retina may vary for different retinal diseases and will need to be investigated.

Repeated treatments—Although the classical definition of stem cells implies they have the potential for unlimited self-renewal, this may not be the case when cells with limited cycling rate, such as EPCs and CD34+ cells, or differentiated cells are used for therapy. Although long-term incorporation of bone marrow CD34+ cells in the retinal vasculature has been observed in animal eyes with acute ischemia reperfusion injury, the majority of injected cells do not remain in the eye. Reported preliminary results of an ongoing phase 1 clinical trial using autologous intravitreal CD34+ cells from bone marrow showed some progression of geographic atrophy in eyes with non-exudative AMD during a 6-month study follow-up although some visual acuity improvement was noted early in the course of the study [162]. Whether repeated cell therapy might further enhance the treatment effect of cell therapy in eyes with retinal disease and slow down progression of the retinal condition is a question that needs to be further explored. In addition, the ideal timing for the repeat treatment also will need to be determined. This may vary with different retinal conditions.

Imaging—Once the stem cells are introduced into the eye, it is difficult to monitor the location and viability of the cells in the eye. Visual acuity and electroretinography have been used to assess the effect of cell therapy on retinal function. However, in vivo retinal imaging may be a better method for monitoring the effect of these cells in the retina. This may be especially important if repeat cell treatments are considered. Based on recent study observations, the intraretinal incorporation of CD34+ cells may be visualized using adaptive-optics optical coherence tomography imaging [162]. Further studies are needed to determine the sensitivity and specificity of using in vivo retinal imaging to study the effect of cell therapy on the retina.

Safety—Although intravitreal autologous bone marrow cell therapy has not been associated with any reported safety issues, we have limited clinical data thus far. Future larger-scale studies with longer follow-up are required to fully establish the ocular and systemic safety profiles of this and other experimental cell therapies.

Other cell-mediated therapy—Stem cell therapy has the potential to be used in synergy with gene therapy or other drug therapy since the stem cell can serve as a vector for delivering selected genes or drugs to the target cells in the retina. The use of concurrent bone marrow stem cell therapy may improve the potential efficacy and specificity of the therapy on the target tissue by improving delivery of the gene or drug to the target cell environment.

Since a paracrine therapeutic effect has been implicated for adult stem cells, an alternative approach utilizing these cells for retinal disease would be to better characterize the cytokines and growth factors derived from these adult stem cells. They might be used alone or in combination with cell therapy as treatment for ischemic or degenerative retinal conditions.

In conclusion, stem cell therapy for retinal diseases is a rapidly developing research area, and autologous use of bone marrow adult stem cells show great promise as a possible treatment that should be explored for various ischemic or degenerative retinal diseases.

References

1. Bourne RR, Stevens GA, White RA, et al.; Vision Loss Expert Group. Causes of vision loss worldwide, 1990–2010: a systematic analysis. Lancet Glob Health. 2013;1:e339–49.
2. Jonas JB, Bourne RR, White RA, et al.; Vision Loss Expert Group of the Global Burden of Disease Study. Visual impairment and blindness due to macular diseases globally: a systematic review and meta-analysis. Am J Ophthalmol. 2014;158:808–15.
3. Bressler NM. Age-related macular degeneration is the leading cause of blindness. JAMA. 2004;291:1900–1.
4. Klein R, Klein BE, Linton KL. Prevalence of age-related maculopathy. The Beaver Dam Eye Study. Ophthalmology. 1992;99:933–43.
5. Mitchell P, Smith W, Attebo K, Wang JJ. Prevalence of age-related maculopathy in Australia. The Blue Mountains Eye Study. Ophthalmology. 1995;102:1450–60.
6. Klaver CC, Assink JJ, van Leeuwen R, et al. Incidence and progression rates of age-related maculopathy: the Rotterdam Study. Invest Ophthalmol Vis Sci. 2001;42:2237–41.
7. Wong TY, Chakravarthy U, Klein R, et al. The natural history and prognosis of neovascular age-related macular degeneration: a systematic review of the literature and meta-analysis. Ophthalmology. 2008;115:116–26.
8. Kawasaki R, Yasuda M, Song SJ, et al. The prevalence of age-related macular degeneration in Asians: a systematic review and meta-analysis. Ophthalmology. 2010;117:921–7.
9. Rudnicka AR, Kapetanakis VV, Jarrar Z, et al. Incidence of late-stage age-related macular degeneration in American whites: systematic review and meta-analysis. Am J Ophthalmol. 2015;160:85–93.
10. Wong WL, Su X, Li X, et al. Global prevalence of age-related macular degeneration and disease burden projection for 2020 and 2040: a systematic review and meta-analysis. Lancet Glob Health. 2014;2:e106–16.
11. Roodhooft JM. Leading causes of blindness worldwide. Bull Soc Belge Ophtalmol. 2002;283:19–25.
12. Prokofyeva E, Zrenner E. Epidemiology of major eye diseases leading to blindness in Europe: a literature review. Ophthalmic Res. 2012;47:171–88.
13. International Diabetes Federation. IDF Diabetes Atlas. 5th ed. Brussels: International Diabetes Federation; 2011.
14. Shaw JE, Sicree RA, Zimmet PZ. Global estimates of the prevalence of diabetes for 2010 and 2030. Diabetes Res Clin Pract. 2010;87:4–14.
15. World Health Organization. Diabetes action now: an initiative of the World Health Organization and the International Diabetes Federation. Geneva: World Health Organization; 2004.
16. Zheng Y, He M, Congdon N. The worldwide epidemic of diabetic retinopathy. Indian J Ophthalmol. 2012;60:428–31.
17. The Eye Disease Case-Control Study Group. Risk factors for central retinal vein occlusion. Arch Ophthalmol. 1996;114:545–54.
18. Hamel C. Retinitis pigmentosa. Orphanet J Rare Dis. 2006;1:40.
19. Blacharski PA. Fundus flavimaculatus. In: Newsome DA, editor. Retinal dystrophies and degenerations. New York: Raven Press; 1988. p. 135–59.
20. Xu L, Wang Y, Li Y, Wang Y, Cui T, Li J, Jonas JB. Causes of blindness and visual impairment in urban and rural areas in

Beijing: the Beijing Eye Study. Ophthalmology. 2006;113:1134.e1–11.

21. Iwase A, Araie M, Tomidokoro A, et al.; Tajimi Study Group. Prevalence and causes of low vision and blindness in a Japanese adult population: the Tajimi Study. Ophthalmology. 2006;113:1354–62.
22. Wu L, Sun X, Zhou X, Weng C. Causes and 3-year-incidence of blindness in Jing-An District, Shanghai, China 2001–2009. BMC Ophthalmol. 2011;11:10.
23. Silva R. Myopic maculopathy: a review. Ophthalmologica. 2012;228:197–213.
24. Fritsche LG, Fariss RN, Stambolian D, Abecasis GR, Curcio CA, Swaroop A. Age-related macular degeneration: genetics and biology coming together. Annu Rev Genomics Hum Genet. 2014;15:151–71.
25. Chakravarthy U, Wong TY, Fletcher A, et al. Clinical risk factors for age-related macular degeneration: a systematic review and meta-analysis. BMC Ophthalmol. 2010;10:31.
26. Tomany SC, Wang JJ, Van Leeuwen R, et al. Risk factors for incident age-related macular degeneration: pooled findings from 3 continents. Ophthalmology. 2004;111:1280–7.
27. Fritsche LG, Chen W, Schu M, et al.; AMD Gene Consortium. Seven new loci associated with age-related macular degeneration. Nat Genet. 2013;45:433–39.
28. Grassmann F, Fritsche LG, Keilhauer CN, Heid IM, Weber BH. Modeling the genetic risk in age-related macular degeneration. PLoS One. 2012;7:e37979.
29. Kahn HA, Leibowitz HM, Ganley JP, et al. The Framingham eye study: I. Outline and major prevalence findings. Am J Epidemiol. 1977;106:17e32.
30. Rosenfeld PJ, Brown DM, Heier JS, et al.; MARINA Study Group. Ranibizumab for neovascular age-related macular degeneration. N Engl J Med. 2006;355:1419–31.
31. Brown DM, Kaiser PK, Michels M, et al.; ANCHOR Study Group. Ranibizumab versus verteporfin for neovascular age-related macular degeneration. N Engl J Med. 2006;355:1432–44.
32. Comparison of Age-related Macular Degeneration Treatments Trials (CATT) Research Group, Martin DF, Maguire MG, et al. Ranibizumab and bevacizumab for treatment of neovascular age-related macular degeneration: two-year results. Ophthalmology. 2012;119:1388–98.
33. Heier JS, Brown DM, Chong V, et al.; VIEW 1 and VIEW 2 Study Groups. Intravitreal aflibercept (VEGF trap-eye) in wet age-related macular degeneration. Ophthalmology. 2012;119:2537–48.
34. Schmidt-Erfurth U, Kaiser PK, Korobelnik JF, et al. Intravitreal aflibercept injection for neovascular age-related macular degeneration: ninety-six-week results of the VIEW studies. Ophthalmology. 2014;121:193–201.
35. Chakravarthy U, Harding SP, Rogers CA, et al.; IVAN study investigators. Alternative treatments to inhibit VEGF in age-related choroidal neovascularisation: 2-year findings of the IVAN randomized controlled trial. Lancet. 2013;382:1258–67.
36. Ciulla TA, Rosenfeld PJ. Anti-vascular endothelial growth factor therapy for neovascular ocular diseases other than age-related macular degeneration. Curr Opin Ophthalmol. 2009;20:166–74.
37. Augood CA, Vingerling JR, de Jong PT, et al. Prevalence of age-related maculopathy in older Europeans: the European Eye Study (EUREYE). Arch Ophthalmol. 2006;124:529–35.
38. Klein R, Klein BE, Knudtson MD, Meuer SM, Swift M, Gangnon RE. Fifteen-year cumulative incidence of age-related macular degeneration: the Beaver Dam Eye Study. Ophthalmology. 2007;114:253–62.
39. Chew EY, Clemons TE, Agrón E, et al.; Age-Related Eye Disease Study Research Group. Long-term effects of vitamins C and E, β-carotene, and zinc on age-related macular degeneration: AREDS report no. 35. Ophthalmology. 2013;120:1604–11.
40. Age-Related Eye Disease Study 2 (AREDS2) Research Group, Chew EY, Clemons TE, Sangiovanni JP, et al. Secondary analyses of the effects of lutein/zeaxanthin on age-related macular degeneration progression: AREDS2 report No. 3. JAMA Ophthalmol. 2014;132:142–9.
41. Wang JJ, Buitendijk GH, Rochtchina E, et al. Genetic susceptibility, dietary antioxidants, and long-term incidence of age-related macular degeneration in two populations. Ophthalmology. 2014;121:667–75.
42. Tarr JM, Kaul K, Chopra M, Kohner EM, Chibber R. Pathophysiology of diabetic retinopathy. ISRN Ophthalmol. 2013;2013:343560.
43. Brown DM, Nguyen QD, Marcus DM, et al.; RIDE and RISE Research Group. Long-term outcomes of ranibizumab therapy for diabetic macular edema: the 36-month results from two phase III trials: RISE and RIDE. Ophthalmology. 2013;120:2013–22.
44. Network DRCR, Wells JA, Glassman AR, Ayala AR, et al. Aflibercept, bevacizumab, or ranibizumab for diabetic macular edema. N Engl J Med. 2015;372:1193–203.
45. Elman MJ, Ayala A, Bressler NM, et al.; Diabetic Retinopathy Clinical Research Network. Intravitreal Ranibizumab for diabetic macular edema with prompt versus deferred laser treatment: 5-year randomized trial results. Ophthalmology. 2015;122:375–81.
46. Do DV, Nguyen QD, Boyer D, et al.; da Vinci Study Group. One-year outcomes of the da Vinci Study of VEGF Trap-Eye in eyes with diabetic macular edema. Ophthalmology. 2012;119:1658–65.
47. Rajendram R, Fraser-Bell S, Kaines A, et al. A 2-year prospective randomized controlled trial of intravitreal bevacizumab or laser therapy (BOLT) in the management of diabetic macular edema: 2-month data: report 3. Arch Ophthalmol. 2012;130:972–9.
48. Kriechbaum K, Prager S, Mylonas G, et al.; Diabetic Retinopathy Research Group. Intravitreal bevacizumab (Avastin) versus triamcinolone (Volon A) for treatment of diabetic macular edema: one-year results. Eye (Lond). 2014;28:9–15.
49. Gillies MC, Lim LL, Campain A, et al. A randomized clinical trial of intravitreal bevacizumab versus intravitreal dexamethasone for diabetic macular edema: the BEVORDEX study. Ophthalmology. 2014;121:2473–81.
50. Callanan DG, Gupta S, Boyer DS, et al.; Ozurdex PLACID Study Group. Dexamethasone intravitreal implant in combination with laser photocoagulation for the treatment of diffuse diabetic macular edema. Ophthalmology. 2013;120:1843–51.
51. Boyer DS, Yoon YH, Belfort R Jr, et al.; Ozurdex MEAD Study Group. Three-year, randomized, sham-controlled trial of dexamethasone intravitreal implant in patients with diabetic macular edema. Ophthalmology. 2014;121:1904–14.
52. Campochiaro PA, Hafiz G, Shah SM, et al.; FAME Study Group. Sustained ocular delivery of fluocinolone acetonide by an intravitreal insert. Ophthalmology. 2010;117:1393–99.
53. Fong DS, Ferris 3rd FL, Davis MD, Chew EY. Causes of severe visual loss in the early treatment diabetic retinopathy study: ETDRS report no. 24. Early Treatment Diabetic Retinopathy Study Research Group. Am J Ophthalmol. 1999;127:137–41.
54. Tzekov R, Arden GB. The electroretinogram in diabetic retinopathy. Surv Ophthalmol. 1999;44:53–60.
55. Rofagha S, Bhisitkul RB, Boyer DS, Sadda SR, Zhang K, SEVEN-UP Study Group. Seven-year outcomes in ranibizumab-treated patients in ANCHOR, MARINA, and HORIZON: a multicenter cohort study (SEVEN-UP). Ophthalmology. 2013;120:2292–9.
56. Meyer CH, Krohne TU, Holz FG. Intraocular pharmacokinetics after a single intravitreal injection of 1.5 mg versus 3.0 mg of bevacizumab in humans. Retina. 2011;31:1877–84.
57. Ahn SJ, Ahn J, Park S, Kim H, et al. Intraocular pharmacokinetics of ranibizumab in vitrectomized versus nonvitrectomized eyes. Invest Ophthalmol Vis Sci. 2014;55:567–73.

58. Krohne TU, Liu Z, Holz FG, Meyer CH. Intraocular pharmacokinetics of ranibizumab following a single intravitreal injection in humans. Am J Ophthalmol. 2012;154:682–6.
59. Stewart MW, Rosenfeld PJ, Penha FM, et al. Pharmacokinetic rationale for dosing every 2 weeks versus 4 weeks with intravitreal ranibizumab, bevacizumab, and aflibercept (vascular endothelial growth factor Trap-eye). Retina. 2012;32:434–57.
60. Kim H, Csaky KG, Gravlin L, et al. Safety and pharmacokinetics of a preservative-free triamcinolone acetonide formulation for intravitreal administration. Retina. 2006;26:523–30.
61. Beer PM, Bakri SJ, Singh RJ, Liu W, Peters 3rd GB, Miller M. Intraocular concentration and pharmacokinetics of triamcinolone acetonide after a single intravitreal injection. Ophthalmology. 2003;110:681–6.
62. Conlon TJ, Deng WT, Erger K, et al. Preclinical potency and safety studies of an AAV2-mediated gene therapy vector for the treatment of MERTK associated retinitis pigmentosa. Hum Gene Ther Clin Dev. 2013;24:23–8.
63. Dinculescu A, Min SH, Deng WT, Li Q, Hauswirth WW. Gene therapy in the rd6 mouse model of retinal degeneration. Adv Exp Med Biol. 2014;801:711–8.
64. Ben M'Barek K, Regent F, Monville C. Use of human pluripotent stem cells to study and treat retinopathies. World J Stem Cells. 2015;7:596–604.
65. Cideciyan AV, Jacobson SG, Beltran WA, et al. Human retinal gene therapy for Leber congenital amaurosis shows advancing retinal degeneration despite enduring visual improvement. Proc Natl Acad Sci U S A. 2013;110:E517–25.
66. Thomson JA, Itskovitz-Eldor J, Shapiro SS, et al. Embryonic stem cell lines derived from human blastocysts. Science. 1998;282:1145–7.
67. Cui L, Guan Y, Qu Z, et al. WNT signaling determines tumorigenicity and function of ESC-derived retinal progenitors. J Clin Invest. 2013;123:1647–61.
68. Melville H, Carpiniello M, Hollis K, Staffaroni A, Golestaneh N. Stem cells: a new paradigm for disease modeling and developing therapies for age-related macular degeneration. J Transl Med. 2013;11:53.
69. Alonso-Alonso ML, Srivastava GK. Current focus of stem cell application in retinal repair. World J Stem Cells. 2015;7:641–8.
70. Booth C, Veys P. T cell depletion in paediatric stem cell transplantation. Clin Exp Immunol. 2013;172:139–47.
71. Schmeisser A, Strasser RH. Phenotypic overlap between hematopoietic cells with suggested angioblastic potential and vascular endothelial cells. J Hematother Stem Cell Res. 2002;11:69–79.
72. Friedenstein AJ, Gorskaja JF, Kulagina NN. Fibroblast precursors in normal and irradiated mouse hematopoietic organs. Exp Hematol. 1976;4:267–74.
73. Ashton BA, Allen TD, Howlett CR, Eaglesom CC, Hattori A, Owen M. Formation of bone and cartilage by marrow stromal cells in diffusion chambers in vivo. Clin Orthop Relat Res. 1980;151:294–307.
74. Bab I, Ashton BA, Gazit D, Marx G, Williamson MC, Owen ME. Kinetics and differentiation of marrow stromal cells in diffusion chambers in vivo. J Cell Sci. 1986;84:139–51.
75. Castro-Malaspina H, Gay RE, Resnick G, et al. Characterization of human bone marrow fibroblast colony-forming cells (CFU-F) and their progeny. Blood. 1980;56:289–301.
76. Campagnoli C, Roberts IA, Kumar S, Bennett PR, Bellantuono I, Fisk NM. Identification of mesenchymal stem/progenitor cells in human first-trimester fetal blood, liver, and bone marrow. Blood. 2001;98:2396–402.
77. Zuk PA, Zhu M, Ashjian P, et al. Human adipose tissue is a source of multipotent stem cells. Mol Biol Cell. 2002;13:4279–95.
78. Chamberlain G, Fox J, Ashton B, Middleton J. Concise review: mesenchymal stem cells: their phenotype, differentiation capacity, immunological features, and potential for homing. Stem Cells. 2007;25:2739–49.
79. Pittenger MF, Mackay AM, Beck SC, et al. Multilineage potential of adult human mesenchymal stem cells. Science. 1999;284: 143–7.
80. Yamanaka S, Takahashi K. Induction of pluripotent stem cells from mouse fibroblast cultures. Tanpakushitsu Kakusan Koso. 2006;51:2346–51.
81. Takahashi K, Tanabe K, Ohnuki M, et al. Induction of pluripotent stem cells from adult human fibroblasts by defined factors. Cell. 2007;131:861–72.
82. Yu J, Vodyanik MA, Smuga-Otto K, et al. Induced pluripotent stem cell lines derived from human somatic cells. Science. 2007;318:1917–20.
83. Park IH, Arora N, Huo H, et al. Disease-specific induced pluripotent stem cells. Cell. 2008;134:877–86.
84. Wernig M, Meissner A, Foreman R, et al. In vitro reprogramming of fibroblasts into a pluripotent ES-cell-like state. Nature. 2007; 448:318–24.
85. Okita K, Ichisaka T, Yamanaka S. Generation of germline-competent induced pluripotent stem cells. Nature. 2007;448:313–7.
86. da Silva ML, Chagastelles PC, Nardi NB. Mesenchymal stem cells reside in virtually all post-natal organs and tissues. J Cell Sci. 2006;119:2204–13.
87. Strioga M, Viswanathan S, Darinskas A, Slaby O, Michalek J. Same or not the same? Comparison of adipose tissue-derived versus bone marrow-derived mesenchymal stem and stromal cells. Stem Cells Dev. 2012;21:2724–52.
88. Bernardo ME, Locatelli F, Fibbe WE. Mesenchymal stromal cells: a novel treatment modality for tissue repair. Ann N Y Acad Sci. 2009;1176:101–17.
89. Hass R, Kasper C, Böhm S, Jacobs R. Different populations and sources of human mesenchymal stem cells (MSC): a comparison of adult and neonatal tissue-derived MSC. Cell Commun Signal. 2011;9:12.
90. Fraser JK, Wulur I, Alfonso Z, Hedrick MH. Fat tissue: an underappreciated source of stem cells for biotechnology. Trends Biotechnol. 2006;24:150–4.
91. Noël D, Caton D, Roche S, et al. Cell specific differences between human adipose-derived and mesenchymal-stromal cells despite similar differentiation potentials. Exp Cell Res. 2008;314: 1575–84.
92. Pachón-Peña G, Yu G, Tucker A, et al. Stromal stem cells from adipose tissue and bone marrow of age-matched female donors display distinct immunophenotypic profiles. J Cell Physiol. 2011;226:843–51.
93. De Ugarte DA, Morizono K, Elbarbary A, et al. Comparison of multi-lineage cells from human adipose tissue and bone marrow. Cells Tissues Organs. 2003;174:101–9.
94. Sakaguchi Y, Sekiya I, Yagishita K, Muneta T. Comparison of human stem cells derived from various mesenchymal tissues: superiority of synovium as a cell source. Arthritis Rheum. 2005;52:2521–9.
95. Kern S, Eichler H, Stoeve J, Klüter H, Bieback K. Comparative analysis of mesenchymal stem cells from bone marrow, umbilical cord blood, or adipose tissue. Stem Cells. 2006;24:1294–301.
96. Rezanejad H, Soheili ZS, Haddad F, et al. In vitro differentiation of adipose-tissue-derived mesenchymal stem cells into neural retinal cells through expression of human PAX6 (5a) gene. Cell Tissue Res. 2014;356:65–75.
97. Vossmerbaeumer U, Ohnesorge S, Kuehl S, et al. Retinal pigment epithelial phenotype induced in human adipose tissue-derived mesenchymal stromal cells. Cytotherapy. 2009;11:177–88.
98. Moviglia GA, Blasetti N, Zarate JO, Pelayes DE. In vitro differentiation of adult adipose mesenchymal stem cells into retinal progenitor cells. Ophthalmic Res. 2012;48(Suppl 1):1–5.

99. Singh AK, Srivastava GK, García-Gutiérrez MT, Pastor JC. Adipose derived mesenchymal stem cells partially rescue mitomycin C treated ARPE19 cells from death in co-culture condition. Histol Histopathol. 2013;28:1577–83.
100. Xia J, Luo M, Ni N, et al. Bone marrow mesenchymal stem cells stimulate proliferation and neuronal differentiation of retinal progenitor cells. PLoS One. 2013;8:e76157.
101. Duan P, Xu H, Zeng Y, Wang Y, Yin ZQ. Human bone marrow stromal cells can differentiate to a retinal pigment epithelial phenotype when co-cultured with pig retinal pigment epithelium using a transwell system. Cell Physiol Biochem. 2013;31:601–13.
102. Huang C, Zhang J, Ao M, et al. Combination of retinal pigment epithelium cell-conditioned medium and photoreceptor outer segments stimulate mesenchymal stem cell differentiation toward a functional retinal pigment epithelium cell phenotype. J Cell Biochem. 2012;113:590–8.
103. Sun X, Chen M, Li J, et al. E13.5 retinal progenitors induce mouse bone marrow mesenchymal stromal cells to differentiate into retinal progenitor-like cells. Cytotherapy. 2011;13:294–303.
104. Tao YX, Xu HW, Zheng QY, FitzGibbon T. Noggin induces human bone marrow-derived mesenchymal stem cells to differentiate into neural and photoreceptor cells. Indian J Exp Biol. 2010;48:444–52.
105. Yang LL, Zhou QJ, Wang Y, Wang YQ. Differentiation of human bone marrow-derived mesenchymal stem cells into neural-like cells by co-culture with retinal pigmented epithelial cells. Int J Ophthalmol. 2010;3:23–7.
106. Johnson TV, DeKorver NW, Levasseur VA, et al. Identification of retinal ganglion cell neuroprotection conferred by platelet-derived growth factor through analysis of the mesenchymal stem cell secretome. Brain. 2014;137:503–19.
107. Sugitani S, Tsuruma K, Ohno Y, et al. The potential neuroprotective effect of human adipose stem cells conditioned medium against light-induced retinal damage. Exp Eye Res. 2013;116:254–64.
108. Tsuruma K, Yamauchi M, Sugitani S, et al. Progranulin, a major secreted protein of mouse adipose-derived stem cells, inhibits light-induced retinal degeneration. Stem Cells Transl Med. 2014;3:42–53.
109. Zhang Y, Wang W. Effects of bone marrow mesenchymal stem cell transplantation on light-damaged retina. Invest Ophthalmol Vis Sci. 2010;51:3742–8.
110. Arnhold S, Absenger Y, Klein H, Addicks K, Schraermeyer U. Transplantation of bone marrow-derived mesenchymal stem cells rescue photoreceptor cells in the dystrophic retina of the rhodopsin knockout mouse. Graefes Arch Clin Exp Ophthalmol. 2007;245:414–22.
111. Li N, Li XR, Yuan JQ. Effects of bone-marrow mesenchymal stem cells transplanted into vitreous cavity of rat injured by ischemia/reperfusion. Graefes Arch Clin Exp Ophthalmol. 2009;247:503–14.
112. Hou HY, Liang HL, Wang YS, et al. A therapeutic strategy for choroidal neovascularization based on recruitment of mesenchymal stem cells to the sites of lesions. Mol Ther. 2010;18:1837–45.
113. Yang J, Zhou W, Zheng W, et al. Effects of myocardial transplantation of marrow mesenchymal stem cells transfected with vascular endothelial growth factor for the improvement of heart function and angiogenesis after myocardial infarction. Cardiology. 2007;107:17–29.
114. Matsumoto R, Omura T, Yoshiyama M, et al. Vascular endothelial growth factor-expressing mesenchymal stem cell transplantation for the treatment of acute myocardial infarction. Arterioscler Thromb Vasc Biol. 2005;25:1168–73.
115. Studeny M, Marini FC, Champlin RE, Zompetta C, Fidler IJ, Andreeff M. Bone marrow-derived mesenchymal stem cells as vehicles for interferon-beta delivery into tumors. Cancer Res. 2002;62:3603–8.
116. Zhang XS, Linkhart TA, Chen ST, et al. Local ex vivo gene therapy with bone marrow stromal cells expressing human BMP4 promotes endosteal bone formation in mice. J Gene Med. 2004;6:4–15.
117. Tang TT, Xu XL, Dai KR, Yu CF, Yue B, Lou JR. Ectopic bone formation of human bone morphogenetic protein-2 gene transfected goat bone marrow-derived mesenchymal stem cells in nude mice. Chin J Traumatol. 2005;8:3–7.
118. Hasharoni A, Zilberman Y, Turgeman G, Helm GA, Liebergall M, Gazit D. Murine spinal fusion induced by engineered mesenchymal stem cells that conditionally express bone morphogenetic protein-2. J Neurosurg Spine. 2005;3:47–52.
119. Park HY, Kim JH, Sun Kim H, Park CK. Stem cell-based delivery of brain-derived neurotrophic factor gene in the rat retina. Brain Res. 2012;1469:10–23.
120. Lee ES, Yu SH, Jang YJ, Hwang DY, Jeon CJ. Transplantation of bone marrow-derived mesenchymal stem cells into the developing mouse eye. Acta Histochem Cytochem. 2011;44:213–21.
121. Lu B, Wang S, Girman S, McGill T, Ragaglia V, Lund R. Human adult bone marrow-derived somatic cells rescue vision in a rodent model of retinal degeneration. Exp Eye Res. 2010;91:449–55.
122. Haddad-Mashadrizeh A, Bahrami AR, Matin MM, et al. Human adipose-derived mesenchymal stem cells can survive and integrate into the adult rat eye following xenotransplantation. Xenotransplantation. 2013;20:165–76.
123. Tzameret A, Sher I, Belkin M, et al. Transplantation of human bone marrow mesenchymal stem cells as a thin subretinal layer ameliorates retinal degeneration in a rat model of retinal dystrophy. Exp Eye Res. 2014;118:135–44.
124. Tomita M, Adachi Y, Yamada H, et al. Bone marrow-derived stem cells can differentiate into retinal cells in injured rat retina. Stem Cells. 2002;20:279–83.
125. Otani A, Dorrell MI, Kinder K, et al. Rescue of retinal degeneration by intravitreally injected adult bone marrow-derived lineage-negative hematopoietic stem cells. J Clin Invest. 2004;114:765–74.
126. Mackie AR, Losordo DW. CD34 positive stem cells in the treatment of hear and vascular disease in human beings. Tex Heart Inst J. 2011;38:474–85.
127. Medina RJ, O'Neill CL, Sweeney M, Guduric-Fuchs J, Gardiner TA, Simpson DA, Stitt AW. Molecular analysis of endothelial progenitor cell (EPC) subtypes reveals two distinct cell populations with different identities. BMC Med Genet. 2010;3:18.
128. Madeddu P. Therapeutic angiogenesis and vasculogenesis for tissue regeneration. Exp Physiol. 2005;90:315–26.
129. Kalka C, Masuda H, Takahashi T, Kalka-Moll WM, Silver M, Kearney M, Li T, Isner JM, Asahara T. Transplantation of ex vivo expanded endothelial progenitor cells for therapeutic neovascularization. Proc Natl Acad Sci U S A. 2000;97:3422–7.
130. Asahara T, Murohara T, Sullivan A, et al. Isolation of putative progenitor endothelial cells for angiogenesis. Science. 1997;275:964–7.
131. Rafii S, Lyden D. Therapeutic stem and progenitor cell transplantation for organ vascularization and regeneration. Nat Med. 2003;9:702–12.
132. Stitt AW, O'Neill CL, O'Doherty MT, Archer DB, Gardiner TA, Medina RJ. Vascular stem cells and ischaemic retinopathies. Prog Retin Eye Res. 2011;30:149–66.
133. Kim SJ, Kim JS, Papadopoulos J, et al. Circulating monocytes expressing CD31: implications for acute and chronic angiogenesis. Am J Pathol. 2009;174:1972–80.

134. Lois N, McCarter RV, O'Neill C, Medina RJ, Stitt AW. Endothelial progenitor cells in diabetic retinopathy. Front Endocrinol (Lausanne). 2014;5:44.
135. Yoder MC, Mead LE, Prater D, et al. Redefining endothelial progenitor cells via clonal analysis and hematopoietic stem/progenitor cell principals. Blood. 2007;109:1801–9.
136. Schuh A, Liehn EA, Sasse A, et al. Transplantation of endothelial progenitor cells improves neovascularization and left ventricular function after myocardial infarction in a rat model. Basic Res Cardiol. 2008;103:69–77.
137. Sasaki K, Heeschen C, Aicher A, et al. Ex vivo pretreatment of bone marrow mononuclear cells with endothelial NO synthase enhancer AVE9488 enhances their functional activity for cell therapy. Proc Natl Acad Sci U S A. 2006;103:14537–41.
138. Nolan DJ, Ciarrocchi A, Mellick AS, et al. Bone marrow-derived endothelial progenitor cells are a major determinant of nascent tumor neovascularization. Genes Dev. 2007;21:1546–58.
139. Asahara T, Takahashi T, Masuda H, et al. VEGF contributes to postnatal neovascularization by mobilizing bone marrow-derived endothelial progenitor cells. EMBO J. 1999;18:3964–72.
140. Bhatwadekar AD, Glenn JV, Curtis TM, Grant MB, Stitt AW, Gardiner TA. Retinal endothelial cell apoptosis stimulates recruitment of endothelial progenitor cells. Invest Ophthalmol Vis Sci. 2009;50:4967–73.
141. Yoon CH, Hur J, Park KW, et al. Synergistic neovascularization by mixed transplantation of early endothelial progenitor cells and late outgrowth endothelial cells: the role of angiogenic cytokines and matrix metalloproteinases. Circulation. 2005;112:1618–27.
142. Llevadot J, Murasawa S, Kureishi Y, et al. HMG-CoA reductase inhibitor mobilizes bone marrow—derived endothelial progenitor cells. J Clin Invest. 2001;108:399–405.
143. Tepper OM, Capla JM, Galiano RD, et al. Adult vasculogenesis occurs through in situ recruitment, proliferation, and tubulization of circulating bone marrow-derived cells. Blood. 2005;105:1068–77.
144. Aicher A, Brenner W, Zuhayra M, et al. Assessment of the tissue distribution of transplanted human endothelial progenitor cells by radioactive labeling. Circulation. 2003;107:2134–9.
145. Park SS, Caballero S, Bauer G, et al. Long-term effects of intravitreal injection of GMP-grade bone-marrow-derived CD34+ cells in NOD-SCID mice with acute ischemia-reperfusion injury. Invest Ophthalmol Vis Sci. 2012;53:986–94.
146. Loomans CJ, de Koning EJ, Staal FJ, Rookmaaker MB, Verseyden C, de Boer HC, Verhaar MC, Braam B, Rabelink TJ, van Zonneveld AJ. Endothelial progenitor cell dysfunction: a novel concept in the pathogenesis of vascular complications of type 1 diabetes. Diabetes. 2004;53:195–9.
147. Ingram DA, Lien IZ, Mead LE, Estes M, Prater DN, Derr-Yellin E, DiMeglio LA, Haneline LS. In vitro hyperglycemia or a diabetic intrauterine environment reduces neonatal endothelial colony-forming cell numbers and function. Diabetes. 2008;57:724–31.
148. Tan K, Lessieur E, Cutler A, et al. Impaired function of circulating CD34(+) CD45(−) cells in patients with proliferative diabetic retinopathy. Exp Eye Res. 2010;91:229–37.
149. Lee IG, Chae SL, Kim JC. Involvement of circulating endothelial progenitor cells and vasculogenic factors in the pathogenesis of diabetic retinopathy. Eye (Lond). 2006;20:546–52.
150. Brunner S, Schernthaner GH, Satler M, et al. Correlation of different circulating endothelial progenitor cells to stages of diabetic retinopathy: first in vivo data. Invest Ophthalmol Vis Sci. 2009;50:392–8.
151. Liu X, Li Y, Liu Y, et al. Endothelial progenitor cells (EPCs) mobilized and activated by neurotrophic factors may contribute to pathologic neovascularization in diabetic retinopathy. Am J Pathol. 2010;176:504–15.
152. Fadini GP, Sartore S, Baesso I, et al. Endothelial progenitor cells and the diabetic paradox. Diabetes Care. 2006;29:714–6.
153. Brunner S, Hoellerl F, Schmid-Kubista KE, et al. Circulating angiopoietic cells and diabetic retinopathy in type 2 diabetes mellitus, with or without macrovascular disease. Invest Ophthalmol Vis Sci. 2011;52:4655–62.
154. Khan ZA, Farhangkhoee H, Chakrabarti S. Towards newer molecular targets for chronic diabetic complications. Curr Vasc Pharmacol. 2006;4:45–57.
155. Sharma NK, Gardiner TA, Archer DB. A morphologic and autoradiographic study of cell death and regeneration in the retinal microvasculature of normal and diabetic rats. Am J Ophthalmol. 1985;100:51–60.
156. Yodoi Y, Sasahara M, Kameda T, Yoshimura N, Otani A. Circulating hematopoietic stem cells in patients with neovascular age-related macular degeneration. Invest Ophthalmol Vis Sci. 2007;48:5464–72.
157. Li Y, Reca RG, Atmaca-Sonmez P, et al. Retinal pigment epithelium damage enhances expression of chemoattractants and migration of bone marrow-derived stem cells. Invest Ophthalmol Vis Sci. 2006;47:1646–52.
158. Sasahara M, Otani A, Yodoi Y, Yoshimura N. Circulating hematopoietic stem cells in patients with idiopathic choroidal neovascularization. Invest Ophthalmol Vis Sci. 2009;50:1575–9.
159. Sasahara M, Otani A, Yodoi Y, Gotoh N, Kameda T, Yoshimura N. Circulating hematopoietic stem cells in patients with choroidal neovascularization secondary to pathologic myopia. Eye (Lond). 2009;23:718–26.
160. Sampat KM, Garg SJ. Complications of intravitreal injections. Curr Opin Ophthalmol. 2010;21:178–83.
161. Recchia FM, Scott IU, Brown GC, Brown MM, Ho AC, Ip MS. Small-gauge pars plana vitrectomy: a report by the American Academy of Ophthalmology. Ophthalmology. 2010;117:1851–7.
162. Park SS, Bauer G, Abedi M, et al. Intravitreal autologous bone marrow CD34+ cell therapy for ischemic and degenerative retinal disorders: preliminary phase 1 clinical trial findings. Invest Ophthalmol Vis Sci. 2014;56:81–9.
163. Siqueira RC, Messias A, Voltarelli JC, Scott IU, Jorge R. Intravitreal injection of autologous bone marrow-derived mononuclear cells for hereditary retinal dystrophy: a phase I trial. Retina. 2011;31:1207–14.
164. Siqueira RC, Messias A, Voltarelli JC, Messias K, Arcieri RS, Jorge R. Resolution of macular oedema associated with retinitis pigmentosa after intravitreal use of autologous BM-derived hematopoietic stem cell transplantation. Bone Marrow Transplant. 2013;48:612–3.
165. Siqueira RC, Messias A, Gurgel VP, Simões BP, Scott IU, Jorge R. Improvement of ischaemic macular oedema after intravitreal injection of autologous bone marrow-derived haematopoietic stem cells. Acta Ophthalmol. 2015;93:e174–6.
166. Jonas JB, Witzens-Harig M, Arseniev L, Ho AD. Intravitreal autologous bone-marrow-derived mononuclear cell transplantation. Acta Ophthalmol. 2010;88:e131.
167. Jakob P, Landmesser U. Current status of cell-based therapy for heart failure. Curr Heart Fail Rep. 2013;10:165–76.

Subretinal Delivery of Cells via the Suprachoroidal Space: Janssen Trial

James S. Baldassarre, Anthony Joseph, Michael Keane, and Jeffrey S. Heier

1 Cell Therapy Overview

Cell therapies are currently in clinical development for the treatment of a number of retinal disorders including age related macular degeneration (both neovascular and atrophic forms), Stargardt disease, and retinitis pigmentosa. The cell types under investigation include retinal pigmented epithelial (RPE) cells derived from embryonic stem cells (ESC) or induced pluripotent stem cells (iPS), human fetal derived neural stem cells (HuCNS-SC), human umbilical cord tissue derived cells, and autologous bone marrow derived cells. There are also preclinical studies investigating neural and photoreceptor cell replacement strategies.

The proposed mechanisms by which these cell therapies may be effective are either by replacement of defective cells or by sustained release of trophic factors that enhance the activity of the patient's own retinal cells. Cell therapies have been delivered to the subretinal space either as a suspension of cells or as an organized cell sheet on a substrate. The surgical delivery of these cells into the subretinal space has conventionally involved a 3-port pars plana vitrectomy, creation of a retinotomy, and transfer of the cells. The safety of the cell therapy encompasses a combination of the surgical safety and the cells, and although the surgical procedure is performed routinely, the safety profile may not be acceptable for the potential benefits [1]. The safety events associated with a vitrectomy include cataracts, retinal detachments, and the development of proliferative vitreoretinopathy (PVR) [2–4]. All of these adverse events have been reported with cell therapies delivered following vitrectomy [5, 6]. As we will discuss, some of these adverse events may be reduced or eliminated if cells can be delivered safely and effectively to the subretinal space without the creation of a retinotomy.

J.S. Baldassarre, M.D. • M. Keane, M.S.
Janssen Cell Therapy, Janssen R&D LLC, Spring House, PA, USA

A. Joseph, M.D. (✉)
Ophthalmic Consultants of Boston, Boston, MA, USA

Vitreoretinal Surgery and Disease, Ophthalmic Consultants of Boston, 800 Washington Street, Suite 600, Boston, MA 02114, USA
e-mail: ajoseph@eyeboston.com

J.S. Heier, M.D.
Ophthalmic Consultants of Boston, Boston, MA, USA

2 Human Umbilical Tissue Cells

Human umbilical tissue derived cells are being developed by Janssen Research & Development, LLC ("Janssen"), a pharmaceutical research company of Johnson & Johnson, for the treatment of retinal degeneration, specifically geographic atrophy secondary to age related macular degeneration. The World Health Organization (WHO) has granted the nonproprietary name palucorcel for human umbilical tissue derived cells.

The umbilical cord tissue is provided to a tissue procurement group following a full term healthy birth. Donors must complete an informed consent, donor eligibility questionnaire, and medical examination that conform to cell and tissue directives for the US, EU, and Japan. The tissue is transferred to a manufacturing facility where further processing conforms to good tissue and good manufacturing practices.

Palucorcel is derived from the enzymatic digestion of the umbilical cord tissue to release the mesenchymal lineage cells from the stroma. Following culture, other cell types from the umbilical cord (such as endothelial, epithelial, and blood derived cells) do not remain attached to the substrate and do not proliferate. The selected cells are banked and cryopreserved before being tested and released for further manufacturing (Fig. 1).

Palucorcel can be expanded to over 40 population doublings and eventually senesce. Although the cells are not clonal, they have the characteristics of a single population of cells when characterized by over 60 cell surface markers. The cells do not spontaneously differentiate in vitro or

S.D. Schwartz et al. (eds.), *Cellular Therapies for Retinal Disease*, DOI 10.1007/978-3-319-49479-1_8

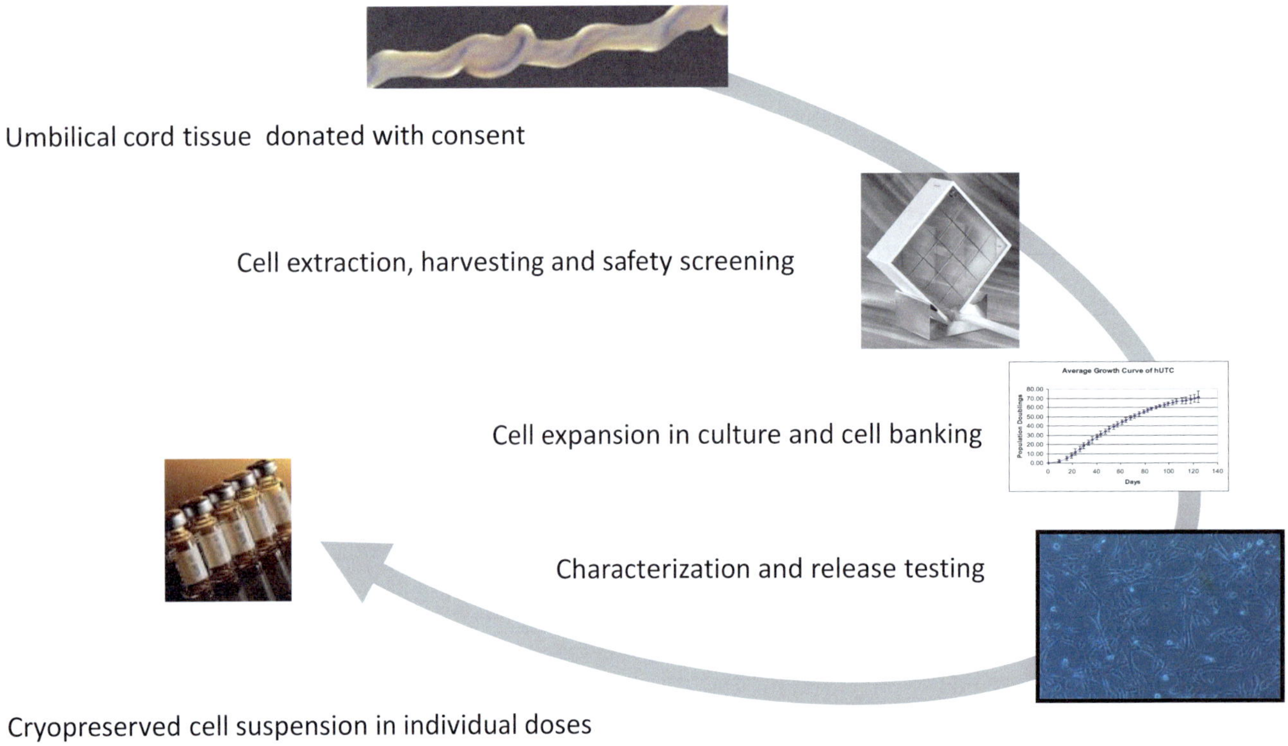

Fig. 1 Human umbilical tissue preparation: Palucorcel is derived from the enzymatic digestion of the umbilical cord tissue to release the mesenchymal lineage cells from the stroma. The selected cells are banked and cryopreserved before being tested and released for further manufacturing

in vivo and do not have the attributes of bone marrow derived stromal (stem) cells. Using a number of different techniques including microarrays, RNA-Seq, and enzyme linked immunosorbent assay (ELISA), it can be demonstrated that the gene expression of the cell types is significantly different.

Palucorcel's proposed mechanism of action is the secretion of trophic factors such as brain derived neurotrophic factor (BDNF), thrombospondins and soluble vascular endothelial growth factor receptor 1 (sVEGFr1), which are naturally expressed and not genetically engineered into the cells. The Royal College of Surgeons (RCS) rat model has been used to demonstrate that palucorcel has the ability to preserve photoreceptors by enhancing the phagocytosis function that was significantly diminished by the mutation in the *mertk* gene. Although palucorcel has activity when injected into the vitreous, the cells can attach to the lens or form membranes that can contract, leading to traction and retinal detachment. The activity of palucorcel is increased when administered to the subretinal space, and the effect can be observed distal to the site of the cell injection site. When palucorcel is injected through the retina into the subretinal potential space, it may also displace RPE, so it would be idea to administer palucorcel to the subretinal space without a retinotomy.

While some cell therapies require immunosuppression, those derived from umbilical tissue do not. Accordingly, palucorcel does not require immune suppression for single or repeated administration to the subretinal space. This has been established through preclinical safety studies in miniature swine where single administration did not induce a detectable immune response [7, 8]. Palucorcel is MHC class I (MHCI) dull and negative for MHC class II (MHCII), but can be activated to increase MHCI and to express MHCII with IFN-gamma stimulation (species specific). Even so, the levels are still relatively low compared to other cell types such as peripheral blood derived mononuclear cells (PBMCs). Using umbilical tissue derived cells from Massachusetts General Hospital MHC-defined miniature swine, we assessed immunogenicity across a full MHC barrier in vivo. Immunogenicity was also assessed by in vitro assays including modified mixed lymphocyte reaction (MLR) and flow cytometry to detect serum alloantibody. The modified MLR assay included viable antigen presenting cells to ensure the assay was reflective of the in vivo situation and to qualify the assays for sensitivity and specificity to ensure positive results were detected. The immune response is influenced by the route of administration as well as the inflammatory status of the site of administration [7, 8].

3 Preliminary Clinical Data

Following completion of appropriate preclinical safety testing, the first human pilot safety study was initiated in patients with advanced retinitis pigmentosa. At this stage, the most widely used and refined technique for the delivery of substrate to the subretinal space was using a transvitreal approach, so accordingly, the delivery of palucorcel was achieved employing a conventional approach. A 3-port, pars plana vitrectomy was performed with induction of a posterior vitreous detachment. Palucorcel was administered as a single dose of cells in phosphate buffered saline (PBS) with a total volume of 100 μL. The preparation was injected over approximately 1–1.5 min into the subretinal space at least one disc diameter from the temporal side of the macula in one eye. The needle tip was placed away from the opening and positioned in an attempt to direct injected cells distal to the retinotomy site in order to minimize the occurrence of cell reflux out of the bleb. At the surgeon's discretion and consistent with local standards of care, a fluid-air exchange could have been completed at the end of the infusion. Subjects received standard postoperative care for pars plana vitrectomy. If the surgeon elected to utilize the fluid-air exchange for retinal tamponade, subjects remained positioned in a face-down or head-bowed position for approximately 36 h following the operation.

Seven subjects were treated. Although the procedure was generally well tolerated, two of seven subjects subsequently developed epiretinal membranes (ERM) with traction and retinal detachment at week 4 follow-up. In one case, ERM peel was performed with subsequent histologic analysis revealing a mixture of host Muller cells, RPE cells, palucorcel, and inflammatory cells [9]. This finding appeared to confirm preclinical data, which suggested that palucorcel in the vitreous have the potential to stimulate ERM formation and retinal detachments due to PVR. Based on this concern, the transvitreal approach to delivery of palucorcel was abandoned.

A second safety study was initiated in subjects with advanced geographic atrophy. In order to minimize the potential for reflux of palucorcel into the vitreous, this study employed an *ab externo* surgical approach with a commercially available subretinal delivery system (iTrack Model 275 microcatheter [iScience Interventional Corporation, Menlo Park, CA]) (Fig. 2). For this approach, an approximately 3 mm long peripheral scleral cutdown was performed 9–12 mm posterior to the limbus, followed by the creation of a choroidotomy. A subretinal bleb was then created using a

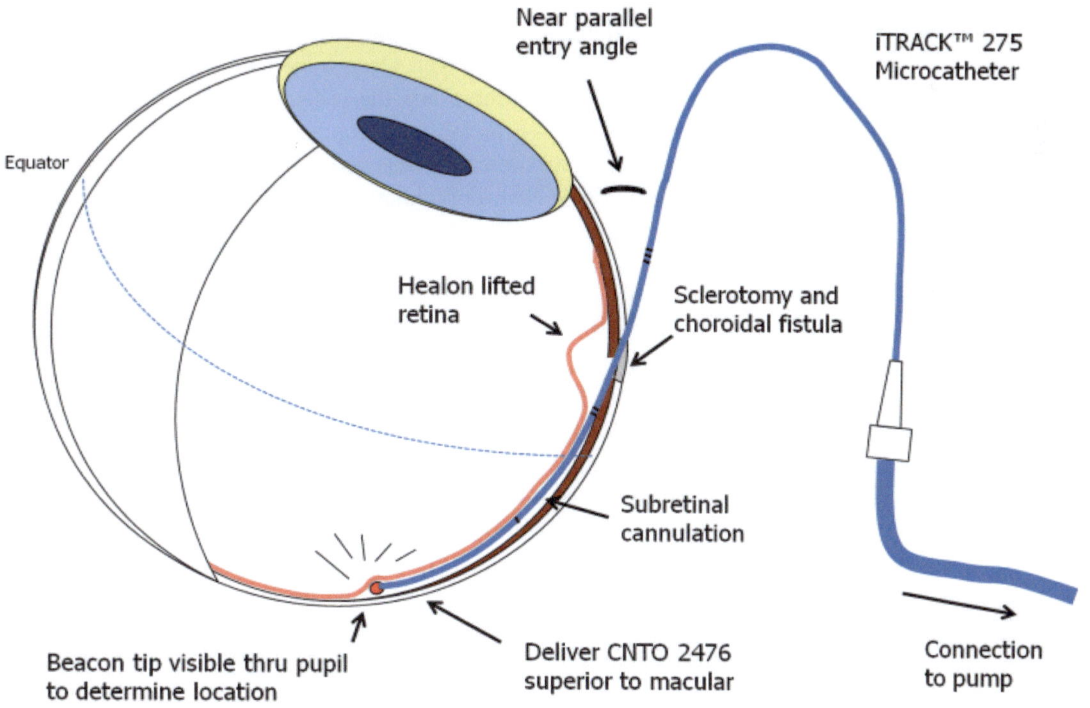

Fig. 2 Ab externo procedure: An approximately 3 mm long peripheral scleral cutdown is performed 9–12 mm posterior to the limbus, followed by the creation of a choroidotomy. A subretinal bleb is then created using a wire-tip microcannula and sodium hyaluronate viscoelastic. The iTrack 275 microcatheter is guided into the subretinal space with visualization to deliver palucorcel in a subretinal fluid bleb adjacent to GA

Table 1 Ocular adverse events: This table summarizes the rate of ocular adverse events in a safety study using an *ab externo* approach to deliver palucorcel with a commercially available subretinal delivery system

Ocular adverse events					
Occurring in ≥10% of subjects					
	6.0×10^4 cells	1.2×10^5 cells	3.0×10^5 cells	5.6×10^5 cells	Total
Subjects treated	12	3	15	3	33
Conjunctival hemorrhage	4 (33.3%)	1 (33.3%)	5 (33.3%)	0	10 (30.3%)
Eye pain	0	1 (33.3%)	3 (20.0%)	0	4 (12.1%)
Retinal detachment	2 (16.7%)	1 (33.3%)	1 (6.7%)	1 (33.3%)	5 (15.2%)
Retinal hemorrhage	1 (8.3%)	1 (33.3%)	1 (6.7%)	2 (66.7%)	5 (15.2%)
Retinal tear	7 (58.3%)	1 (33.3%)	4 (26.7%)	0	12 (36.4%)
Visual acuity reduced	2 (16.7%)	1 (33.3%)	1 (6.7%)	0	4 (12.1%)

wire-tip microcannula and sodium hyaluronate viscoelastic (Healon, Abbott Medical Optics, Abbott Park, IL). The iTrack 275 microcatheter was guided into the subretinal space with visualization to deliver palucorcel in a subretinal fluid bleb adjacent to GA. Intraocular visualization was attempted with indirect ophthalmoscopy but the procedure was subsequently amended to allow the use of intraocular endoscopy (Endo Optiks E2 System, Beaver Visitec, Waltham, MA). If the surgeon suspected that an inadvertent retinal perforation had occurred, the area could be treated prophylactically with peripheral cryopexy or laser retinopexy. If a retinal perforation posterior to the equator was observed, the investigator was not to inject palucorcel. For retinal perforations observed anterior to the equator, injection of the cells was based on the investigator's clinical judgment.

The study was planned to enroll approximately 45 eligible subjects; however, owing to the length and difficulty of the procedure, the study was terminated after 33 subjects had been treated. The primary difficulty was related to the inability to visualize the site of the choroidotomy and subretinal bleb, leading to an unacceptable incidence of retinal tears (43.5%). Most perforations occurred at the site of the entry bleb, but additional perforations occurred along the subretinal track of the cannula. Retinal perforations allowed reflux of palucorcel into the vitreous. Two of 33 subjects (6.1%) developed PVR, with an overall rate of retinal detachments of 15% (Table 1). Despite the limitations of the delivery procedure, a number of subjects experienced improvements in vision; at the month 12 follow-up, 24% of subjects had a documented improvement in BCVA of ≥15 letters on a standard ETDRS eye chart [10].

Janssen concluded that palucorcel were sufficiently safe and well tolerated when delivered to the subretinal space without a perforation, but that the delivery procedure as designed was unsuitable for further development. Given the safety of the cells and preliminary evidence of clinical benefit, the sponsor believed that further exploration was warranted. Thus, the program was suspended pending development of a safer and more efficient delivery procedure.

4 A Novel Surgical Approach to the Subretinal Space

Following the suspension of the Phase 1 clinical trial, Janssen set forth to develop de novo a delivery procedure and a device that would mitigate the risk of retinal perforations, retinal detachments, and PVR. The primary consideration was elimination of any retinotomy or retinal perforation that might allow egress of cells into the vitreous. As mentioned earlier, the most commonly used route for subretinal injection is vitrectomy followed by a direct transvitreal approach to the posterior retina. Following vitrectomy, a subretinal cannula is introduced through a port. As the tip penetrates the retina, BSS is infused to hydrodissect the retina from the RPE layer. Once a bleb is raised, the therapeutic agent is infused and the cannula is withdrawn. Although simple in principle, this approach has a number of important limitations. The injection cannulas are typically very simple bevel cut polyimide tubes of 35–39 Ga. Surgical approaches are not well standardized. The direction and size of the expanding subretinal bleb is difficult to predict or control. And because the insertion point in the pars plana creates a fulcrum, small motions by the surgeon may create continual enlargement of the retinotomy and retinal tears. The enlarging retinotomy increases the likelihood of reflux of fluid back into the vitreous cavity. This in turn prevents accurate and complete dosing of the therapeutic agent, and post injection air-fluid exchanges may be necessary to remove misplaced product. After considering a number of potential approaches to mitigating these issues (Fig. 3), Janssen developed a novel approach to subretinal injection, using the suprachoroidal space as a conduit to the posterior pole.

5 The Suprachoroidal Space

The suprachoroidal space refers to the potential space between the sclera and the choroid with its anterior boundary at the scleral spur and its posterior boundary at the connections

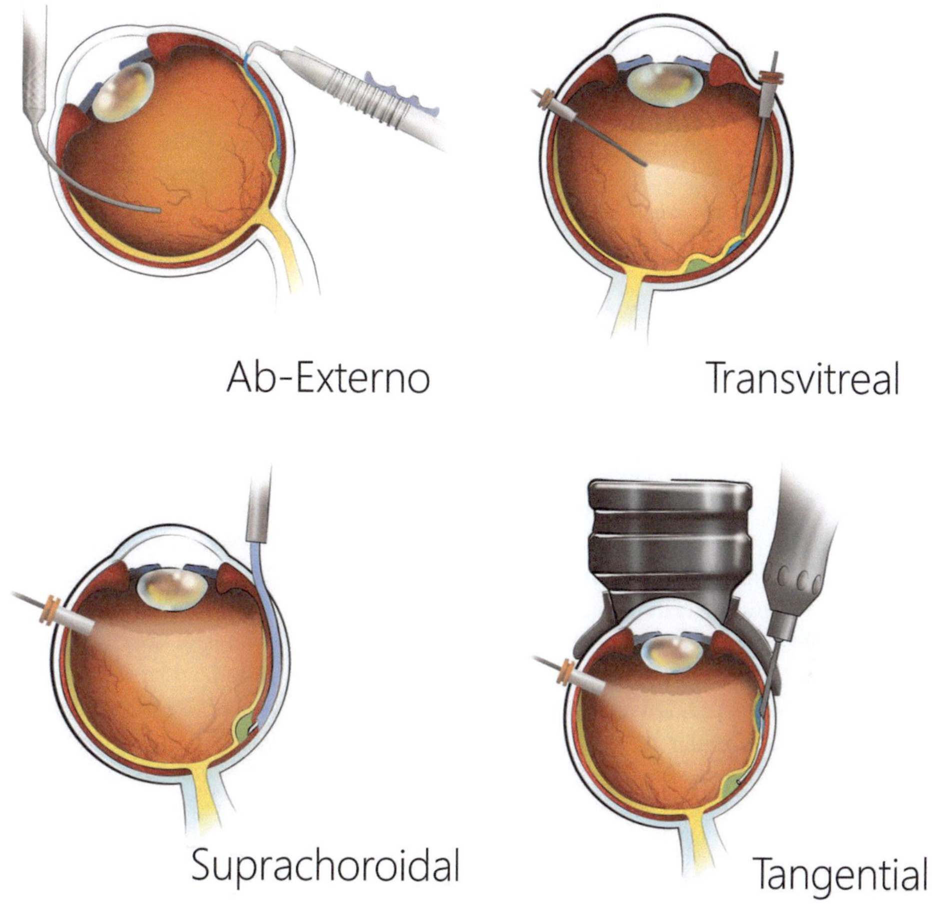

Fig. 3 Surgical options: The ab externo approach accesses the subretinal space through a scleral cutdown and choroidotomy—a microcatheter is subsequently guided in the subretinal space to the site of cell delivery. The tranvitreal approach accesses the subretinal space using a conventional pars plana vitrectomy technique with the creation of a retinotomy at the site of cell delivery. The suprachoroidal approach accesses the subretinal space by introducing a cannula into the suprachoroidal space through a sclerotomy and subsequently advancing a needle from the cannula through the choroid into the subretinal space at the site of cell delivery. The tangential approach is similar to the ab externo approach whereby the subretinal space is directly accessed through a sclerotomy and choroidotomy, and a small cannula is guided in the subretinal space to the site of cell delivery

of the short posterior ciliary vessels to the choroid. Focal equatorial connections exist where venous blood exits the globe at the vortex ampullae [11]. Before more recent advances in imaging, the anatomy of the suprachoroidal space was studied primarily in cadaver eyes. Krohn and Bertelsen used corrosion casting techniques by injecting methyl methacrylate into the suprachoroidal space of pig [12] as well as human eyes [13] to highlight uveoscleral drainage routes. They continued their work using light microscopy and gelatine injections to show drainage through perivascular and perineural spaces as well as through pre-formed channels into scleral veins [14]. As these studies suggest, a major area of clinical interest in the suprachoroidal space involves its role as a route for aqueous humor outflow from the eye and the potential that might play in the surgical management of glaucoma [15–17].

As advances in posterior segment imaging, including optical coherence tomography (OCT), have transformed the management of many retinal disease processes by allowing high resolution in vivo imaging of retinal anatomy, so have these advances improved our understanding of the suprachoroidal space. Techniques such as enhanced depth imaging (EDI)-OCT offer improved imaging of the layers beneath the retina [18], and further advances have even allowed visualization of the suprachoroidal space in vivo as a hyporeflective band on EDI-OCT [19] as well as swept source OCT [20]. Further research into imaging findings of the suprachoroidal space in specific diseases may offer insight into future management of those conditions [21, 22].

Just as advances in imaging revolutionized treatment of many retinal diseases, so did the rapidly growing utilization of intravitreal injections [23]. While this technique has proven to be a very effective means of administering drug therapies to the retina, researchers have also explored the suprachoroidal space as an alternative route to deliver treatments to the retina and choroid. Surgical microcannulation systems have been shown to be safe and reproducible in a monkey model [11], although further examination suggests a more rapid clearance of bevacizumab from the suprachoroidal space compared with traditional intravitreal administration [24]. Microneedle techniques have also been explored in rabbit [25] and porcine [26] models, with recent completion of two clinical trials by Clearside Biomedical Inc. evaluating this system for administration of suprachoroidal triamcinolone acetonide and aflibercept for macular edema after vein occlusion as well as the safety and efficacy of suprachoroidal triamcinolone for macular edema due to noninfectious uveitis (NCT02303184 and NCT01789320) [27]. It should be noted that the microcannulation and microneedle techniques described here were designed only to deliver pharmacologic agents to the suprachoroidal space by passing just through the sclera. This is in contrast to the surgical techniques under development by Janssen that use suprachoidal space as pathway to access the subretinal space for delivery of cell therapy.

While the utilization of the suprachoroidal space as a means for drug delivery in the treatment of retinal disease has been explored, its utilization in retina surgery has been limited. As mentioned earlier, its role in aqueous humor outflow has made the suprachoroidal space a target for glaucoma surgery. Conversely, vitreoretinal surgeons only rarely access the space when choroidal effusions require drainage; otherwise it has primarily been explored for relatively novel surgical procedures. Mittl and Tiwari initially described the injection of sodium hyaluronate into the suprachoroidal space in rabbit eyes to create a buckling effect [28]. El Rayes further described using this technique in humans to repair retinal detachments [29, 30] as well treat myopic foveoschisis and myopic macular holes [31]. There has also been recent work to develop a retinal prosthesis that is implantable using a suprachoroidal approach to more easily access the subretinal space [32, 33] with the first in-human trials of such a prosthesis showing a safety benefit to this approach in 3 patients [34]. It seems that accessing the suprachoroidal space holds promise in both vitreoretinal surgery and the treatment of vitreoretinal disease, although further technical progress must be made for its use to become widespread.

6 Suprachoroidal Approach to the Subretinal Space

The subretinal space is the potential space between the pigment epithelium and the photoreceptors. A separation can be created between these two layers by introducing a needle between them while infusing fluid. The proposed suprachoroidal approach to the subretinal space requires introducing a microneedle from the suprachoroidal space (posterior), perforating the choroid, Bruch's membrane, and the RPE layer to reach the subretinal space (anterior). Balanced salt solution is infused as the needle approaches the subretinal space, and a subretinal bleb appears when the two tissue planes separate.

6.1 Surgical Technique

Following placement of ports for fluid and illumination, the conjunctiva is taken down in the superior temporal quadrant. An inked marker is applied to indicate the location of stay sutures and sclerotomy. Two suture loops are placed to guide the introduction of the delivery cannula and to maintain the appropriate angle of cannula introduction during manipulation; the tails of these sutures are left long to control the rotation of the eye. Using a crescent blade, a 3 mm wide sclerotomy is made 6 mm posterior to the limbus. The choroid is visualized and any remaining scleral fibers are cleared using a Sinsky hook or Westcott scissors. Concurrently, the delivery cannula is primed with balanced salt solution and a suspension of palucorcel; the primed delivery cannula is attached to a "third arm" that mounts on the wrist rest.

Once the operator is satisfied with the sclerotomy, the delivery cannula and third arm are positioned carefully to align the cannula tip with the sclerotomy. The third arm is then fixed to the wrist rest. The tip of the cannula is slipped under the suture loops to ensure it remains tangential to the globe. The cannula will slide easily into the suprachoroidal space, and is advanced approximately 5 mm.

The operator then utilizes the chandelier and a noncontact wide-field viewing system for an internal view. Under direct visualization, the cannula is advanced towards the area of atrophy, stopping approximately 1–2 disc areas short of the atrophy border. The microneedle is then extended by turning a screw mechanism at the rear of the delivery cannula. BSS is infused as the needle is extended. When the needle perforates the RPE layer, a bleb forms in the subretinal space, indicating that the needle is in the correct location. The operator must ensure that there is no retinal perforation. If a perforation is detected, the procedure is terminated. After ensuring the integrity of the bleb, the operator turns a valve to engage the cell suspension, and then injects 50 μL of cells into the subretinal space. The cells are preceded by an air bubble. The operator carefully watches the bleb to ensure no egress of cells. The needle is then retracted, the cannula is withdrawn, and the surgical incisions are closed using standard techniques (Fig. 4).

7 Subretinal Access Kit

A subretinal access kit with a delivery cannula is used to facilitate delivery of palucorcel cells to the targeted perimacular subretinal delivery site via a suprachoroidal approach as described above. The kit comprises a subretinal injection cannula (Fig. 5), a specific ophthalmic marker, two 1-mL syringes, and tubing set. All of these components are single-patient-use disposables.

The subretinal injection cannula comprises a plastic body connected to flexible cannula shaft with a working length of approximately 50 mm. The distal end of the cannula measures 1.6 mm wide by 0.6 mm thick and houses a 35 gauge nitinol needle, which can be extended by a screw mechanism inside the plastic body. The exit angle of the needle is

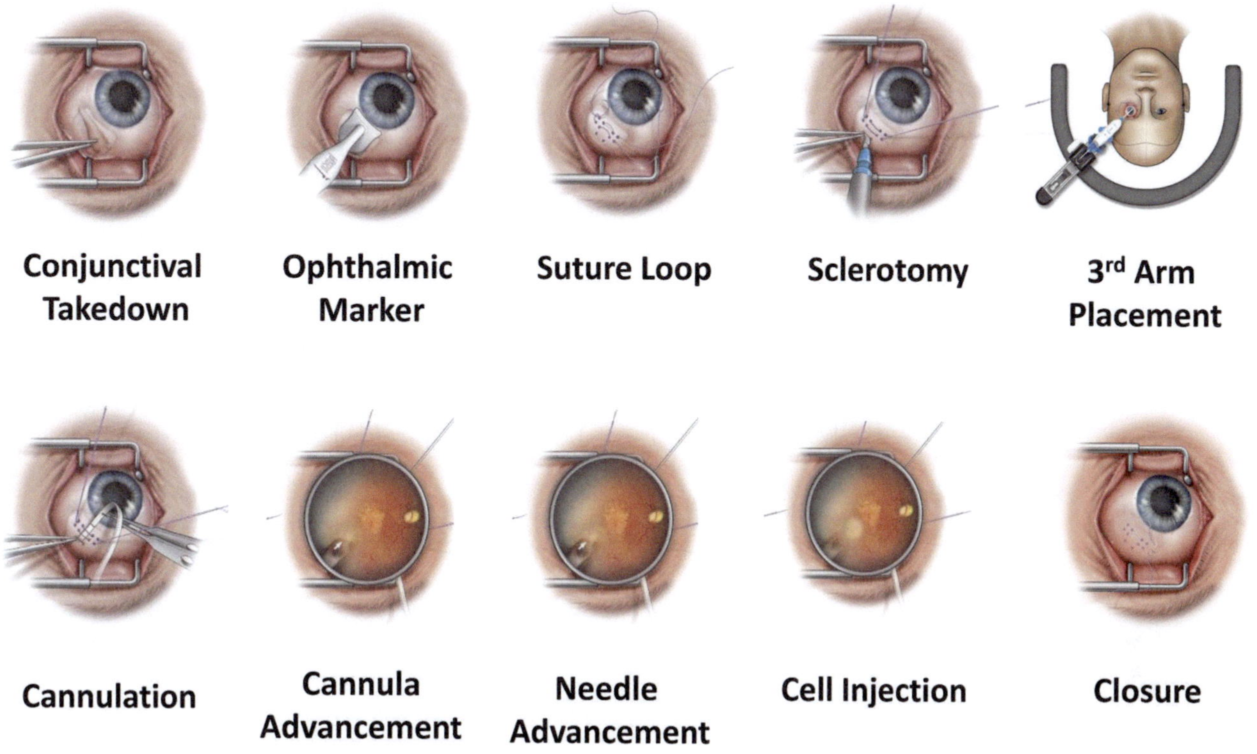

Fig. 4 Surgical procedure: This diagram illustrates the basic steps of the suprachoroidal approach to cell delivery with creation of a sclerotomy, introduction of a cannula into the suprachoroidal space, advancement of the cannula under direct visualization to the desired delivery site, advancement of the needle through the choroid into the subretinal space, and delivery of cells

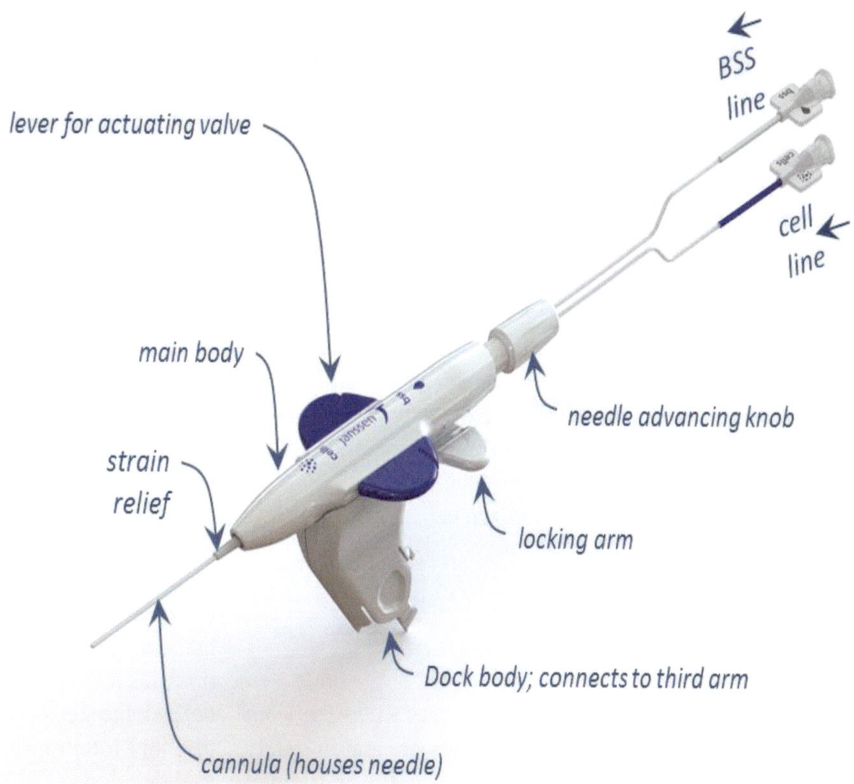

Fig. 5 Subretinal cannula: The subretinal injection cannula comprises a plastic body connected to flexible cannula shaft with a working length of approximately 50 mm. The distal end of the cannula measures 1.6 mm wide by 0.6 mm thick and houses a 35 gauge nitinol needle, which can be extended by a screw mechanism inside the plastic body. The device houses two separate fluid lines, one each for BSS and for cell suspension, and these are selected by an operator-controlled valve. A slide mechanism allows for small adjustments in the positioning of the plastic body relative to the patient

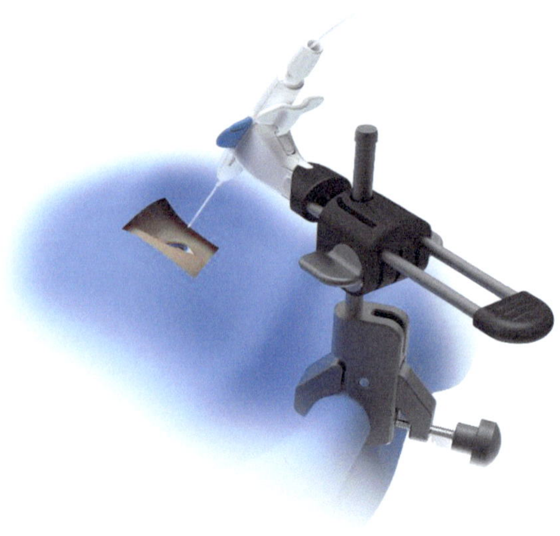

Fig. 6 Third arm accessory: A reusable "third arm" accessory interfaces with the subretinal injection cannula body and attaches to standard operating room table equipment

controlled by a metal ramp. The device houses two separate fluid lines, one each for BSS and for cell suspension, and these are selected by an operator-controlled valve. A slide mechanism allows for small adjustments in the positioning of the plastic body relative to the patient.

A reusable "third arm" accessory (Fig. 6), which interfaces with the subretinal injection cannula body and attaches to standard operating room table equipment, is also provided to aid in the surgical procedure.

In addition to the above palucorcel product specific devices, the following ancillary devices may also be employed:

- A standard operating microscope, lighting system, and video system and standard surgical instruments (e.g., eyelid speculum, sutures, electrocautery pen, vitrectomy lens) may be used.
- A balanced salt solution (BSS Plus®) will be used to separate tissues to visualize and verify proper subretinal access.
- A commercially available pressure regulator may be used to automatically depress the syringe plunger to infuse BSS Plus.
- Heating block for thawing of frozen cells.

8 Procedure Training Program

To ensure safe use of the subretinal access kit, significant efforts were made to ensure that surgeons were appropriately trained to perform the procedure as designed. The training program follows an advanced Learning Continuum that was initially developed to support surgeon education and skill development at Ethicon, a medical device division of Johnson & Johnson. It follows a thorough adult learning process, incorporating procedure and disease state awareness and in-depth knowledge development provided through online training modules. The learner then transitions to the Skills Development stage, where they engage in device and procedure hands-on learning, animate and inanimate tissue labs (porcine) led by surgeon faculty, device assembly practice, live technical demonstrations, and procedure reviews. The learning process continues into the live surgical procedure, with the availability of an interactive web-based Surgeon Expert Remote Observation (SERO) for live surgeon faculty support during surgery for surgeons less familiar with the procedure.

After the surgeon begins performing the procedure, there are a variety of tools available for the surgeon to refresh their knowledge of the procedure including surgical videos, live communication with other surgeons and procedure training faculty, live telesurgery, and the ability to attend refresher skills development labs if desired.

9 Special Considerations with Cell Therapy

Cell therapies are unique in that they are viable products with the capacity to proliferate and sustain a therapeutic benefit through multiple mechanisms. They also have the potential to transmit pathogens from the donor or from the manufacturing process as there is no ability to sterilize the final product. Stringent controls need to be in place for donor eligibility, consent, and manufacture according to Good Tissue Practices (GTPs) in the US or cell and tissue directives in the EU. These products are complex and must be manufactured according to Good Manufacturing Practices (GMPs) to obtain a consistent quality product.

Shipping and handling of cell therapies is not standardized and is unlikely to go through a pharmacy or institutional cell therapy group. Cell therapies are either made to order and shipped in medium at 2–10 °C or ambient conditions or cryopreserved at below −120 °C in the vapor phase of a dry shipper. As each product is unique with respect to how it is shipped, the sponsor of the clinical study or the manufacturer of the product is responsible for training the clinical staff to handle and use the cell therapy appropriately in order to ensure quality. As cell therapy in retinal diseases matures, the surgical procedures and the devices to deliver the cells will improve. In addition, the infrastructure and training of staff to store, handle, and deliver cells will develop.

10 Conclusions and Next Steps

The use of the suprachoroidal space to access the subretinal space is a useful innovation which has the potential to change the way that therapeutic agents are delivered to the retina. This approach may improve the accuracy of targeting drug delivery, as the cannula is advanced under direct visualization and can be repositioned prior to injection. There is also the possibility for improved precision in the volume and dose of the therapeutic agent, as there should be no leak anteriorly to the vitreous, and any leakage posteriorly into the choroid may be limited by the elasticity of Bruch's Membrane. Finally, because there is no egress of therapeutic agent into the vitreous, this procedure might offer an improved safety profile by reducing the risk for adverse reactions related to pharmacologic agents in the vitreous. Even so, challenges remain as this technique utilizes a novel device and nonstandard surgical techniques including the creation of a sizeable sclerotomy. Future safety data will offer insight into whether the risk of retinal tears and detachments are reduced with this technique and whether new concerns arise.

This device and procedure described here are not yet cleared for clinical use. The system has completed small-scale preclinical validation studies and is now being tested in conjunction with clinical trials of palucorcel for the treatment of geographic atrophy due to AMD. Given the experience of the clinical trial collaborators and the device's specific design to access the subretinal space, there is optimism surrounding these trials. Furthermore, the experience provides the opportunity to uncover areas for improvement as Janssen and the clinical collaborators have already identified a number of areas for improvements in safety, usability, and efficiency. With continued refinement, this delivery system and approach to the subretinal space holds promise for wider use in vitreoretinal surgery and the treatment of retinal disease.

Conflict of Interest James Baldassarre is an employee of Janssen R&D.

Michael Keane is an employee of Janssen R&D.

Jeffrey S. Heier, MD, has received research support from and served as a scientific consultant to Janssen R&D, Acucela, Astellas, and Genentech/Roche.

References

1. Stein JD, et al. Adverse events after pars plana vitrectomy among medicare beneficiaries. Arch Ophthalmol. 2009;127(12):1656–63.
2. Sadaka A, Giuliari GP. Proliferative vitreoretinopathy: current and emerging treatments. Clin Ophthalmol. 2012;6:1325–33.
3. Wang LC, et al. Assessment of retinal pigment epithelial cells in epiretinal membrane formation. J Chin Med Assoc. 2015;78(6):370–3.
4. Vedantham V, Ramasamy K. Pigmented epiretinal membranes caused by RPE migration: OCT-based observational case reports. Indian J Ophthalmol. 2007;55(2):148–9.
5. Schwartz SD, et al. Human embryonic stem cell-derived retinal pigment epithelium in patients with age-related macular degeneration and Stargardt's macular dystrophy: follow-up of two open-label phase 1/2 studies. Lancet. 2015;385(9967):509–16.
6. Song WK, et al. Treatment of macular degeneration using embryonic stem cell-derived retinal pigment epithelium: preliminary results in Asian patients. Stem Cell Reports. 2015;4(5):860–72.
7. Lutton BV, et al. Approaches to avoid immune responses induced by repeated subcutaneous injections of allogeneic umbilical cord tissue-derived cells. Transplantation. 2010;90(5):494–501.
8. Cho PS, et al. Immunogenicity of umbilical cord tissue derived cells. Blood. 2008;111(1):430–8.
9. Spencer R, et al. Case study of epiretinal membrane in a subject with retinitis pigmentosa who received palucorcel (CNTO 2476) via transvitreal subretinal delivery. Submitted, 2016.
10. Ho, A.C., T.S. Chang, and M. Samuel. A novel cell-based therapy administered subretinally to eyes with geographic atrophy secondary to age-related macular degeneration. Submitted, 2016.
11. Olsen TW, et al. Cannulation of the suprachoroidal space: a novel drug delivery methodology to the posterior segment. Am J Ophthalmol. 2006;142(5):777–87.
12. Krohn J, Bertelsen T. Corrosion casts of the suprachoroidal space and uveoscleral drainage routes in the pig eye. Acta Ophthalmol Scand. 1997;75(1):28–31.
13. Krohn J, Bertelsen T. Corrosion casts of the suprachoroidal space and uveoscleral drainage routes in the human eye. Acta Ophthalmol Scand. 1997;75(1):32–5.
14. Krohn J, Bertelsen T. Light microscopy of uveoscleral drainage routes after gelatine injections into the suprachoroidal space. Acta Ophthalmol Scand. 1998;76(5):521–7.
15. Bailey AK, Sarkisian Jr SR, Vold SD. Ab interno approach to the suprachoroidal space. J Cataract Refract Surg. 2014;40(8):1291–4.
16. Patrianakos TD. Anatomic and physiologic rationale to be applied in accessing the suprachoroidal space for management of glaucoma. J Cataract Refract Surg. 2014;40(8):1285–90.
17. Gigon A, Shaarawy T. The suprachoroidal route in glaucoma surgery. J Curr Glaucoma Pract. 2016;10(1):13–20.
18. Spaide RF, Koizumi H, Pozzoni MC. Enhanced depth imaging spectral-domain optical coherence tomography. Am J Ophthalmol. 2008;146(4):496–500.
19. Yiu G, et al. Characterization of the choroid-scleral junction and suprachoroidal layer in healthy individuals on enhanced-depth imaging optical coherence tomography. JAMA Ophthalmol. 2014;132(2):174–81.
20. Michalewska Z, et al. Suprachoroidal layer and suprachoroidal space delineating the outer margin of the choroid in swept-source optical coherence tomography. Retina. 2015;35(2):244–9.
21. Spaide RF, Ryan Jr EH. Loculation of fluid in the posterior choroid in eyes with central serous chorioretinopathy. Am J Ophthalmol. 2015;160(6):1211–6.
22. Kim JH, et al. Imaging suprachoroidal layer in exudative age-related macular degeneration. Curr Eye Res. 2016;41(5):715–20.
23. Peyman GA, Lad EM, Moshfeghi DM. Intravitreal injection of therapeutic agents. Retina. 2009;29(7):875–912.
24. Olsen TW, et al. Pharmacokinetics of pars plana intravitreal injections versus microcannula suprachoroidal injections of bevacizumab in a porcine model. Invest Ophthalmol Vis Sci. 2011;52(7):4749–56.
25. Patel SR, et al. Targeted administration into the suprachoroidal space using a microneedle for drug delivery to the posterior segment of the eye. Invest Ophthalmol Vis Sci. 2012;53(8):4433–41.
26. Gilger BC, et al. Treatment of acute posterior uveitis in a porcine model by injection of triamcinolone acetonide into the suprachoroidal

space using microneedles. Invest Ophthalmol Vis Sci. 2013;54(4): 2483–92.
27. Pearce W, Hsu J, Yeh S. Advances in drug delivery to the posterior segment. Curr Opin Ophthalmol. 2015;26(3):233–9.
28. Mittl RN, Tiwari R. Suprachoroidal injection of sodium hyaluronate as an 'internal' buckling procedure. Ophthalmic Res. 1987;19(5):255–60.
29. El Rayes EN, Elborgy E. Suprachoroidal buckling: technique and indications. J Ophthalmic Vis Res. 2013;8(4):393–9.
30. El Rayes EN, Oshima Y. Suprachoroidal buckling for retinal detachment. Retina. 2013;33(5):1073–5.
31. El Rayes EN. Suprachoroidal buckling. Dev Ophthalmol. 2014;54:135–46.
32. Saunders AL, et al. Development of a surgical procedure for implantation of a prototype suprachoroidal retinal prosthesis. Clin Exp Ophthalmol. 2014;42(7):665–74.
33. Villalobos J, et al. Cortical activation following chronic passive implantation of a wide-field suprachoroidal retinal prosthesis. J Neural Eng. 2014;11(4):046017.
34. Ayton LN, et al. First-in-human trial of a novel suprachoroidal retinal prosthesis. PLoS One. 2014;9(12):e115239.

Part III

Disease Modeling Using Induced Pluripotent Stem Cells

"Disease in a Dish" Modeling of Retinal Diseases

Huy V. Nguyen and Stephen H. Tsang

1 Introduction

Human vision is vital for nearly every major activity of daily living. Dysfunction of one of the responsible cell types, the retinal pigmented epithelium (RPE), leads to retinal disease and consequently severe visual impairment and blindness. RPE cells exist as a monolayer located between the retina and Bruch's membrane and is essential for photoreceptor function and survival. Retinal disorders such as age-related macular degeneration (AMD) and retinitis pigmentosa (RP) result in progressive loss of RPE. The adult retina does not intrinsically regenerate, so RPE degeneration may ultimately lead to blindness [1]. Since there is no cure for RPE degenerative diseases, the disease burden of these conditions are expected to continue to rise. Projected to affect 196 million people in 2020, AMD alone accounts for 8.7% of all blindness worldwide and is the most common cause of blindness in developed countries [2]. Currently, nine million Americans have been diagnosed with AMD, and its incidence is expected to double within a decade, affecting 20% of Americans between the ages of 65 and 75 years [3].

Cell transplantation into the human retina has the potential to provide treatment for diseases like AMD and RP with significant RPE loss, with the goal being stabilizing or even restoring vision. Since these diseases spare the inner retina and optic nerve, retinal transplantation has focused on replacement of the photoreceptors and RPE. Retinal stem cells have been shown to be efficient at integrating into the degenerating host retina in a mouse model of RP [4]. In 2015, Schwartz et al. showed for the first time in a clinical trial that RPE derived from human embryonic stem cells (ESCs) could be transplanted into human eyes with encouraging results [5].

Induced pluripotent stem (iPS) cells, which result from reprogramming of adult somatic cells into a pluripotent state with the transcription factors *Oct3/4*, *Sox2*, *Klf4*, and *c-Myc*, offer the possibility of generating patient-specific cell lines in vitro [6]. Generating iPS cells (iPSCs) from readily available cells, such as skin fibroblasts, eliminates the need for fertilized embryos and opens the possibility for creating personalized stem cells from diseased patients. As a platform to study patient-specific targeted disease cells, iPSCs have exciting potential in regenerative medicine and human disease modeling. As one example, after human embryonic stem cells were shown to be able to produce 3-D optic vesicle-like structures displaying a precise apical-basal orientation [7], human iPSCs were used to also create optic vesicle-like structures which self-assembled into rudimentary, multilayered retinal tissue [8]. Similarly, human iPSCs have been used to model primary open-angle glaucoma (POAG). The optineurin E50K mutation is a mutation associated with POAG, and human iPSCs have been created with the E50K mutation to study the molecular and cellular characterization of POAG onset [9]. Human iPSC modeling has also suggested that normal-tension glaucoma via TBK1 gene duplication is due to increased levels of LC3-II, a key marker of autophagy [10].

Specifically, iPSC-based therapies hold great promise for treating retinal degenerative diseases. Compared to other organ systems, the eye is an immune-privileged tissue, which affords allografts a level of protection from rejection. Imaging techniques to monitor the progress and success of therapies are noninvasive and do not pose significant risks as compared to techniques which rely on radiation. With their

H.V. Nguyen, M.D.
Massachusetts Eye and Ear Infirmary, Boston, MA 02114, USA
e-mail: hnvguyen10@gmail.com

S.H. Tsang, M.D., Ph.D. (✉)
Jonas Children's Vision Care, and Bernard & Shirlee Brown Glaucoma Laboratory, Department of Ophthalmology, Columbia University Medical Center, New York, NY, USA

Edward S Harkness Eye Institute, New York-Presbyterian Hospital, New York, NY, USA

Department of Pathology & Cell Biology, Institute of Human Nutrition, College of Physicians and Surgeons, Columbia University, New York, NY, USA
e-mail: sht2@cumc.columbia.edu

S.D. Schwartz et al. (eds.), *Cellular Therapies for Retinal Disease*, DOI 10.1007/978-3-319-49479-1_9

application to the retina, iPSC technology facilitates investigations of pathophysiological mechanisms of genetic mutations and testing of gene therapy vectors on RPE-based disease models. Indeed, iPS-derived RPE (iPS-RPE) can be reproducibly isolated and closely monitored both morphologically and functionally before experiments, effectively minimizing variability in the timing of differentiation. Figure 1 depicts a culture of iPS-RPE differentiated from cells from a standard human donor which exhibits the typical hexagonal monolayer. In addition, the RPE, unlike many other human cell types, has a well-described culture standard, which ensures proper controls [4, 11].

The in vitro phenotypes of disease-specific iPS-derived cells can be used to bridge the gap between the clinical phenotype and molecular or cellular mechanisms, creating new strategies for the development of novel therapeutic agents [12]. Human iPSC-based disease models can prove that a disease is caused by a genetic mutation, hypothesize potential treatment options before using more expensive animal models [13], and assist in the development of novel treatments for clinical trials [14–16].

2 Scientific Basis for Intervention: iPSC Disease Modeling and iPS-Derived RPE Cells for Cell Therapy

The eye is an ideal site for stem cell therapies. First, it is considered an immune-privileged organ since the inflammatory responses of the eye differ significantly from those in other tissues. Comprising of tight junctions of the epithelial cell layer of the uveal track, the endothelial cells of the inner retina capillaries, and the avascular cornea, the blood-ocular barrier allows for the creation of a microenvironment with

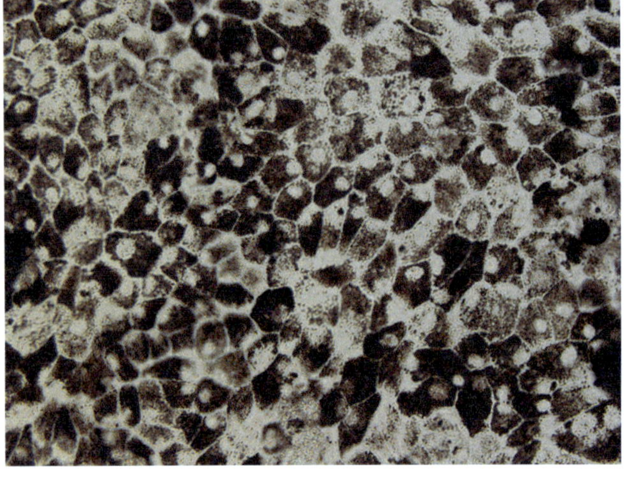

Fig. 1 Light microscopy of cultured human iPS-derived RPE cells from a wild-type control donor. The typical hexagonal structure of the RPE monolayer is clearly seen

anti-inflammatory properties [17]. As its own microenvironment, the eye has no direct lymphatic drainage and so any foreign transplanted material cannot reach regional lymph nodes [18]. Therefore, placement of foreign antigen into the anterior chamber, vitreous, or subretinal space will induce a systemic tolerance to the antigen. Second, ocular imaging techniques are safe and noninvasive, so that treatment progress to stem cell therapies can be effectively monitored. Ocular imaging techniques such as ocular coherence tomography allow for imaging of tissue microstructure in situ with micron-scale resolution. Third, in the case of serious complications, the eye as a unit can be removed, due to its relative isolation from other body systems. Stem cells in turn are an appealing option for retinal cell replacement due to their pluripotency and potentially unlimited capacity for self-renewal. Currently, there are two leading options for stem cells in retinal transplantation: (1) ESCs, which can be isolated from developing embryos 4–5 days after fertilization, and (2) iPSCs, which can be created from adult cells by the transduction of transcription factors [6]. However, due to the technical concerns with using ESCs, iPSCs have largely been favored for retinal transplantation.

iPSCs in particular offer a compelling alternative approach for stem cell therapy. When derived from the transplant recipient, autologous iPS-derived cells reduce the risk of post-transplant rejection and potentially obviate the need for immunosuppression after transplantation. The well-described iPSC culture standards also aid in the development of functional testing and optimization studies. Likewise, RPE transplantation into the retina poses fewer challenges than other kinds of cell transplantation since the generation, isolation, culture, manipulation, and testing of RPE cells has been well described [19, 20]. RPE monolayers exist in an easily identifiable hexagonal structure and can be isolated and transferred to a variety of substrates without the need for synaptic integration. Additionally, the quality of the RPE on a dish can be measured qualitatively using established morphological features and quantitatively using measurements of transepithelial resistance.

Studies on RPE replacement therapies using pluripotent stem cells have progressed rapidly. A multicenter trial focusing on the treatment of dry macular degeneration and Stargardt macular dystrophy showed that purified human ESC-derived RPE can be subretinally injected into patients with good safety results [21]. However, the blood-retina barrier breaks down in a diseased RPE state due to faulty tight junctions, so the retina may become more pro-inflammatory [22, 23]. Therefore, cells transplanted into a diseased retina are likely to be at a higher risk for rejection, so autologous iPSC transplantation may represent an alternative to ESC-derived therapy for curing degenerative retinal diseases.

Human iPS-derived RPE (iPS-RPE) experiments have largely been confined to animal models. In 2009, Carr et al.

performed subretinal injections of dissociated human iPS-RPE into Royal College of Surgeons (RCS) rats and observed restoration of RPE phagocytotic function, as measured by intracellular RHO staining, and long-term preservation of visual function, as measured by optokinetic head-tracking [24]. Another model is the RPE-specific protein 65 kDa (RPE65) mutant mouse model, which is used to study Leber congenital amaurosis (LCA) and RP since the RPE65 defect leads to a faulty isomerase which can no longer convert the chromophore necessary for rhodopsin to detect light [25]. In a 2012 study, dissociated human iPS-RPE was injected into the subretinal space of a RPE65 mutant mouse model and produced integration of the transplant with host RPE, as well as a modest improvement of visual function as measured by electroretinogram (ERG) [11]. The *Mfrprd6/Mfrprd6* (*rd6*) mouse, which has a frameshift mutation in the Membrane Frizzled-Related Protein (*Mfrp*) gene, is another widely used model. The resulting MFRP protein, an RPE-specific membrane receptor of unknown function, is abnormal and the mice exhibit progressive retinal degeneration, making the model a preclinical and progressive model of RP [26]. In a recent study by Li et al., subretinal injections of AAV-packaged wild-type *Mfrp* into *rd6* mice showed improvement in visual function and RPE cell layer thickness. The authors also showed that AAV-*MFRP* therapy was able to restore the damaged structure of the iPS-RPE monolayer in affected patients in vitro (see Fig. 2, reproduced with permission) [27].

The most advantageous aspect of iPSC-based therapy is the potential of autologous transplantation, which intends to address the problem of immune rejection. Despite the assumption that these autologous cells should not provoke an immune response in the recipient from whom the cells were derived, there have been conflicting reports that raise some concern of the immunogenicity of iPSCs. In a recent study, teratomas originating from subcutaneous injection of murine-derived iPSCs were found to have abnormal gene expression in some cells, which elicited a T-cell dependent immune response in syngeneic mice [28]. However, when Guha et al. transplanted various types of murine iPS-derived cells to a site under the kidney capsule of B6 mice, they found no evidence of immune response to the iPSCs, no increased T-cell proliferation in vitro, no rejection of syngeneic iPSC-derived cells after transplantation, and no antigen-specific secondary immune response [29]. Findings by Liu et al. in 2013 suggest that iPSC immunogenicity increases with ex vivo differentiation, as the authors observed immune responses after transplantation of differentiated iPS-derived cardiomyocytes but no response when transplanting undifferentiated iPSCs [30]. In contrast, Morizane et al. performed a direct comparison between autologous and allogeneic transplantation of iPS-derived neural cells in brains of nonhuman primates and found that the autologous transplantation of iPS-derived neurons caused only a minimal immune response in the brain, while the allografts elicited an acquired immune response [31]. Moreover, a higher number of dopaminergic neurons survived in autografted iPS-derived cells, which further support their use. Taken together, these findings reveals that different cell types derived from iPSCs might have distinctive immunogenicities in their syngeneic hosts. For the development of human iPS-based cell therapy, there remains still a challenge to evaluate the immunogenicity of human iPS-derived cells in an autologous human immune system. Despite promising early results, the first approved transplantation trial using human iPSC-derived RPE for treatment of AMD in Japan has been placed on hold for safety concerns [32].

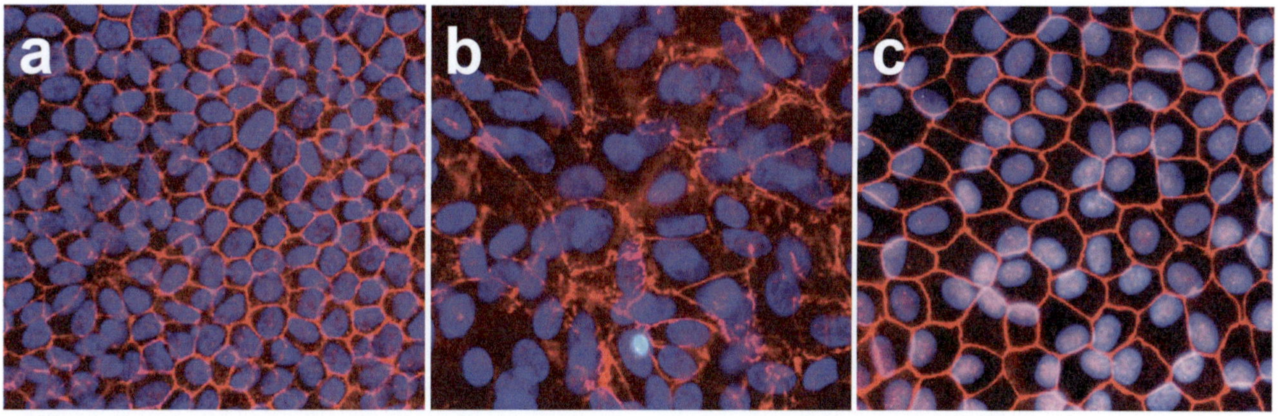

Fig. 2 (**a**) Immunohistochemistry staining for zonula occludens (ZO)-1 (*red*) and nuclear DAPI staining (*blue*) confirms that iPS-RPE forms a monolayer. Hexagonal red structure indicates the presence of ZO-1 on wild-type RPE. (**b**) Cellular morphology is lost in RPE derived from MFRP-deficient patient, but (**c**) hexagonal pattern of ZO-1 expression is restored following AAV-*MFRP* vector treatment of RPE

3 Preclinical Data

3.1 Progress of RPE Disease Modeling Using iPSCs

Human iPS cells are useful for modeling RPE disorders since they can be isolated, expanded, re-seeded, and closely monitored both morphologically and functionally prior to testing [33]. Phenotypes of patient-specific iPS cells may differ from a mouse model with the same mutation [27], underscoring the necessity for multiple models of human genetic diseases. Since differences in phenotypic expression can be observed among species with the same genetic mutation, it is important to study patient-specific cell lines as a complement to mouse models.

The first "disease in a dish" model of retinal disease using patient-specific iPS cells was introduced by Singh et al. and focused on Best vitelliform macular dystrophy (BVMD) [14]. Caused by a defect in the RPE gene BEST1, which results in the subretinal accumulation of photoreceptor waste products, BVMD is characterized by central vision loss due to photoreceptor death. Singh et al. created iPS-RPE from affected patients and compared them with those created from unaffected siblings. From their model, they concluded that the pathophysiology of the disease included delayed rhodopsin degradation after photoreceptor outer segment feeding, as evidenced by disrupted fluid flux and increased accumulation of autofluorescent material [14]. This human iPSC model of BVMD possessed functional deficiencies consistent with the clinical features of the disease and was used to characterize clinically relevant disease phenotypes for BVMD. Specifically, the findings suggest that BEST1 plays a crucial role in photoreceptor outer segment turnover and homeostasis, and that BVMD results from a disruption of this pathway.

iPS-derived RPE cells have also recently been used to model and study the pathophysiology of AMD. Therapeutic interventions such as anti-vascular endothelial growth factor agents have helped many patients suffering from choroidal neovascular membrane growth in wet AMD, but for those patients who have lost photoreceptors, RPE, and choroidal endothelial cells, cell replacement therapy is still being pursued. To this end, the genetic causes and pathogenic mechanisms involved in AMD must be understood. While genome-wide association studies (GWAS) have identified risk alleles for the disease, such as the ARMS2 and HTRA1 genes, how these alleles lead to pathology is still unclear since there is currently a lack of appropriate models for AMD. Autopsy eyes from end-stage patients already possess terminal changes and cannot be used to determine how abnormal gene expression can lead to RPE pathology, and mice do not have maculae so they cannot serve as appropriate models. To bypass these obstacles, Yang et al. created a model for AMD by obtaining patient-specific iPS-derived RPE and pharmacologically accelerating the aging process with treatment of bisretinoid N-retinylidine-N-ethanolamine (A2E) and blue light [13]. From a proteome screen of multiple A2E-aged patient-specific iPS-RPE lines, impaired superoxide dismutase 2 (SOD2) function was identified as a high risk factor for developing AMD. Using their iPS model, the researchers concluded that the ARMS2/HTRA1 risk alleles decreased SOD2 defense, making RPE more susceptible to oxidative damage and thus contributing to AMD pathogenesis.

3.2 Development of Gene Therapy on Patient-Specific iPSCs

Gene targeting refers to the use of homologous recombination in specific gene modification. In 2007, gene targeting was first successfully used to correct iPSCs when Hanna et al. converted mouse tail fibroblasts into iPSCs and used the loxP and HSV selection cassettes to correct the β-globin gene in a mouse model of sickle cell anemia [34]. Gene-corrected patient-specific iPSCs offer a unique approach to autologous therapies, which have the potential to treat a wide range of acquired and inherited diseases. However, gene targeting in human pluripotent stem cells has remained exceedingly difficult [35].

3.2.1 Viral Vector-Mediated Targeting

One approach to carry genes into affected iPSCs is to use viral vectors, given their high transduction efficiency. Adenovirus, a double-stranded DNA virus, has an approximate genome size of 30 kb and can be used with high homologous recombination efficiency in gene targeting via the helper-dependent adenoviral vector (HDAdV). Similarly, adeno-associated virus (AAV), a single-stranded DNA virus with an approximate genome size of 4.7 kb, has been successfully used as a transfer vector to carry missing genes into cells, including pluripotent stem cells. Several clinical trials have already shown that direct injection of *RPE65*-carrying AAV vectors can restore vision to patients suffering from *RPE65*-associated LCA [36, 37]. Vasireddy et al. published the first study which successfully transduced iPSCs developed from a patient with choroideremia with AAV subtype 2 (AAV2) [38]. Choroideremia is an inherited disorder due to loss of the *CHM* gene and the resulting Rab Escort Protein 1 (REP-1), leading to degeneration of the choroid and retina and blindness by the second decade of life. Research moving towards clinical trials has been stymied due to a lack of an animal model with similar functional and morphological features as the human retina, since the knockout of the murine

Chm is lethal. The authors developed a preclinical model of choroideremia using iPSCs and successfully transduced wild-type human *Chm* cDNA into these cells using AAV2-mediated therapy. They observed a functional restoration of REP-1 enzymatic activity and protein trafficking, showing that their gene therapy was successful and that iPSCs can be used as a preclinical model for choroideremia [38]. However, complications to their construction and production as well as the stringent biosafety precautions are required to prevent viral vectors from being widely used [39].

3.2.2 Genome Editing with Engineered Nucleases

The development of the engineered nuclease genome editing tools such as zinc finger nucleases (ZFNs), transcription activator-like effector nucleases (TALENs), and the clustered regularly interspaced short palindromic repeats (CRISPR)-Cas system have facilitated gene targeting in human iPSCs [40]. These tools use double-strand break induction and subsequent homology-directed repair to edit the mutations in the patients' genomic DNA, so that the corrected gene will remain under the normal endogenous promoters and enhancers. Thus, compared to conventional viral-mediated gene replacement, gene editing using ZFNs, TALENs, or the CRISPR system can avoid genetic expression in inappropriate cell types as well as incorrect levels of expression [41].

The zinc finger DNA recognition domain has been used to bind to a wide range of DNA sequences. ZFNs were developed to specifically create a double-stranded break by conjugating zinc finger domains with a sequence-independent nuclease domain from the FokI restriction enzyme. ZFN-mediated genome editing has been applied many times to human iPS cells and represents a powerful tool to modify genomic sequences [42, 43]. However, ZFNs are not more broadly used owing to their bias for higher GC content in target sequences, cytotoxicity due to off-target effects, and context-dependent cleavage activity. An alternative DNA binding domain, the transcription activator-like effector (TALE), has also been used in gene editing since it can be quickly engineered to bind any desired DNA sequence. Conjugating the TALE binding domain with the FokI nuclease creates the TALEN system, which has also been shown to be effective in gene targeting of human iPS cells [44, 45].

The most recent gene editing technique, CRISPR/Cas9, can edit DNA with much improved accuracy. Just one protein, the CRISPR-associated 9 (Cas9) protein, can be used with two small RNAs, crRNA (CRISPR RNA) and tracrRNA (*trans* crRNA), to mediate specific double-stranded breaks. crRNA and tracrRNA can even be combined into one 20 nucleotide sgRNA (single guide RNA), which retains DNA cleavage activity and simplifies expression in mammalian cells. Following a double-stranded break, homology-directed repair occurs using the exogenous repair sequence as template. In this fashion, the wild-type gene and protein expression and function may be restored as the mutation is corrected in diseased patient cells.

The CRISPR-Cas9 system has several advantages over its gene-editing predecessors, ZFNs and TALENs, for enhancing gene targeting efficiency. Most CRISPR-Cas9 subtypes target DNA directly (see Fig. 3), suggesting the possibility of engineered, RNA-directed gene editing systems. This usage of easily generated RNA guides avoids the need for repeated protein design, which sets CRISPR-Cas9 apart from ZFNs and TALENs, which use protein-based DNA targeting motifs. Over the past several years, several groups have already demonstrated the effectiveness of using the CRISPR-Cas9 system in gene targeting of human pluripotent cells. In 2013, Mali et al. targeted the endogenous AAVS1 locus in human iPSCs to achieve CRISPR-Cas9-mediated homology-directed repair of fibroblast-derived iPSCs [46]. Recently, Hou et al. developed a CRISPR-Cas9 system from *N. meningitides* to generate accurately targeted clones in human iPSCs with increased efficiency as compared to TALENs [47]. Several other groups have already demonstrated that CRISPR-Cas9 can successfully be used on gene targeting in iPSCs [48, 49]. However, there remains several concerns with CRISPR-Cas technology in human genome editing, primarily off-target DNA cleavage [50]. However, recent experiments showed that "nickases," or enzymes that cleave only the single-strand noncomplementary to the guide RNA, can increase the specificity and safety of the CRISPR-Cas9 system [51]. Single-stranded nicks are repaired by the base excision repair pathway, maintaining genome integrity. Double-stranded breaks with minimal off-target modification can be created by delivering the Cas9-nickase guided by a pair of sgRNAs targeting opposite strands of the target locus. Continued improvements to the CRISPR-Cas9 system as a gene targeting tool raise its potential as being a key player for gene therapy in human pluripotent stem cells.

3.3 Gene Therapy on Patient-Specific iPSC-Derived RPE Cells

With the aim of correcting genetic defects in retinal disease, gene therapy has been attempted not only on patient-specific iPS cells, but also RPE cells derived from these cell lines. A proof of concept study was performed by Cereso et al. which used a hybrid vector comprised of AAV2 and AAV5 (AAV2/5) to mediate gene therapy to the RPE derived from iPS created from a choroideremia patient [52]. The authors successfully developed a human iPS-derived retinal cell model of choroideremia, performed gene therapy on the iPS-RPE, and showed that AAV2/5-mediated therapy could potentially restore RPE phenotype. Working with MFRP, Li

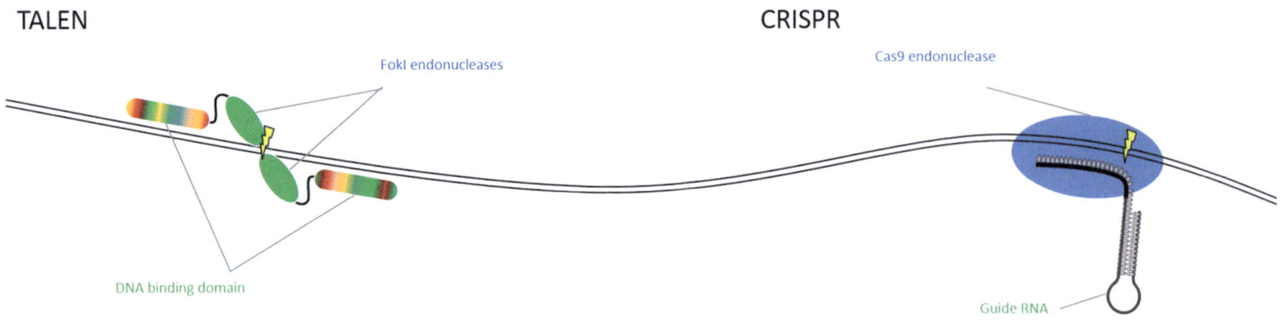

Fig. 3 TALEN and CRISPR-Cas9. The TALEN (transcription activator-like effector nuclease) uses custom-engineered DNA binding domains conjugated with FokI endonucleases to effect double-stranded breaks. The CRISPR (clustered regularly interspaced short palindromic repeat) system uses a protein component Cas9 (CRISPR-associated 9) and the sgRNA (single guide RNA) to generate a double-stranded break at a target site

et al. also showed that patient-specific iPS-RPE could be a recipient for gene therapy [27]. The researchers applied the AAV8 vector expressing human MFRP to iPS-RPE from patients with MFRP mutations and confirmed that gene therapy led to restoration of RPE phenotype, specifically with regard to actin organization. These studies suggest that gene therapy using AAV vectors can be applied to RPE created from patient-specific iPS for retinal diseases without previous models, and that these diseases may be potential targets for additional gene therapy trials.

4 Translational Strategies

Translational strategies involving patient-specific pluripotent stem cells span both scientific and economic domains. Progress has already been made; in 2010, the first phase I clinical trial to use human embryonic stem cells was conducted. In that study, oligodendrocyte progenitors were injected into the site of spinal cord damage in five patients (ClinicalTrials.gov ID: NCT01217008), with the patients still being monitored for adverse outcomes. However, few therapies have translated successfully from the "proof-of-concept" stage in animal models to patient treatments. Some diseases might be so rare that it is not economically feasible to develop specific treatments. To address this, autologous patient-derived pluripotent stem cells are being used to test the molecular efficacy of a drug or response to gene therapy. Dosages of therapies need to be comparable in the culture dish and in vivo, and a range of tolerated and effective doses must also be determined. Improvements still must be made in reprogramming, differentiation, and cell characterization protocols of patient-specific pluripotent stem cells before employing this technology in further clinical transplantation trials.

In moving from animal models to human trials, potential safety issues must be carefully addressed. The use of potent oncogenic transgenes such as c-myc and Klf4 in the reprogramming process as outlined by Yamanaka is one area of concern [6]. If these transgenes are not silenced or are reactivated after reprogramming, genomic instability may result and not only confound results of disease modeling studies but also cause tumor formation after transplantation. To this end, iPS reprogramming protocols are still being optimized. An alternative reprogramming protocol by Yu et al. obviates the use of oncogenic transgenes by using a combination of Oct4, Sox2, Nanog, and Lin28 [53]. The methodology for generating iPSCs has markedly improved and now integration-free iPSCs, without transgene insertion in the host genome, can be obtained using plasmid vectors, RNA viruses, or mature microRNAs [54–58]. Integration-free iPSCs appear ideal since exogenous genes integrated in the host genome may affect the genetic properties of the iPSCs generated and thus modify the resulting cellular phenotypes of differentiated progeny. Additional studies are also required to ensure that the risk of rejection is significantly reduced in patient-specific iPSCs, given that immune rejection when certain tissues derived from iPSCs were transplanted into syngeneic murine hosts have been reported [28].

5 Human Data

5.1 Transplantation of ES and iPSC-Derived RPE Cells

Considerable attention has been paid to the potential of human pluripotent stem cells as a source for regenerative medicine, disease modeling, and drug testing. In particular, the limitations in existing treatments for retinal disease, including AMD, have led to attention being given to alternative approaches in which damaged RPE is replaced by healthy RPE. Human embryonic stem cells have already been used in a 2011 clinical trial investigating Stargardt macular dystrophy (SMD) and dry AMD [21]. RPE was generated from the MA09 human ES cell line and a cell suspension was injected into the subretinal space in immunosuppressed patients. In the follow-up report in 2015, none of the patients experienced serious adverse effects from the transplantation, no evidence of teratoma formation or adverse proliferation has been observed, and roughly half of the patients showed some improvement in visual acuity at 6 months post-transplantation [5].

Compared to allogenic transplantation with ES-derived RPE, autologous transplantation with iPSC-derived RPE minimizes immune rejection. In September 2014, in a landmark trial, patient-specific iPSC-derived RPE cells were transplanted for the first time into a human patient with wet AMD by Dr. Masayo Takahashi's group at RIKEN CDB in Japan. Clearance for the phase I human trial was given after Takahashi et al. showed that transplantation of iPSC-derived RPE did not provoke an immune reaction nor lead to tumor growth in monkeys or mice [59]. Autologous iPSCs were created from the patient's skin cells and then differentiated into RPE so that they would grow in a monolayer without the use of synthetic scaffolds or matrices. To achieve this, iPSC-RPE were seeded onto type I collagen gel on a Transwell insert. After the RPE reached confluence, collagenase was applied to dissolve the collagen gel and leave a sheet of RPE. A 1.3 millimeter by 3.0 millimeter cut of this sheet was then grafted into the patient's retina following excision of her existing damaged RPE. This marks the first clinical trial on humans using iPSC-derived somatic cells, although the trial is currently on hold due to safety concerns. In September 2015, human trials for transplantation of iPS-derived RPE into patients with wet AMD featuring RPE tear were commenced at Moorfields Eye Hospital in London.

Preliminary clinical data for the use of RPE derived from human pluripotent stem cells appears favorable with respect to safety. Improvements in visual acuity were observed, but further research needs to be done to elucidate how much of this improvement can be attributed to therapeutic intervention. The feasibility of using iPSCs from patients to treat their blindness is still being established, but these human trials hold great potential for the advancement of translational medicine in retinal disease.

6 Future Directions

Patient-specific iPSCs have been shown to not only complement animal models of human disease, but also function as an excellent model in their own right. Patient-specific cell lines created from somatic cells from patients with inherited eye diseases can (1) provide a window for testing the efficacy of gene or drug-based therapies, (2) elucidate previously unknown mechanisms and pathways of disease, (3) demonstrate the pathogenicity of unusual mutations in individual patients, and (4) enable researchers to optimize parameters for successful cell replacement therapy in vitro. Skin-derived iPSCs can be used to investigate the function or dysfunction of a mutant gene product in tissues such as retina that are inaccessible to molecular analysis in living patients [60]. Finally, gene therapy tools such as ZFNs, TALENs, and the CRISPR-Cas system are rapidly improving the prospects of restoring the function of diseased RPE from patients with inherited retinal diseases. These patient-specific iPS-RPE, after undergoing gene therapy, can be optimized to become transplantable retinal cells, with the goal of restoring sight to patients with no other therapeutic options.

ESCs are still the gold standard for in vitro pluripotency. However, transplantation of human ES cell-derived products in human trials are allogenic and have been conducted without matching HLA types, so applications are mainly limited to the immune-privileged tissues such as the eyes and spinal cord. This necessitates development of iPSC technology, although a significant concern of using iPSCs in development of therapies is still whether they are truly equivalent to ESCs. For example, key differences between iPSCs and ESCs in transcribed genes, epigenetic landscape, differentiation potential, mutational load, and premature senescence has been described [61]. If iPSCs cannot closely replicate ESCs, the results from studies using iPSCs must be interpreted with this in mind. Significant differences between iPSCs and ESCs may hinder the translation of study results from an in vitro iPSC-based disease model to human disease.

A further step likely to accelerate the integration of iPS technology in regenerative medicine is the development of industry and biotechnology collaboration in order to develop large-scale stem cell production [62]. In this way, availability of iPS-based technology will increase, making them more widespread in investigative and translational studies in the future. Large-scale cell production is crucial to the success of future studies, and how to increase production without compromising cell quality and purity will become vital. Another key issue will be the cell transplantation stage. Terminally differentiated cells, such as RPE, are functional but do not proliferate. Progenitor cells such as iPSCs may be able to continuously self-renew in vivo, but risks such as tumorigenicity remain. The best cell types for transplantation must still be determined for treatment of retinal disease.

Patient-specific iPS-derived cells offer the hope of slowing progression or improving visual function for patients with currently untreatable retinal diseases. In addition to curing blindness, stem cell transplantation in the eye can also be seen as a model system for investigating cell-based treatments for other degenerative disorders of the CNS.

7 Conclusions

iPSCs have revolutionized the field of precision medicine. As long as a differential protocol is available, an immortal population of patient-specific pluripotent cells, which can theoretically differentiate into any cell type in the body, can be created. This technology, when applied to retinal cells, has the promise to make significant contributions to our understanding of the most pressing blinding diseases of our time. Stem cells also allow for the development of gene therapies for exceedingly rare retinal conditions which currently have little to no funding for research, with the hope that disease mechanisms and potential treatments may be elucidated. iPSCs can be generated from adult somatic cells, thus avoiding the ethical considerations involved with using embryonic stem cells. New gene targeting methods such as the CRISPR-Cas9 system are important tools for modeling retinal disease when used with the growing potential of patient-specific iPSCs. However, further efforts are needed to make use of the full potential of iPSC-mediated therapy to benefit human health. Until then, iPSCs will continue to be a sustainable method to model disease as gene therapies, drug therapies, and transplantable retinal cells continue to be developed for inherited retinal disorders.

References

1. Rosenfeld PJ, Brown DM, Heier JS, et al. Ranibizumab for neovascular age-related macular degeneration. N Engl J Med. 2006;355(14):1419–31.
2. Wong WL, Su X, Li X, et al. Global prevalence of age-related macular degeneration and disease burden projection for 2020 and 2040: a systematic review and meta-analysis. Lancet Glob Health. 2014;2(2):e106–16.
3. Phillips CO, Higginbotham EJ. Multivitamin supplements, ageing, and loss of vision: seeing through the shadows. Arch Intern Med. 2009;169(13):1180–2.
4. Wang N-K, Tosi J, Kasanuki JM, et al. Transplantation of reprogrammed embryonic stem cells improves visual function in a mouse model for retinitis pigmentosa. Transplantation. 2010;89(8):911–9.
5. Schwartz SD, Regillo CD, Lam BL, et al. Human embryonic stem cell-derived retinal pigment epithelium in patients with age-related macular degeneration and Stargardt's macular dystrophy: follow-up of two open-label phase 1/2 studies. Lancet. 2015;385(9967):509–16.
6. Takahashi K, Tanabe K, Ohnuki M, et al. Induction of pluripotent stem cells from adult human fibroblasts by defined factors. Cell. 2007;131(5):861–72.
7. Nakano T, Ando S, Takata N, et al. Self-formation of optic cups and storable stratified neural retina from human ESCs. Cell Stem Cell. 2012;10(6):771–85.
8. Phillips MJ, Wallace KA, Dickerson SJ, et al. Blood-derived human iPS cells generate optic vesicle-like structures with the capacity to form retinal laminae and develop synapses. Invest Ophthalmol Vis Sci. 2012;53(4):2007–19.
9. Minegishi Y, Iejima D, Kobayashi H, et al. Enhanced optineurin E50K-TBK1 interaction evokes protein insolubility and initiates familial primary open-angle glaucoma. Hum Mol Genet. 2013;22(17):3559–67.
10. Tucker BA, Solivan-Timpe F, Roos BR, et al. Duplication of TBK1 stimulates autophagy in iPSC-derived retinal cells from a patient with normal tension glaucoma. J Stem Cell Res Ther. 2014;3(5):161.
11. Li Y, Tsai Y-T, Hsu C-W, et al. Long-term safety and efficacy of human-induced pluripotent stem cell (iPS) grafts in a preclinical model of retinitis pigmentosa. Mol Med. 2012;18(9):1312–9.
12. Tsuji O, Miura K, Okada Y, et al. Therapeutic potential of appropriately evaluated safe-induced pluripotent stem cells for spinal cord injury. Proc Natl Acad Sci U S A. 2010;107(28):12704–9.
13. Yang J, Li Y, Chan L, et al. Validation of genome-wide association study (GWAS)-identified disease risk alleles with patient-specific stem cell lines. Hum Mol Genet. 2014;23(13):3445–55.
14. Singh R, Shen W, Kuai D, et al. iPS cell modeling of best disease: insights into the pathophysiology of an inherited macular degeneration. Hum Mol Genet. 2013;22(3):593–607.
15. Jin Z-B, Okamoto S, Osakada F, et al. Modeling retinal degeneration using patient-specific induced pluripotent stem cells. PLoS One. 2011;6(2):e17084.
16. Lustremant C, Habeler W, Plancheron A, et al. Human induced pluripotent stem cells as a tool to model a form of Leber congenital amaurosis. Cell Reprogram. 2013;15(3):233–46.
17. Campbell M, Humphries P. The blood-retina barrier: tight junctions and barrier modulation. Adv Exp Med Biol. 2012;763:70–84.
18. Stein-Streilein J. Immune regulation and the eye. Trends Immunol. 2008;29(11):548–54.
19. Idelson M, Alper R, Obolensky A, et al. Directed differentiation of human embryonic stem cells into functional retinal pigment epithelium cells. Cell Stem Cell. 2009;5(4):396–408.
20. Sonoda S, Spee C, Barron E, Ryan SJ, Kannan R, Hinton DR. A protocol for the culture and differentiation of highly polarized human retinal pigment epithelial cells. Nat Protoc. 2009;4(5):662–73.
21. Schwartz SD, Hubschman J-P, Heilwell G, et al. Embryonic stem cell trials for macular degeneration: a preliminary report. Lancet. 2012;379(9817):713–20.
22. Rutar M, Provis JM, Valter K. Brief exposure to damaging light causes focal recruitment of macrophages, and long-term destabilization of photoreceptors in the albino rat retina. Curr Eye Res. 2010;35(7):631–43.
23. Chinnery HR, McLenachan S, Humphries T, et al. Accumulation of murine subretinal macrophages: effects of age, pigmentation and CX3CR1. Neurobiol Aging. 2012;33(8):1769–76.
24. Carr A-J, Vugler AA, Hikita ST, et al. Protective effects of human iPS-derived retinal pigment epithelium cell transplantation in the retinal dystrophic rat. PLoS One. 2009;4(12):e8152.
25. Jin M, Li S, Moghrabi WN, Sun H, Travis GH. Rpe65 is the retinoid isomerase in bovine retinal pigment epithelium. Cell. 2005;122(3):449–59.
26. Kameya S, Hawes NL, Chang B, Heckenlively JR, Naggert JK, Nishina PM. Mfrp, a gene encoding a frizzled related protein, is mutated in the mouse retinal degeneration 6. Hum Mol Genet. 2002;11(16):1879–86.
27. Li Y, Wu W-H, Hsu C-W, et al. Gene therapy in patient-specific stem cell lines and a preclinical model of retinitis pigmentosa with membrane frizzled-related protein defects. Mol Ther. 2014;22(9):1688–97.
28. Zhao T, Zhang Z-N, Rong Z, Xu Y. Immunogenicity of induced pluripotent stem cells. Nature. 2011;474(7350):212–5.
29. Guha P, Morgan JW, Mostoslavsky G, Rodrigues NP, Boyd AS. Lack of immune response to differentiated cells derived from syngeneic induced pluripotent stem cells. Cell Stem Cell. 2013;12(4):407–12.
30. Liu Z, Wen X, Wang H, et al. Molecular imaging of induced pluripotent stem cell immunogenicity with in vivo development in ischemic myocardium. PLoS One. 2013;8(6):e66369.

31. Morizane A, Doi D, Kikuchi T, et al. Direct comparison of autologous and allogeneic transplantation of iPSC-derived neural cells in the brain of a non-human primate. Stem Cell Reports. 2013;1(4): 283–92.
32. Cyranoski D. Stem cells cruise to clinic. Nature. 2013;28:413.
33. Okamoto S, Takahashi M. Induction of retinal pigment epithelial cells from monkey iPS cells. Invest Ophthalmol Vis Sci. 2011;52(12):8785–90.
34. Hanna J, Wernig M, Markoulaki S, et al. Treatment of sickle cell anemia mouse model with iPS cells generated from autologous skin. Science. 2007;318(5858):1920–3.
35. Zwaka TP, Thomson JA. Homologous recombination in human embryonic stem cells. Nat Biotechnol. 2003;21:319–21.
36. Bainbridge JWB, Smith AJ, Barker SS, et al. Effect of gene therapy on visual function in Leber's congenital amaurosis. N Engl J Med. 2008;358(21):2231–9.
37. Maguire AM, High KA, Auricchio A, et al. Age-dependent effects of RPE65 gene therapy for Leber's congenital amaurosis: a phase 1 dose-escalation trial. Lancet. 2009;374(9701):1597–605.
38. Vasireddy V, Mills JA, Gaddameedi R, et al. AAV-mediated gene therapy for choroideremia: preclinical studies in personalized models. PLoS One. 2013;8(5):e61396.
39. Li M, Suzuki K, Kim NY, Liu G-H, Izpisua Belmonte JC. A cut above the rest: targeted genome editing technologies in human pluripotent stem cells. J Biol Chem. 2014;289(8):4594–9.
40. Zou J, Maeder ML, Mali P, et al. Gene targeting of a disease-related gene in human induced pluripotent stem and embryonic stem cells. Cell Stem Cell. 2009;5(1):97–110.
41. Tucker BA, Mullins RF, Stone EM. Stem cells for investigation and treatment of inherited retinal disease. Hum Mol Genet. 2014;23(R1):R9–R16.
42. Wang Y, Zhang WY, Hu S, et al. Genome editing of human embryonic stem cells and induced pluripotent stem cells with zinc finger nucleases for cellular imaging. Circ Res. 2012;111(12):1494–503.
43. Ramalingam S, London V, Kandavelou K, et al. Generation and genetic engineering of human induced pluripotent stem cells using designed zinc finger nucleases. Stem Cells Dev. 2013;22(4):595–610.
44. Hockemeyer D, Wang H, Kiani S, et al. Genetic engineering of human pluripotent cells using TALE nucleases. Nat Biotechnol. 2011;29(8):731–4.
45. Ding Q, Lee Y-K, Schaefer EAK, et al. A TALEN genome-editing system for generating human stem cell-based disease models. Cell Stem Cell. 2013;12(2):238–51.
46. Mali P, Yang L, Esvelt KM, et al. RNA-guided human genome engineering via Cas9. Science. 2013;339(6121):823–6.
47. Hou Z, Zhang Y, Propson NE, et al. Efficient genome engineering in human pluripotent stem cells using Cas9 from Neisseria meningitidis. Proc Natl Acad Sci U S A. 2013;110(39):15644–9.
48. Li HL, Fujimoto N, Sasakawa N, et al. Precise correction of the dystrophin gene in duchenne muscular dystrophy patient induced pluripotent stem cells by TALEN and CRISPR-Cas9. Stem Cell Reports. 2015;4(1):143–54.
49. Horii T, Tamura D, Morita S, Kimura M, Hatada I. Generation of an ICF syndrome model by efficient genome editing of human induced pluripotent stem cells using the CRISPR system. Int J Mol Sci. 2013;14(10):19774–81.
50. Fu Y, Foden JA, Khayter C, et al. High-frequency off-target mutagenesis induced by CRISPR-Cas nucleases in human cells. Nat Biotechnol. 2013;31(9):822–6.
51. Ran FA, Hsu PD, Lin C-Y, et al. Double nicking by RNA-guided CRISPR Cas9 for enhanced genome editing specificity. Cell. 2013;154(6):1380–9.
52. Cereso N, Pequignot MO, Robert L, et al. Proof of concept for AAV2/5-mediated gene therapy in iPSC-derived retinal pigment epithelium of a choroideremia patient. Mol Ther Methods Clin Dev. 2014;1:14011.
53. Yu J, Vodyanik MA, Smuga-Otto K, et al. Induced pluripotent stem cell lines derived from human somatic cells. Science. 2007;318(5858):1917–20.
54. Ban H, Nishishita N, Fusaki N, et al. Efficient generation of transgene-free human induced pluripotent stem cells (iPSCs) by temperature-sensitive Sendai virus vectors. Proc Natl Acad Sci U S A. 2011;108(34):14234–9.
55. Warren L, Manos PD, Ahfeldt T, et al. Highly efficient reprogramming to pluripotency and directed differentiation of human cells with synthetic modified mRNA. Cell Stem Cell. 2010;7(5):618–30.
56. Yu J, Hu K, Smuga-Otto K, et al. Human induced pluripotent stem cells free of vector and transgene sequences. Science. 2009;324(5928):797–801.
57. Miyoshi N, Ishii H, Nagano H, et al. Reprogramming of mouse and human cells to pluripotency using mature microRNAs. Cell Stem Cell. 2011;8(6):633–8.
58. Kim D, Kim C-H, Moon J-I, et al. Generation of human induced pluripotent stem cells by direct delivery of reprogramming proteins. Cell Stem Cell. 2009;4(6):472–6.
59. Kamao H, Mandai M, Okamoto S, et al. Characterization of human induced pluripotent stem cell-derived retinal pigment epithelium cell sheets aiming for clinical application. Stem Cell Reports. 2014;2(2):205–18.
60. Tucker BA, Scheetz TE, Mullins RF, et al. Exome sequencing and analysis of induced pluripotent stem cells identify the cilia-related gene male germ cell-associated kinase (MAK) as a cause of retinitis pigmentosa. Proc Natl Acad Sci U S A. 2011;108(34):E569–76.
61. Bilic J, Izpisua Belmonte JC. Concise review: induced pluripotent stem cells versus embryonic stem cells: close enough or yet too far apart? Stem Cells. 2012;30(1):33–41.
62. Borooah S, Phillips MJ, Bilican B, et al. Using human induced pluripotent stem cells to treat retinal disease. Prog Retin Eye Res. 2013;37:163–81.

Retinal Organoids: An Emerging Technology for Retinal Disease Research and Therapy

Jennifer G. Aparicio, Dominic W.H. Shayler, and David Cobrinik

Blinding diseases afflict millions of people worldwide at great personal and societal cost [1]. Many forms of vision loss, such as age-related macular degeneration (AMD), retinitis pigmentosa (RP), or glaucoma, involve the death of retinal cells critical for vision [2, 3]. Current therapeutic strategies can slow disease progression but new strategies aiming to preserve and restore vision are needed to better treat these disorders. Cell replacement using PSC-derived cells is one of several promising approaches being pursued to restore vision and improve quality of life [4–6].

J.G. Aparicio, Ph.D. (✉)
Division of Ophthalmology, Department of Surgery, The Vision Center, Children's Hospital Los Angeles, 4650 Sunset Blvd, MS 163, Los Angeles, CA 90027, USA
e-mail: japaricio@chla.usc.edu

D.W.H. Shayler, B.A.
Division of Ophthalmology, Department of Surgery, The Vision Center, Children's Hospital Los Angeles, 4650 Sunset Blvd, MS 163, Los Angeles, CA 90027, USA

Development, Stem Cells, and Regenerative Medicine Program, Keck School of Medicine, University of Southern California, Los Angeles, CA, USA
e-mail: dshayler@chla.usc.edu

D. Cobrinik, M.D., Ph.D.
Division of Ophthalmology, Department of Surgery, The Vision Center, Children's Hospital Los Angeles, 4650 Sunset Blvd, MS 163, Los Angeles, CA 90027, USA

Development, Stem Cells, and Regenerative Medicine Program, Keck School of Medicine, University of Southern California, Los Angeles, CA, USA

The Saban Research Institute, Children's Hospital Los Angeles, Los Angeles, CA, USA

Department of Biochemistry & Molecular Medicine, Keck School of Medicine, University of Southern California, Los Angeles, CA, USA

USC Roski Eye Institute, Keck School of Medicine, University of Southern California, Los Angeles, CA, USA
e-mail: dcobrinik@chla.usc.edu

PSCs are a renewable source of human cells that can be differentiated in vitro to any tissue, including retina [7–10]. They can be genetically manipulated and thus permit gene repair [10–12]. They can be produced to match patient genetic backgrounds and thus their differentiated progeny can theoretically avoid immune rejection [13, 14]. For these reasons, there is great optimism that PSC-derived cells and tissue can be used effectively for therapeutic purposes.

Recent years have witnessed a major milestone as PSCs were used to generate self-organizing three-dimensional structures with properties remarkably similar to developing human retina [15–19]. With these methodologies cells pass through the same developmental stages as cells differentiating in vivo, and cells develop in the context of a tissue that preserves cell-cell contacts and intercellular signaling of retinal cells in vivo. All of these factors increase the likelihood that the molecular nature of the cells (the expression and localization of proteins and noncoding RNAs and the epigenetic marks governing chromatin structure) will set the stage for appropriate cellular behavior, whether used for disease modeling, regenerative therapeutics, human retina development studies, drug screening, or toxicity assessments. Thus, the ability to generate three-dimensional retinal tissue is providing previously unachievable opportunities to model human retinal development and disease, and has enhanced the probability that stem cell-derived tissues and cells will contribute to effective treatment of eye disease.

This chapter will discuss PSC-derived retina, including current culture methods, their capacity for recapitulation of human retinal development, and some potential uses. Limitations of PSC-derived retina to further understanding of retinal disease pathogenesis and to develop novel retinal disease therapies will also be considered. While many terms have been utilized to describe the various forms of PSC-derived tissue, we adopt "retinal organoid" except where referring to a specific methodology that used another term.

S.D. Schwartz et al. (eds.), *Cellular Therapies for Retinal Disease*, DOI 10.1007/978-3-319-49479-1_10

Abbreviations

AMD	Age-related macular degeneration
BMP	Bone morphogenetic protein
ECM	Extracellular matrix
FBS	Fetal bovine serum
GFP	Green fluorescent protein
hESC	Human embryonic stem cell
hiPSC	Human iPSC
ILM	Inner limiting membrane
INL	Inner nuclear layer
IPL	Inner plexiform layer
iPSC	Induced pluripotent stem cell
mESC	Mouse embryonic stem cell
NR	Neural retina
ONL	Outer nuclear layer
OPL	Outer plexiform layer
OV	Optic vesicle
PSC	Pluripotent stem cell
RP	Retinitis pigmentosa
RPE	Retinal pigment epithelium

1 The Need for Human Retinal Tissue

Excitement surrounding PSCs stems from the fact that they are a renewable source that enables production of *human* cells and tissues. While animals are invaluable assets to basic and clinical research, humans are sufficiently distinct in biological detail from common laboratory animals or even other primates such that diseases can behave differently and drugs can have variable effects [10, 20–24]. Only human tissue can be used to decipher the aspects of human biology that cause the uniquely human response. While fetal tissue has also been considered as a tissue source for degenerative conditions, it carries ethical, reproducibility, and logistical issues that are far too complex for routine use.

Indeed, even within the realm of human tissue, a particular individual's genetic background can influence disease progression or drug response. Induced pluripotent stem cell (iPSC) technology, in which somatic cells are reprogrammed to pluripotency [8, 25], makes it possible to study how genetic makeup influences biological responses. With respect to regenerative medicine, only human cells can be expected to behave in a fully integrated manner with appropriate cell-cell and paracrine signaling via species-matched ligand-receptor pairs, and only human iPSC (hiPSC)-derived tissue has the prospect of avoiding immune rejection without immune suppression (although some types of autologous iPSC-derived tissues can also generate immune responses) [14]. Fulfilling the potential of PSC-based therapies for retinal diseases depends upon the ability to produce biologically authentic human retinal cells.

2 Production of Mouse Retina in Three-Dimensional Cultures

In April 2011, the cover of Nature was graced by a picture of a spectacular accomplishment, the in vitro production of an embryonic optic cup from mouse embryonic stem cells (mESCs) (Fig. 1) [9]. To achieve this, Yoshiki Sasai's group in Japan had coaxed these cells toward a neuroectoderm fate, specifically to that of the "eye field" cells that populate a region of the neural plate from which neuroectodermal-derived eye structures (neural retina (NR) and retinal pigment epithelium (RPE)) develop (Fig. 2). Within the eye field multiple co-expressed transcription factors (Pax6, Rx, Six3, Six6, Lhx2) seal the fate of eye field cells to that of optic vesicle tissue [15, 30, 31]. In vitro the eye field neuroepithelium proved to have remarkable self-organizing ability (Fig. 3a–i). Because the mESCs had been engineered to produce green fluorescent protein (GFP) along with the eye field transcription factor Rx, eye field induction produced green fluorescent neuroepithelial tissue that evaginated from an aggregate of neuroectoderm and subsequently invaginated (Fig. 3). These self-driven morphogenetic events formed a bi-layered tissue composed of embryonic NR expressing the retinal progenitor marker Chx10 juxtaposed to an immature RPE expressing its markers Mitf and Pax6 (Fig. 3h–i). Interestingly, invagination depended on activated actomyosin within the RPE domain to maintain tissue rigidity, while simultaneous inactivation of actomyosin in the NR reduced tissue rigidity. This coupled with rapid cell proliferation within the NR and apical constriction in the "hinge" region between the two domains caused invagination of the NR to form the bi-layered tissue or cup. Notably, the morphogenetic movements observed in vitro accurately replicated those naturally occurring in the developing embryo (Fig. 2) [27, 30, 32].

Optic tissue morphogenesis was accompanied by other biological properties that mimic developmental events in vivo. The tissue has the characteristic apical-basal polarity of neuroepithelium, such that the exterior of the budding vesicle is the basal surface (Figs. 2 and 3j), identified by laminin expression, while the apical surface, strongly expressing N-cadherin, forms internally, equivalent to the ventricular surface in vivo. As the tissue invaginates, the apical surface of the neural retina comes into close association with that of the RPE, an important relationship for proper development of photoreceptor outer segments [33, 34]. The retinal progenitor cells in the NR are elongated and extend cytoplasmic processes to both the apical and basal surfaces while their nuclei migrate between these surfaces. Mitosis occurs on the apical surface and the nuclei move basally for G1 and S phase and return toward the apical surface during G2. This phenomenon, termed interkinetic nuclear migration, is

Fig. 1 The cover of Nature in April 2011 shows an optic cup-like structure generated in vitro in a three-dimensional floating tissue that was differentiated from mESCs. Reprinted by permission from Macmillan Publishers Ltd: *Nature,* doi:10.1038/nature09941, 2011

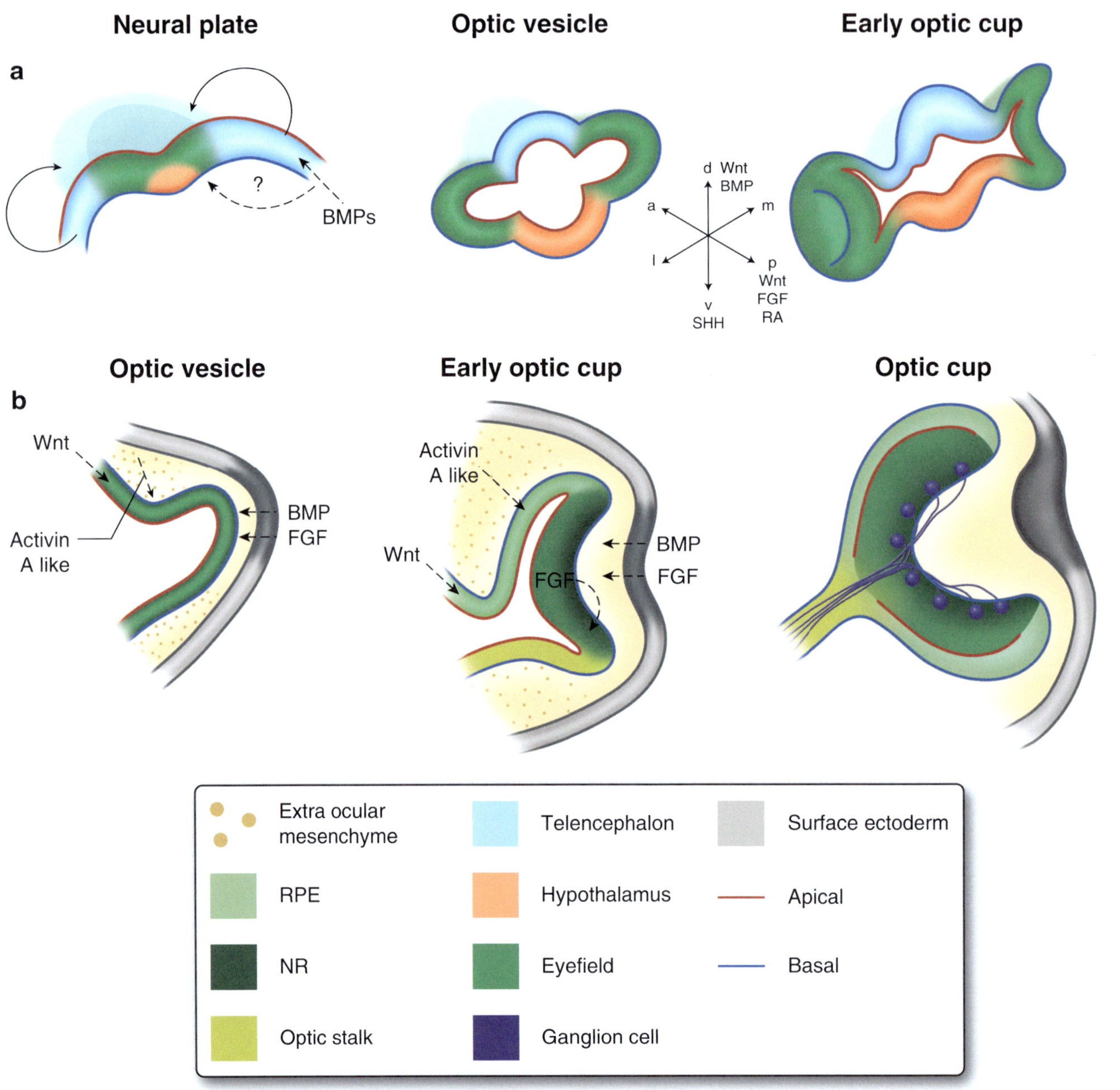

Fig. 2 Schematic of vertebrate early eye development and formation of an optic cup in vivo. (**a**) A three-dimensional representation of early embryonic eye development. The eye field specified region of neuroectodermal tissue evaginates after neural tube formation (tissue folding indicated by *arrows* in "neural plate") to form optic vesicles. An optic cup is created after the tissue invaginates bringing together early neural retina and RPE. *a* anterior, *p* posterior, *m* medial, *l* lateral, *d* dorsal, *v* ventral [26]. (**b**) A two-dimensional representation of optic vesicle formation that includes the ocular mesenchyme and overlying surface ectoderm. Signaling molecules that can influence developmental fate in the in vitro differentiation methods are shown along with **possible** in vivo sources based on reports from other vertebrates [27–29]. The source of Wnt is suggested by the in vitro experiments of Eiraku et al. 2011 [9]. At the optic cup stage retinal ganglion cells differentiating in the basal layer of the retina send axons through the optic nerve head forming the optic nerve

characteristic of proliferating neural progenitor cells in developing neuroepithelium and may maintain tissue integrity and ensure production of proper numbers of each cell type [35–37]. Given time in culture the progenitor cells generate all major cell types of the NR: ganglion cells, rod and cone photoreceptors, amacrine cells, horizontal cells, bipolar cells, and Müller glia (Fig. 4). Neurogenesis occurs in the temporal order seen in vivo with ganglion cells first to exit the cell cycle and the peak of photoreceptor and bipolar cell cycle exit delayed by 6 and 8 days, respectively (Fig. 4m–p). Spatially the young retinal cells stratify into layers with photoreceptors at the apical surface, bipolar and horizontal cells below the photoreceptors, and amacrine and ganglion cells populating a basal layer (Fig. 4). Neurites extending from

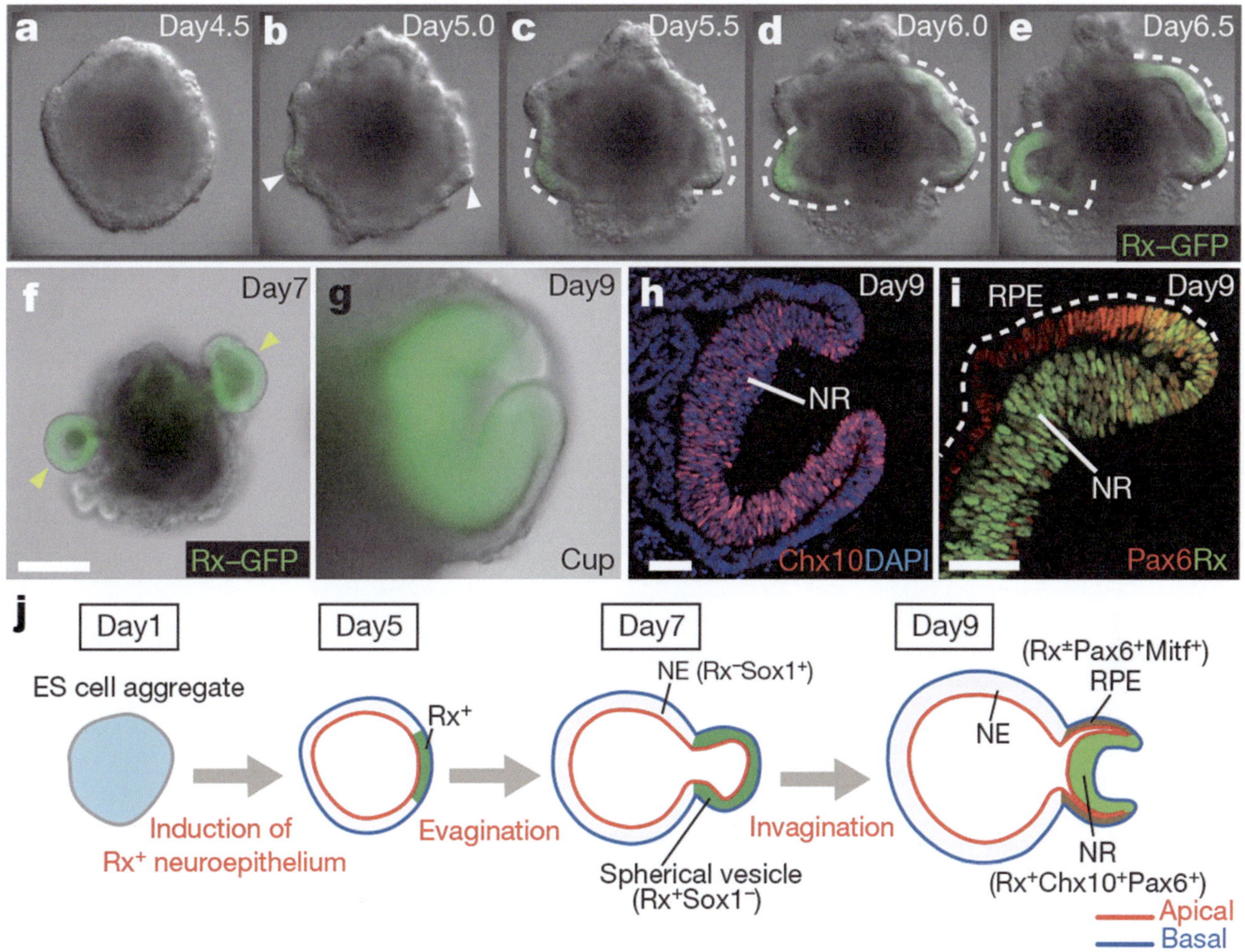

Fig. 3 Self-formation of optic cup-like structures in vitro from three-dimensional culture of mESCs. (**a**–**f**) Maturation of mESC aggregates in floating culture shows the emergence of the eye field (Rx-GFP) from a portion of a thin neuroepithelium and evagination of an optic vesicle-like structure. (**g**–**i**) By day 9 the NR and RPE tissues are closely associated due to invagination of the NR to form a cup. (**j**) Schematic of mouse optic-cup self-formation in vitro. *NE* neuroepithelium. Scale bars: 200 mm (**b**, **e**–**f**, **k**), 50 mm (**h**, **i**). Adapted by permission from Macmillan Publishers Ltd: *Nature,* doi:10.1038/nature09941, 2011

cells appear to form synapses in plexiform-like layers as staining for synaptic proteins is evident [9]. Thus, this seminal paper beautifully demonstrated the dramatic potential of PSC-derived tissue to replicate complex developmental processes in vitro and suggested that PSC-derived retina tissues and cells can have the verisimilitude needed to positively impact disease research and clinical progress.

3 Production of Human Retina in Three-Dimensional Cultures

Multiple publications have demonstrated that three-dimensional retinal tissue can form from human embryonic stem cells (hESCs) as well as from hiPSCs [15–19, 38–40]. These studies build upon previous two-dimensional differentiation strategies and other three-dimensional neural differentiation methods (reviewed in [4]). In all of the approaches, the human PSCs differentiate through the same stages of embryonic development as in mouse three-dimensional cultures and as in normal human development. At the stage of optic cup formation, the cultures can be readily manipulated to yield NR or RPE or both, reflecting the common embryonic eye field origin of NR and RPE (Fig. 2). In the NR, a polarized neuroepithelium contains retinal progenitor cells that undergo interkinetic nuclear migration [17, 18] and give rise to the major retinal cell types. With time, the cells form a stratified retinal tissue and exhibit appropriate morphological differentiation with neurites, synapses, and rudimentary photoreceptor outer segments. Different methodologies showcase distinct aspects of the inherent potential in PSC differentiation and each has its advantages and disadvantages (Table 1). The main methodologies are outlined as follows.

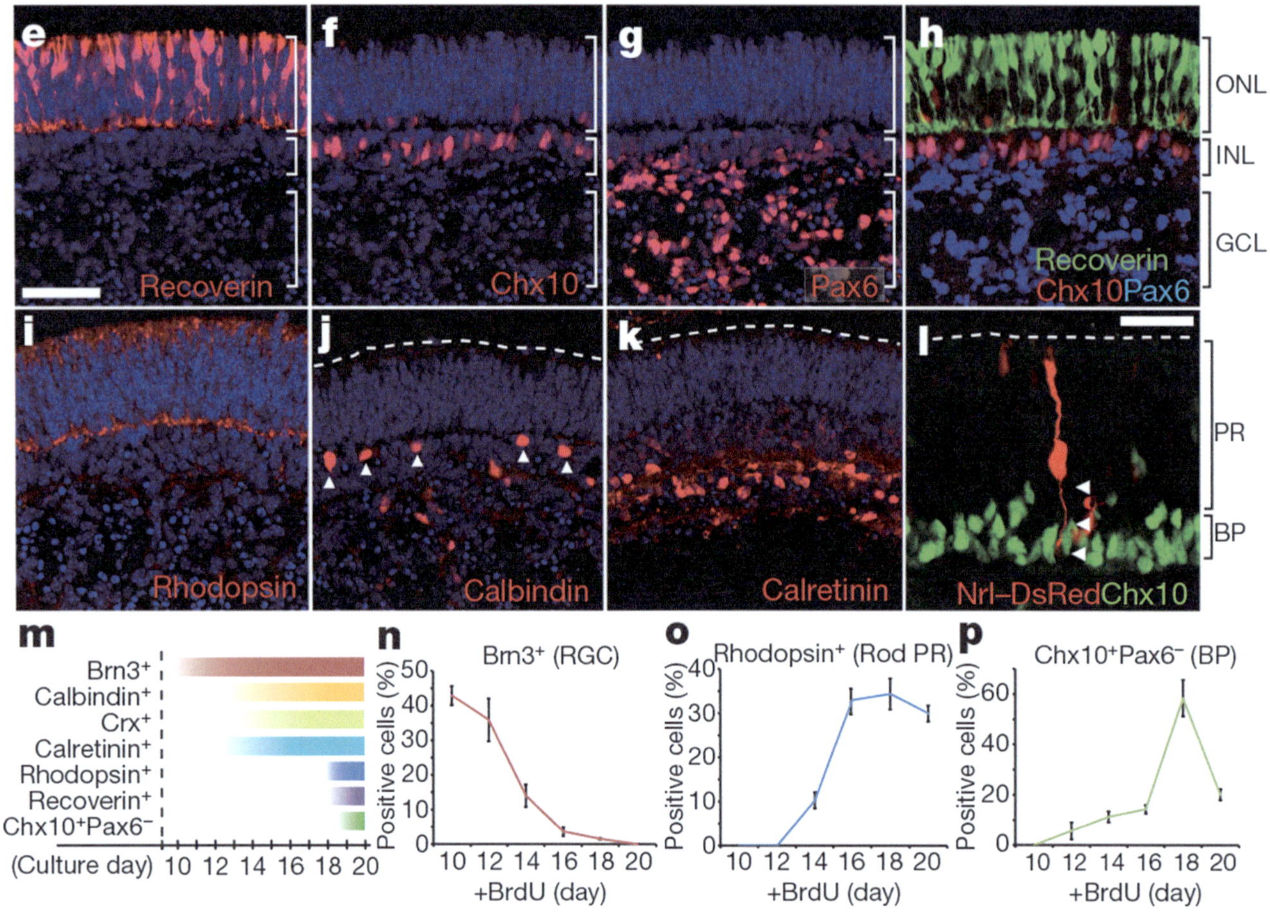

Fig. 4 Generation of stratified neural retina tissue from mESC-derived invaginated epithelia. (**e–k**) Neural retina markers in cross sections on day 24 of differentiation. Recoverin, photoreceptors; Chx10, Pax6-, bipolar cells; Pax6, ganglion, and amacrine cells; Rhodopsin, rod photoreceptors; calretinin, amacrine, and ganglion cells; calbindin, horizontal cells. Not shown is CRALBP staining of Müller glia. Note the RPE tissue has been removed for long-term culture. (**l**) pNrl-DsRed2 was electroporated (apically) on day 16, and analyzed on day 24 and shows the morphology of a rod PR. *GCL* ganglion cell layer, *INL* inner nuclear layer, *ONL* outer nuclear layer, *PR* photoreceptors, *BP* bipolar cells. (**m**) Temporal expression profile of neural retina markers. (**n–p**) BrdU was incorporated on each day indicated in the plots and the BrdU-retaining cell types were analyzed on day 24. Scale bars: 50 mm (**e–k**), 20 mm (**l**). Error bars represent s.e.m. Adapted by permission from Macmillan Publishers Ltd: *Nature,* doi:10.1038/nature09941, 2011

3.1 Production of "Optic Vesicles" and "Retinal Cups"

In 2014, Zhong et al. reported production of tissue with light-detecting photoreceptors. This remarkable feat was methodologically built upon the foundations laid by Meyer et al. 2009, 2011 and Philips et al. 2012 [15, 16, 18, 38] initiated prior to the initial report of mouse ESC-derived retinal tissue.

Meyer et al. demonstrated that differentiation of human PSCs to a retinal fate involved progression from neuroectoderm, to eye field, to optic vesicle, and finally to optic cup-like stages as shown by expression of key markers of each developmental stage in a time frame similar to that of normal human development [15]. In this protocol hESC colonies are cultured in suspension in the presence of canonical Wnt and bone morphogenetic protein (BMP) inhibitors to promote efficient anterior neuroectoderm formation (Table 1) [26, 41–44]. On day 6 the Wnt and BMP inhibitors are removed and neuroectodermal aggregates are plated onto laminin to foster formation of neural rosettes that subsequently develop into densely packed neural colonies. These visually identifiable structures are dislodged from the laminin substrate on day 16 for continued culture as three-dimensional floating neuroaggregates. Ensuing work noted that by day 25 retinal neuroaggregates, termed "optic vesicles" (OVs), have a characteristic phase-bright appearance that enabled distinction in culture from darker appearing forebrain structures [16, 38]. By day 35 the OVs were bi-layered with an apical neuroblastic layer containing retinal progenitors and an interior or

Table 1 Summary of published methods for differentiation of human PSCs to neural retina and associated structures

Method	Meyer et al. [15, 38], Phillips et al. [16]	Zhong et al. [18]	Reichman et al. [40]	Nakano et al. [17]	Kuwahara et al. [19]	Mellough et al. [39]
Features of Tissue Produced	NR or RPE; rosettes form in NR by day 50	NR	Rapid differentiation of NR and RPE; rosettes form in NR by day 42	Bi-layered optic cup of NR and RPE or NR	NR, RPE, or NR with ciliary margin	NR, RPE, lens and cornea, in general varying structure
Pluripotent cell lines used	WA09 [a]IMR90–4 [a]6–9-12 T [a]iPSC-12 [a]iPSC-12.4 [a]TiPSC-5	[a]IMR90–4 [a]KA.1 [a]C-BiPSC6.2	[a]hiPSC-2	KhES-1	KhES-1	WA09
Anterior neuroectoderm induction (days)	DKK and Noggin (2–4)	None	None	Canonical Wnt inhibitor (IWR1) (1–12)	none	IGF (0–37)
Eye field induction	None	Matrigel®	None	Matrigel® Canonical Wnt inhibitor (IWR1)	BMP	IGF
Retina induction	None	None	FGF2	Shh pathway activation (SAG) and FBS	FBS	IGF
RPE induction	Activin A	None	None	Canonical Wnt activator (CHIR99021)	Canonical Wnt activator (CHIR99021) and FGF inhibition (SU5402)	None
Undefined differentiation components	[b]KSR	[b]KSR Matrigel® FBS	None (?)	Matrigel® FBS	[b]KSR FBS	[b]KSR
Manual manipulation	Excision from tissue culture plate	Excision from tissue culture plate	Excision from tissue culture plate	Excision of NR from floating aggregate	No	No
Adherent growth stage	Yes	Yes	Yes	No	No	No
Polarity of organoid	Apical exterior	Apical exterior	Apical exterior	Basal exterior until excision on day 18	Apical exterior	Basal exterior
Primary Media components (days)	DMEM/F12, 20% [b]KSR (0–4)	mTeSR1 (Stem Cell Technologies) (0–3 decreasing%)	ReproStem (ReproCell) (0–2)	GMEM 20% [b]KSR (0–12)	IMDM/F12, 10% [b]KSR, lipid concentrate (0–18)	DMEM/F12, [b]KSR, B27 (0–37 KSR decreasing from 20–10%)
Primary Media components (days)	DMEM/F12 N2 (4–16)	DMEM/F12 N2 (1–3 increasing %, 3–16,100%)	DMEM/F12 N2 (2 onward)	GMEM, 20% [b]KSR FBS (12–18)	DMEM/F12 N2 FBS (18 onward)	DMEM/F12 N2 B27 (37–90)
Primary Media components (days)	DMEM/F12 (3:1) B27(w/o RA) (16- ~50)	DMEM/F12 (3:1) B27(w/o RA) (16–42)	–	DMEM/F12N2 FBS (18 onward)	–	–
Primary Media components (days)	–	DMEM/F12 (3:1) B27(w/o RA) FBS (42 onward)	–	–	–	–

[a]Indicates these are hiPSC lines
[b]KSR knockout serum replacement

basal layer of retinal ganglion cells. With continued culture most OVs lost their laminar appearance and instead formed rosettes that became increasingly disorganized. A few maintained their laminar structure and at day 72 exhibited a photoreceptor layer that had developed apical to the neuroblastic layer. This work also demonstrated that RPE can be produced at the expense of NR by adding Activin A starting at day 20, likely mimicking the role of extraocular mesenchyme in vivo (Fig. 2) [27, 45]. An important feature of this method is that the differentiation media after day 4 is chemically defined, a prerequisite for preparation of cells or tissue for clinical use. The media prior to day 4 is ESC media for which there now exists several chemically defined options that presumably may substitute.

The most recent iteration of the methodology described above introduced multiple modifications that include deviating from the chemically defined differentiation media [18]. A prominent change is the addition of fetal bovine serum (FBS) to the media after day 42 enabling orderly retina maturation significantly beyond that previously achieved. In addition, Matrigel® substitutes for laminin as the extracellular matrix (ECM) substrate that promotes neural development and may also contribute to eye field induction [17, 26]. hiPSCs differentiated using this methodology form a remarkable in vitro human embryonic retina-like structure, which they termed a "retinal cup" (RC). The RC exhibits a spatiotemporal differentiation pattern similar to that occurring in the human retina in vivo such that seven major retinal cell types are generated. There is no associated RPE, but these RCs exhibit an outer nuclear layer (ONL) comprising mostly rods, but also both S and L/M cones, as well as a developing outer plexiform layer (OPL) that by 21 weeks of culture expresses proteins required for synapse formation. Formation of inner retinal layers are less clear, but horizontal cells appear to align beneath the ONL and amacrine cells are largely observed more internally to the horizontal cells. Retinal ganglion cells appear to form their own layer among neuronal processes reminiscent of a nerve fiber layer, and Müller cell cytoplasmic processes span all layers. Long-term culture, on the order of 27 weeks, permits maturation of photoreceptors to a state where ultrastructure examination by electron microscopy reveals an outer limiting membrane, inner segments, basal bodies, connecting cilia, and the small stacks of membranous discs critical for proper assembly of the phototransduction apparatus (Fig. 5). A number of phototransduction proteins are localized within rudimentary outer segments including opsins, the retina-specific guanylate cyclase, and subunits of transducin, phosphodiesterase, and the cGMP-gated channel. Many of these features were also identified in rudimentary photoreceptors of PSC-derived retinal organoids produced by a different methodology [39]. Promisingly, a subset of photoreceptors produced by Zhong et al. exhibited a reduced inward current following light stimulation, a response consistent with the closure of cGMP channels that follows light absorption by opsins in vivo. Nevertheless these photoreceptors lack the distinctive outer segment morphology of mature photoreceptors in vivo, a structure integral to many aspects of a rod's response to light. Thus, further characterization is required to evaluate whether they efficiently mature to fully functional cells. However, given the intimate relationship that photoreceptors have with the RPE and the requirement of RPE for proper maturation in vivo, complete differentiation seems unlikely when they are bathed in media rather than swaddled by their natural RPE caretakers [33, 34, 46–50].

3.2 Production of "Optic Cups" and "Neural Retina"

The Sasai group has published two methods to produce human embryonic retina from hESCs [17, 19]. Each uses distinct media and produces unique structures not observed in the Meyer et al. and Zhong et al. approaches (Table 1). The first, reported in Nakano et al. 2012, optimized their mouse optic cup methodology described by Eiraku et al. 2011 for human retina production. The method can produce either NR comprising only retina tissue, or "optic cups" comprising retina plus RPE in an optic cup configuration similar to the mouse work described above. The difference between producing NR or optic cup is simply the addition of a canonical Wnt pathway activator on days 15 through 18. This is the time of optic vesicle formation, where evagination of the eye field from the neuroepithelial aggregate can be observed, analogous to what occurs in the mouse cultures (Fig. 3c–f). The activation of the Wnt pathway tips the balance of some cells toward an RPE fate and induces the invagination behavior required for optic cup formation (see Fig. 2b). In contrast, addition of the Wnt activator days 18 through 21 causes all the cells to convert to RPE, consistent with mouse studies that have demonstrated that Wnt signaling is required for RPE differentiation in vivo [51–53]. While the authors reported that 21–24% of cultures formed NR plus RPE using the optic cup conditions, long-term optic cup structures were not shown and the authors stated that NR cultures were more efficient and hence preferred. Thus, whether the optic cup methodology enables long-term growth of RPE adjacent to NR is unclear.

The Nakano et al. protocol is capable of producing beautiful stratified NR similar to that of Zhong et al. The eye field is induced by adding Matrigel®, as was previously demonstrated in mouse tissue [54]. Matrigel®, a basement membrane secretion from mouse tumor cells rich in ECM proteins like laminin, comprises up to 1850 different proteins whose variation between different lots can influence the reproducibility of in vitro retina production [19, 55]. The Sasai lab's more recent work, Kuwahara et al., overcomes this issue, producing equally stunning stratified neural retina without Matrigel® [19]. Instead, they used BMP4 to induce the eye field (defined by RX protein expression) at day 6 of differentiation, after the hESCs have transitioned from the epiblast state to neuroectoderm. The exact role of BMP is not understood [19]. BMPs are expressed in surface ectoderm and anterior neuroectoderm, and in general have a dorsalizing function with roles in patterning the anterior neural tissue, but not a clearly described function for eye

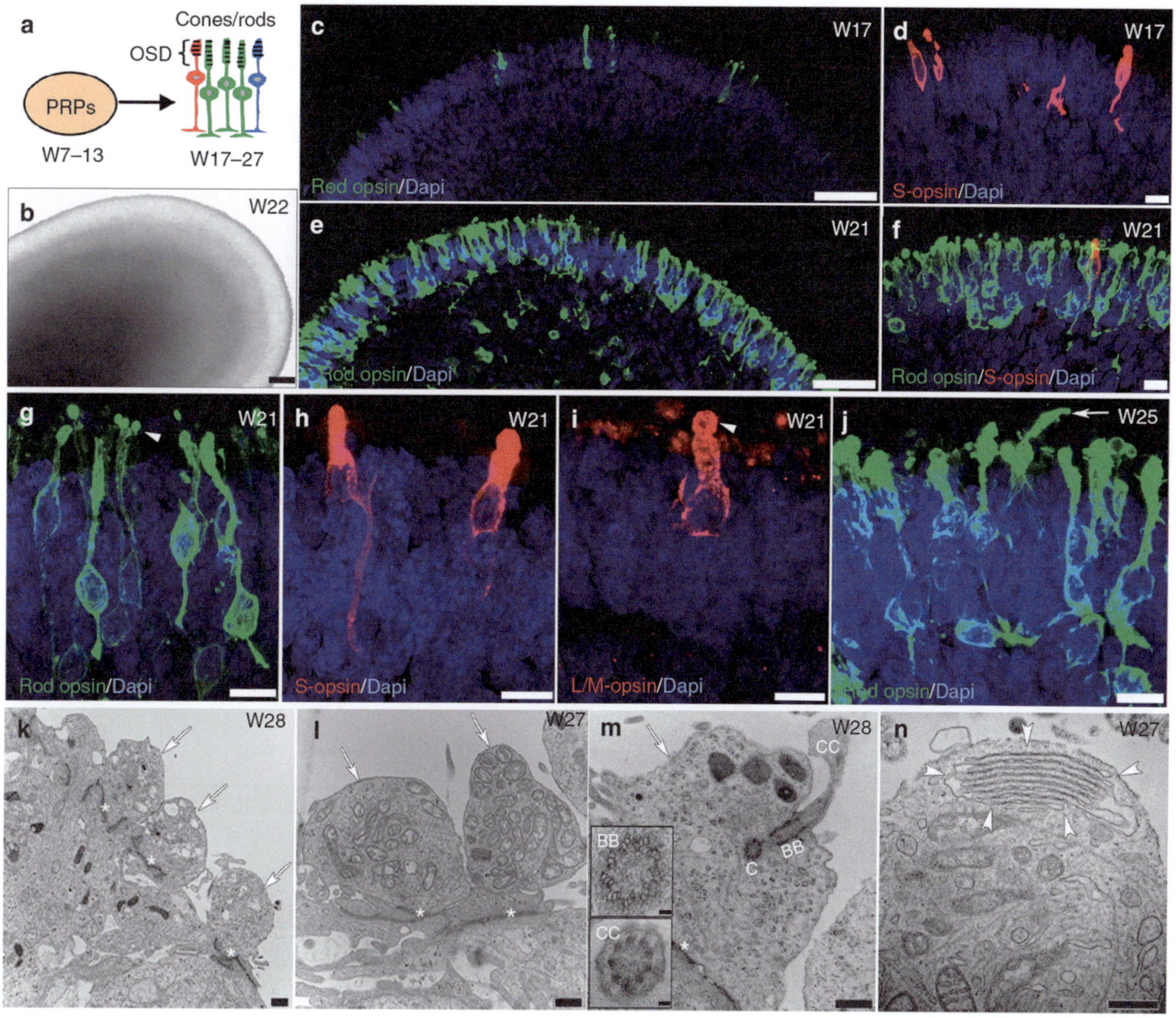

Fig. 5 Human iPSC-derived rod and cone photoreceptors achieved an advanced level of differentiation including outer-segment disc formation. (**a**) During normal development photoreceptor precursors (PRPs) differentiate into rods, L/M- and S-cones (in *green*, *red,* and *blue* color, respectively). (**b**) A week 22 RC that has been exposed to 1-mM RA in week 10–week 14. (**c–j**) Under this condition, photoreceptors expressing high levels of rhodopsin in the entire cell body were first observed by week 17 (**c**), significantly increasing in number and forming large patches by week 21 (**e**, **f**). S-opsin expression could be observed in some rhodopsin-negative photoreceptors (**d**, **f**). High-magnification images of week 21 RCs showing rods (**g**), S- (**h**) and L/M- (**i**) cones with a morphology and a topological organization similar to those of the in vivo retina, including structures reminiscent of short, nascent outer segments (*arrowheads*). By week 25, elongated structures resembling more developed outer segments were rarely observed (**j**, *arrow*). (**k–n**), Transmission electron microscopic analysis revealed the presence of an outer limiting membrane (*Asterisk*), inner segments (*arrows*), basal bodies (*BB*), connecting cilia (*CC*), and stacks of outer-segment discs (demarcated by *arrowheads*). BB and CC presented the photoreceptor-specific microtubule arrangement consisting of $9 \times 3 + 0$ and $9 \times 2 + 0$, respectively (insers in **m**). *C* centriole. Scale bars, 50 mm (**b**, **c**, **e**); 10 mm (**d**, **f** and **g–j**); 0.5 mm (**k–n**); 0.05 mm (**m** insets). Reprinted by permission from Macmillan Publishers Ltd: *Nature Communications,* doi:10.1038/ncomms5047, 2014

field formation [26, 28, 43, 56, 57]. In chick, post-eye field formation, overlying surface ectoderm secretes BMP, which is required for retina formation (Fig. 2) [27, 58], but in zebrafish BMPs suppress eye field formation [59]. Whether there are non-conserved BMP functions or the addition of BMP4 in the in vitro artificial environment achieves some level of signaling cross talk required for eye field induction is not known.

In the Kuwahara et al. study [19] once again, the NR fate is readily switched to that of RPE (defined by MITF expression). However, in this case, Wnt pathway activation was combined with fibroblast growth factor (FGF) inhibition at day 18, taking advantage of evidence that FGF promotes neural retina [27, 60–62]. Interestingly, transient Wnt activation and FGF inhibition (on days 18–24) followed by continued culture in the NR differentiation medium converts many of these cells back to NR, leaving only specific domains with RPE characteristics (Fig. 6). In this process, which was termed "induction-reversal culture," the secondarily produced NR is not part of a bi-layered optic cup but rather shows a distinctive tapering of the epithelium as it approaches the RPE-like domain (Fig. 6). The thinning tissue is less differentiated, analogous to the peripheral retina developing in vivo. In a domain adjacent to the RPE-like region, the tapered NR tissue exhibits ciliary margin characteristics, as it contains SSEA1+ neurosphere forming stem cells, yielding a NR with the potential to study the properties of these unique cells.

Two studies have utilized these methods to demonstrate properties of the retinal ganglion cells, particularly the functionality of retinal ganglion cell axons. Tanaka et al. tweaked the Nakano et al. protocol to grow retinal organoids in floating culture [63]. Then, to permit outgrowth of retinal ganglion cell axons, the organoids were placed on a two-dimensional laminin substrate. Organoids between day 26 and day 29 proved optimal to yield robust axon outgrowth, producing axons expressing tau, neurofilament proteins, and Tuj1, and exhibiting enhanced survival in the presence of brain-derived neurotrophic factor [64, 65]. These axons also exhibited the functional properties of sodium-dependent

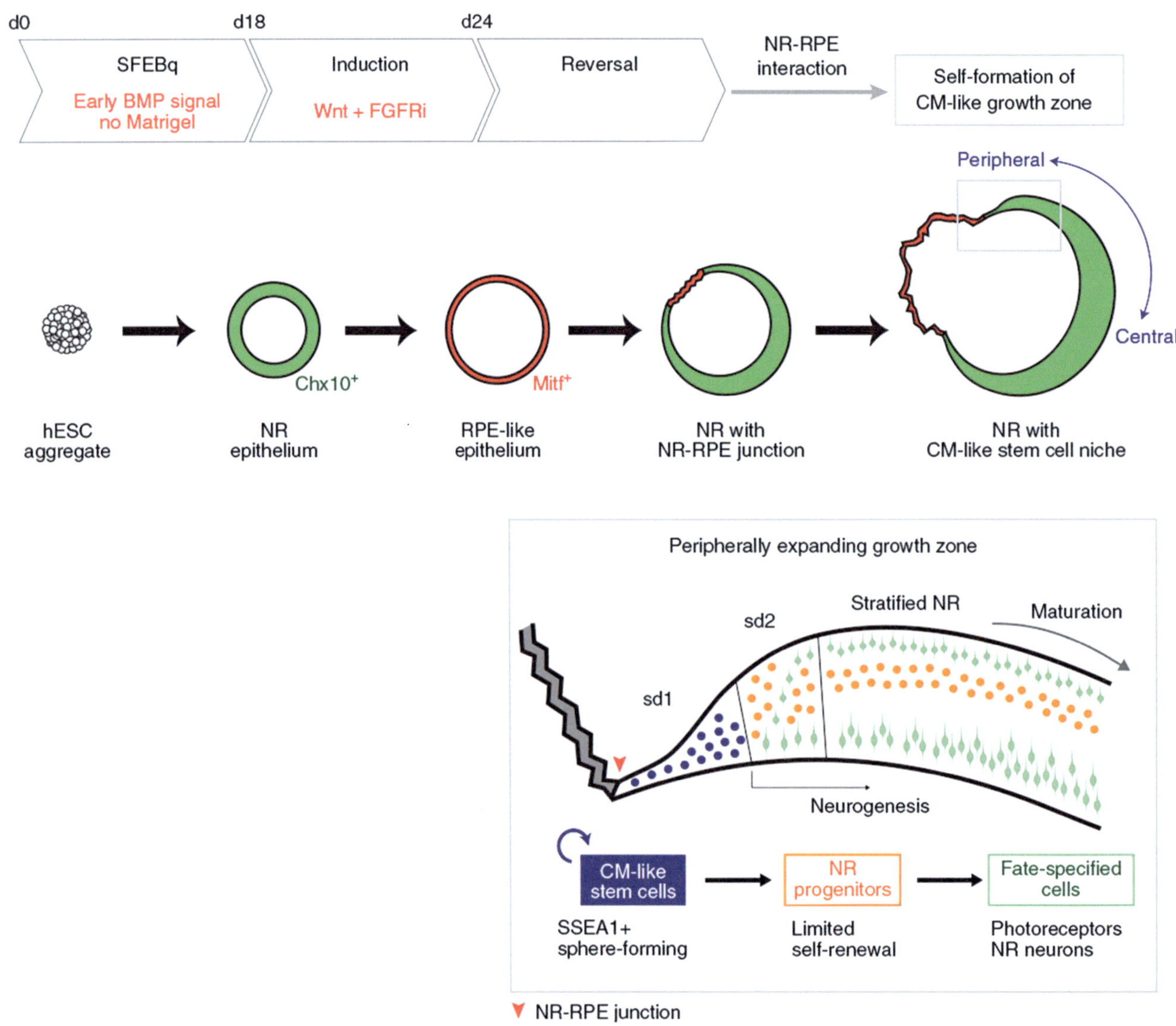

Fig. 6 Self-organization of growth zone with stem cell niche from hESCs. Schematic of self-organization culture of hESCs involving selective NR differentiation with transient BMP4 treatment, induction to RPE and reversal to NR and self-formation of ciliary margin-like tissue at the NR–RPE boundary. Inset, schematic of a self-forming ciliary margin-like zone at the margin of hESC-derived NR with SSEA1+ sphere-forming stem cells (*blue*), which produces NR progenitors. The CM-like growth zone functions in a similar manner with the progress zone in the developing limb, with regard to self-renewal and tissue extension. Reprinted by permission from Macmillan Publishers Ltd: *Nature Communications*, doi:10.1038/ncomms7286, 2015

action potential generation and antero- and retrograde axonal transport dependent upon intact microtubules. Using the Kuwahara protocol, Maekawa et al. optimized media conditions for neurite outgrowth and demonstrated that retinal ganglion cell axons could extend up to 4 mm in vitro [66].

3.3 Expedient Neural Retina Formation

Reichman et al. have shown in an hiPSC line that NR and RPE can develop within 14 days [40]. By allowing a confluent plate of PSCs to spontaneously differentiate and changing to a neuronal supportive media on day 2, retinoepithelial structures develop adjacent to pigmented regions by day 14 [40]. Excision of the NR-like structures and continued culture floating in a NR favorable media produces a retinal organoid that by day 21 has many ganglion cells in a basal layer, CRX-positive photoreceptors accumulating at the apical surface, and inner retinal neurons populating the epithelium. (Note RPE can be excised from the dish adjacent to the NR and expanded in culture as well.) By day 42, stratification has given way to internal rosettes where photoreceptors develop rudimentary outer segment structures similar to that seen by others [18]. By avoiding Matrigel® and using defined media, this protocol permits rapid generation of early embryonic retina; however, it remains unclear whether this method will be effective for many different PSC lines.

3.4 Summary

Thus, from human PSCs it is possible to generate tissue that evaginates and invaginates to form a bi-layered optic cup composed of both NR and RPE; tissue that forms NR capable of producing all major retinal cell types with the spatiotemporal characteristics of developing human fetal retina; tissue that contains a ciliary margin enriched in stem cells; tissue that produces light transducing photoreceptors; tissue that generates cells with synapse-forming neurites; and tissue that produces retinal ganglion cells capable of extending axons competent in axonal transport and action potential generation.

Since the cells differentiating in these cultures proceed through the same stages of development as cells in vivo and differentiation occurs in a three-dimensional tissue microenvironment replicating tissue architecture, one anticipates that the progression of gene expression and epigenetic changes generates cells with molecular characteristics very close to those in vivo. If this is true, then they are likely to respond to environmental signals in a similar manner as native cells, whether in the context of transplantation, a drug or toxicology screen, or an experimental manipulation to understand development. In this regard, cells within PSC-derived retinas might behave more authentically than cells produced from two-dimensional culture or by transdifferentiation protocols that lack the cell-cell contacts and the normal progression from the embryonic state. Indeed, a comparison of epigenetic modifications of in vivo-generated and two-dimensional in vitro-generated pancreatic cells demonstrated many similarities but also stark differences in the distribution of repressive chromatin marks, suggesting that the in vitro differentiated cells might exhibit altered behavior [67, 68]. That cells differentiated in two-dimensional platforms differ from their three-dimensional counterparts is widely supported in literature pertaining to retina and other fields [6, 69–74].

Finally, it should be noted that OCT4- or NANOG-expressing cells were not detected in three-dimensional NR, suggesting that all PSCs either differentiate or undergo cell death [17]. This is an important characteristic that can prevent accidental outcomes due to undifferentiated cells with aberrant developmental behavior and tumorigenic potential. However, caution is warranted for SSEA1+ cells that can be produced as in Kuwahara et al. 2015 [19], since SSEA1+ cells proved to be tumorigenic upon transplant [75], but whether these two SSEA1+ cell populations behave similarly is not clear. Still, the evidence leaves little doubt that three-dimensional retinal differentiation methods offer tremendous potential to mimic or substitute for retinal cells or tissue produced in vivo.

4 Strengths and Limitations of Current Retinal Organoid Culture Methods

Harnessing the promise of three-dimensional retinal differentiation methods requires progress on several fronts. Ultimately, the purpose for which one intends to use a three-dimensional retina culture defines the characteristics required of the cultures, and hence the challenges faced. Some of the current general or method-specific strengths and limitations are discussed below.

4.1 Scalability

One might imagine that retinal production will become an automated, high-throughput, high-quality process as has recently been achieved for iPSC production [76]. Minimizing variability and achieving scalability requires that human intervention is minimized. In the Meyer, Phillips, Zhong, and Reichman et al. studies differentiation proceeds though a stage of adherent culture from which retina-destined tissue is manually excised [15, 16, 18, 38, 40]. This step is avoided in the Sasai lab methods since cell aggregates are cultured in suspension throughout the entire differentiation process [17, 19]. However, production of NR in suspension culture with

Matrigel® (as in Nakano et al.) promotes complex structures from which retina tissue must be manually excised [17]. The Kuwahara protocol, on the other hand, tends to produce single round spheres or crescents in which the entire structure represents NR or RPE, requiring no "clean up." Thus, of these protocols, Kuwahara minimizes the need for human interference and appears to be the most easily scalable option.

The ability to produce large quantities of tissue is important for feasibility, whether tissue will be used as an experimental system to study human development, a tissue to screen for toxicology studies, a source of cells for transplantation or any other imagined purpose. Currently, in the 96-well plate format used in Nakano et al. and Kuwahara et al., one person can initiate hundreds of PSC aggregates in several hours. By day 90 these NR aggregates contain on the order of 150,000 cells (Aparicio et al., unpublished data). If 23% of the cells are photoreceptors [77], 500 aggregates or about five 96-well plates would generate approximately 17 million photoreceptors. While a number that might be required, for example, for transplantation, is not yet clear, scale up from five 96-well plates is feasible and hence it is reasonable to expect to produce enough tissue for practical use.

4.2 Efficiency and reproducibility

Efficiency and reproducibility of NR production will be critical for minimizing cost and time, and maximizing results. The rough calculation of number of photoreceptors above assumes that NR is produced in 100% of culture wells. The efficiency of the methodologies discussed above range from 21 to 95%, with the Kuwahara method (without ciliary margin) being the most efficient. Reproducibility in part depends upon consistently produced chemically defined reagents. The undefined natures and lot-to-lot variabilities of Matrigel® and FBS are likely to impede consistent differentiation. Kuwahara's method bypasses Matrigel® completely, but still depends upon FBS for long-term differentiation (from day 18 or day 24). Thus, there is a need to identify the required FBS components. This is also a prerequisite for good manufacturing practice reagent production necessary for clinical use.

Efficient differentiation of hiPSCs to retina is required if they are to fulfill their envisioned role as an autologous source of tissue for disease modeling, drug or toxicology screening, and transplantation repair. Although iPSCs reproduce the genetic background of the cell donors from which they derive, some iPSC epigenetic marks can be maintained from the source cells [78, 79]. This epigenetic memory can alter or bias differentiation potential and makes them more difficult to efficiently differentiate than ESCs [80–84]. Thus, to be widely useful, a hiPSC differentiation methodology must redirect any inherent bias. To date, several protocols have generated retina from hiPSCs with 62–85% efficiency, but it is not clear how applicable these methodologies will be for all hiPSC lines [16, 18, 40]. Differentiation of hiPSCs to retina using the Sasai lab protocols has not been reported in the literature. In our experience working with hiPSC lines derived from peripheral blood cells, none of the methods discussed above yielded >6% retinal tissue production for any of the 14 cell lines tested (Aparicio et al. unpublished data), demonstrating the need for better methodology. In the future, requirements for a universally applicable differentiation methodology may be reduced as techniques for producing high quality iPSCs are continually improving [85, 86]. Additionally the newly discovered ability to culture human PSCs in the naive (versus primed) state might make hiPSC differentiation less problematic since the naïve state exhibits a more primitive epigenetic pattern [87, 88].

4.3 Cultured Retina Maturation

The maturation limitations of retinal organoid cultures have to be carefully weighed against the desired use. To date, these cultures do not mature to a state resembling the human adult retina. All cell types are produced, and thus the embryonic period encompassing neurogenesis appears to be well represented. The extent to which different types of inner neurons are generated and whether neuronal circuitry is established with appropriate connections remains unexplored. Over time, ganglion and amacrine cells gradually disappear and complete morphological and functional differentiation of photoreceptors is very rare [18]. Thus, the organoids currently produced do not well represent later stages of fetal retina development—roughly post mid-gestation. This might in part be due to the lack of contributions from other maturing structures present in vivo. There is no associated developing lens, ciliary body, optic nerve head, or structurally intact RPE. Molecules secreted by these structures normally contribute to retina development and function. For example, the inner limiting membrane (ILM) is formed by extracellular matrix proteins secreted from the lens and ciliary body [89, 90]. In addition, astrocytes do not migrate into the retina through the optic nerve head and retinal vascularization does not occur [30, 91, 92]. The lack of a route for retinal ganglion cells to extend their axons and find appropriate targets may contribute to their cell death [93]. The lack of RPE adjacent to photoreceptors likely precludes proper maturation [33, 46–50].

Disease modeling optimally requires the ability to mature tissue to the developmental stage in which the disease phenotype is manifested. Thus, modeling diseases that exhibit a phenotype in early retina development, or investigating cellular processes that underlie adult disease phenotypes may be appropriate. However, more sophisticated culture that better mimics a later developing or adult retina will be required for robust modeling of adult retinal diseases. A similar argument applies for drug screening or toxicology studies; human PSC-derived retina might be useful to assess fetal but not adult effects. On the other hand, maturation obtained by current culture methods seems suitable for cell replacement purposes since both animal and human studies indicate that immature cells and tissue survives and restores vision better than their more mature counterparts [94–96].

For studying human fetal retina development, these cultures offer an unprecedented opportunity to study early developmental stages, such as eye field formation or neurogenesis. The ability to readily grow tissue, to genetically manipulate tissue, to virally infect tissue, and to expose tissue to small molecules such as pathway activators and inhibitors suggests it will be a powerful experimental system. The value is already apparent from several recent publications. In Kaewkhaw et al. hESCs were engineered to express GFP coincident with CRX, enabling isolation of photoreceptor precursors from different culture ages and analysis of the transcriptome at the different stages [77]. This has generated the first molecular database of human photoreceptor development. In a similar vein, the Sasai lab has utilized their three-dimensional mouse culture system and the ability to manipulate NR versus RPE formation to create a transcriptome database of early NR and early RPE, providing detailed information about differential gene expression at this stage of differentiation [97]. Using mESC-derived tissue, La Torre et al. elucidated the importance of the GIPCI-PI3K-Akt1 signaling pathway in eye field formation [98].

4.4 Validation of Retinal Organoids as a Model

The preceding paragraphs illustrate how three-dimensional retinal cultures can replicate many aspects of early retinal development. This in itself is significant validation. However, how authentic are the transcriptome databases? Do they accurately represent the in vivo processes? A comparison to human fetal tissue is the only way to answer such questions. Defining the similarity of each human PSC-derived retinal cell type to its fetal counterpart by isolation, RNA sequencing, and epigenetic analyses is required to establish whether the appearance of cell type-specific and maturation stage-specific transcriptional programs resembles that occurring in vivo. These studies will be important for in vitro developmental, disease modeling, drug or toxicology studies, and cell transplantation since a molecular validation is required to understand the degree to which normal biology is replicated. Cells produced in vitro are unlikely to precisely mimic fetal cells due to adaption to their media environment. Accordingly, validation studies may help to identify factors that could be altered in the media to better mimic the in vivo state.

Quality control is another aspect of validation. It will be important to establish noninvasive methods to assess maturity and health of growing retinal organoids, such as by assaying secreted factors or by noninvasive imaging [99, 100].

4.5 Summary

While there are a number of concerns about current retinal organoid culture systems, it is important to recognize that the field is only about 4 years old and that culture methods will continue to develop. No doubt improvements in efficiency and reproducibility will be made. Media and other aspects of the culture environment will be altered to overcome at least some current limitations of maturation. Molecular validation of the system will highlight biological processes that are not reproduced and likely will provide clues to factors that can be changed to overcome current limitations.

5 Use of PSC-Derived Retinal Organoids for Cell Replacement Therapy

One of the most promising applications of human PSC-derived retinas is their use as a tissue source for retinal cell replacement therapies for blinding diseases such as RP, AMD, and glaucoma [101–107].

5.1 Photoreceptor Replacement

RP and AMD patients lose vision when photoreceptors degenerate. In RP and other photoreceptor dystrophies, vision loss can result from a failure to produce one or more types of photoreceptors, from production of abnormal nonfunctional photoreceptors, or from production of abnormal photoreceptors that initially function but degenerate at varied rates (often over many decades) [108]. Thus, the principal challenge of cell replacement is to transplant and functionally integrate cells that have the capacity to develop into mature photoreceptors within dysfunctional retinal tissue. In dry AMD, the RPE layer is dysfunctional or lost, further complicating the challenge of sustaining a viable layer of transplanted photoreceptors [109]. Thus, replacement of both RPE and photoreceptors may be a more promising cell replacement strategy [101].

Currently, restoration of photoreceptors is being pursued via two approaches: (1) transplantation of dissociated cells and (2) transplantation of "sheets" of embryonic retinal tissue. Each of these shows promise to improve vision in animal models and, in the latter case, in a human clinical trial as well [4, 95]. While retinal cells from human PSC-derived retinal organoids in principle should serve both strategies well, this remains to be shown. However, encouraging results have been obtained with mESC-derived photoreceptors. Some highlights of the striking progress in restoring vision through cell replacement are presented below. Further details may be found in recent reviews [4, 110].

5.1.1 Transplantation of Dissociated Cells

The first compelling evidence that dissociated retinal cells can functionally integrate into a diseased retina came from a landmark paper by MacLaren et al. in 2006 [94]. The authors found that GFP-labeled retinal cells from post-natal day 1 (P1) mice integrated into the developing retina and developed photoreceptor morphological features, had correct orientation relative to the ONL, expressed rod proteins such as phosducin and rhodopsin, produced functional ribbon synapses with the synaptic protein bassoon interfacing with protein kinase C-positive bipolar cells, and responded to stimulation of a rod glutamate receptor. Integration also succeeded using GFP+ photoreceptor precursors from *Nrl–gfp* transgenic mice and was maximal when cells were derived from P1 to P7 mice, a developmental stage when rods functionally integrate into the developing normal retina. Integration of dissociated cells required photoreceptor-committed precursors and was not effective either with undifferentiated retinal progenitor cells or with mature rods. Moreover, integration was similar either using recipient mice of the same age as the donor or using recipient adults, and occurred in *rds*, *rd*, and *rho−/−* retinal degeneration mouse models. In the *rho−/−* model, the transplanted cells restored light-evoked extracellular field potentials as well as light-induced pupil constriction. The authors presciently concluded that their work "might facilitate the generation of appropriate cells for transplantation into humans from either embryonic or adult-derived stem cells" [94].

The landmark study of MacLaren et al. [94] provided a proof-of-principle that photoreceptor precursors can functionally integrate into degenerating retinae; however, many challenges remained. One of the first issues was whether cone as well as rod photoreceptor precursors could integrate into and function within a damaged retina, and this was demonstrated in a mouse model of Leber congenital amaurosis [111]. Cone transplantations were more recently further characterized [112] and have so far been less efficient than rod replacement [4]. An additional concern was whether the synaptic connections and evoked potentials observed by MacLaren et al. also provided functional vision. Indeed, rod precursor transplantation restored scotopic (night) vision to *Gnat1−/−* mice, which entirely lack rod function but retain normal ONL morphology and cone-mediated photopic vision. Transplanted mice had evidence of visual stimuli projection to higher visual areas of the cortex, head tracking, and water maze navigation under scotopic conditions [113]. Finally, the capacity to integrate and provide visual responses was extended to many additional retinal degeneration models [114–116] and parameters influencing photoreceptor precursor integration efficiency and survival are being actively investigated [115, 117–122].

Given the astounding progress in restoring vision by syngeneic transplantation of normal mouse photoreceptor precursors to mouse retinal degeneration models, a critical next step is to identify optimal sources of retinal photoreceptor precursors for human retinal cell therapies. To date, only one study using cells from mouse retinal organoids has been reported [123], whereas studies with human PSC-derived cells have been limited to those from two-dimensional culture [124, 125]. In the former, retinal organoids cultured following the Eiraku et al. 2011 protocol were produced from mESCs and transduced with a virus expressing a rhodopsin promoter driven *GFP* (Rhop.GFP+) [123]. Rhop.GFP+ cells were subsequently transplanted to the subretinal space of *Gnat1−/−* mice and integrated into the ONL. Cells from younger cultures (day 26 and 29) integrated at higher efficiency than those from older cultures, confirming the requirement for photoreceptor precursors of a precise developmental stage for optimal transplantation. Integrated cells formed ribbon synapses with appropriate synaptic protein expression, indicating connection to existing retinal circuitry as well as outer segment maturation. Integrated cells also had similar response to pharmacological agonists and antagonists as the host retinal cells, demonstrating that the in vitro-derived cells can perform similar to native cells. Integration efficiency was lower in Peripherin-2 null mutants (*Prph2rd2/rd2*) and in *rhodopsin* knockouts (*Rho−/−*) than in *Gnat1−/−* retinae, highlighting a need to tailor the integration strategy to each retinal disease.

One critical issue in the development of PSC-derived photoreceptor replacement therapy relates to the number of correctly integrated cells that are needed for functional vision. In previous work, scotopic ERG responses were observed in *Gnat1−/−* mice after transduction of a *Gnat1* transgene into approximately 150,000 rods but not after transduction into ~60,000 rods, suggesting that >60,000 rods are needed for an ERG response [113]. In keeping with this, no ERG responses were detected after integration of up to ~26,000 wild-type photoreceptors in *Gnat1−/−* retinae. However, integration of as few as ~10,000 cell stimulated primary visual cortex activity and optokinetic head tracking, and integration of ~35,000 substantially improved water maze navigation despite a lack of ERG response [113]. In the Gonzalez-Cordero et al. study performed by the same group, far fewer cells integrated from three-dimensional cultures compared with native mouse cells, with only ~0.3% of 200,000 transplanted, or ~600 integrated rods per *Gnat1−/−* eye [123]. This suggested that the number of integrated photoreceptors was insufficient for functional visual recovery. The reason for the diminished number of integrated PSC-derived rod precursors is unclear, but could relate to the transplantation of Rhop.GFP+ expressing cells at a more mature stage than is optimal, an inherently decreased transplantation efficiency of PSC-derived retinal cells, a lower post-transplantation survival rate of the in vitro produced cells, or a greater heterogeneity in the photoreceptor precursors from PSC-derived retinal organoids compared to native cells. Whatever the case, the data suggests that there may be differences between the two cell populations.

These studies indicate that PSC-derived photoreceptor precursors can exhibit functional properties upon transplantation. Human PSC-derived retinal organoids are anticipated to be a suitable source of human replacement cells for photoreceptors since they produce both rods and cones (S and L/M cones) [17, 18, 77]. The level of maturity achieved with current culture methods should be suitable as well, since the rod marker NRL is expressed at about day 90 and rhodopsin and other phototransduction proteins are expressed at around day 125 [17, 18, 77], suggesting that the age required for successful transplantation of immature rod precursors will be within this range [94, 123].

5.1.2 Retinal Sheet Transplantation

A body of work spanning at least 14 years has described results of transplanting retinal sheets, or pieces of intact immature embryonic retina, with and without adjacent RPE, from fetal rat or human into several rat models exhibiting severe photoreceptor degeneration [95, 126–131]. Following surgical transplantation into the subretinal space, transplanted tissue that contains RPCs expands and differentiates into all major retinal cell types except ganglion cells, forming multiple types of synapses [132–137]. The engrafted tissue yields a continuous ONL with mature photoreceptor outer segments contacting host or donor RPE, and produces its own inner nuclear layer (INL) as well as inner and outer plexiform layers [132, 133, 135, 137–140]. Interestingly, the result is a retina with two INLs and two inner plexiform layers (IPL), one donor and one host. While this retina has an unusual architecture, it exhibits a number of positive characteristics. Phototransduction was demonstrated with characteristic shifts in cellular localization of phototransduction proteins following light stimulation [141]. The donor tissue extended processes that exhibit both pre- and postsynaptic markers into the host inner nuclear and IPL [136, 139]. Trans-synaptic tracing from the superior colliculus demonstrated chains of functional synapses from host ganglion cells to transplant inner neurons and photoreceptors [130, 142]. Importantly, optokinetic head tracking significantly improved, and recordings from the superior colliculus detected visual responses in areas corresponding topographically to the transplanted area in up to 80% of transplanted rats [126–129, 136, 139, 140, 142, 143]. Finally, in a phase II clinical trial six RP patients and four AMD patients received fetal retinal sheet transplants that included RPE [144]. The results appeared encouraging in that at 1 year 3/6 RP eyes and 4/4 AMD eyes exhibited improved visual acuity. In the most promising case vision improved in the treated eye of an RP patient from 20/800 to 20/160 and remained stable for at least 5 years while vision in the untreated eye deteriorated [95, 144]. While this study lacked an untreated control group, as well as sham surgeries to rule out placebo effects, the overall results of retinal sheet transplantation in both rats presented above and humans suggest that further investigation is warranted.

Human PSC-derived retinal organoid tissue may be an adequate substitute for fetal retinal sheet therapy since current evidence suggests that the developmental state of embryonic tissue required for these transplantation studies (8–14 weeks postconception) is well replicated. This time period in vivo is characterized by rapid genesis of photoreceptor precursors, as well as retinal progenitor cell proliferation and genesis of other retinal neurons [95, 145, 146]. The equivalent maturation time in culture is approximately day 30 through 70, when retinal progenitor cells are proliferative and producing photoreceptor [77] and other neuronal precursors.

Whether human PSC-derived photoreceptors in retinal organoids will adequately mature after retinal sheet transplantation is beginning to be explored. A first study of mouse PSC-derived retinal sheet transplantation showed encouraging results [147]. In this study, mouse PSC-derived retinal organoids were differentiated to various stages of maturity, from day 11, when the NR is populated with only retinal progenitor cells, through day 24 following the genesis of all photoreceptor precursors. The organoids were cut into smaller pieces and transplanted as retinal sheets into 6-week-old *rd1* mice exhibiting advanced retinal degeneration including complete loss of photoreceptors. The grafted tissue from all ages continued to develop, producing ONLs populated with photoreceptors elaborating outer segments [147]. The transplants exhibited several different morphologies depending on their ages, with the best graft-to-host synaptic connections formed from relatively immature 11–17 day cultures. However, the functionality of these transplants awaits future assessment.

A more recent study tested the transplantation of human PSC-derived retinal sheets in nude rats and more critically in a newly developed primate model of ONL degeneration [148]. In both species, human organoid-derived sheets, differentiated for 50–60 days and transplanted subretinally, expanded and matured forming rosette structures adjacent to the host INL. The grafted tissue was shown to express mature photoreceptor markers as well as to form structurally mature outer segments. In primates, the synapse protein RIBEYE was expressed between graft photoreceptors and both host and graft bipolar cells. In addition, host bipolar cells extended dendrites toward and contacted grafted photoreceptors. While focal ERG yielded no positive signal, this study demonstrated that transplanted hESC-derived tissue can form mature photoreceptors as well as establish synaptic interactions with a host retina, significant steps toward achieving effective transplantation in humans.

The above studies make clear that transplanting dissociated or intact sheets of PSC-derived retinal cells are plausible routes to treating photoreceptor degeneration. Indeed, a recent workshop has addressed the state of the field and identified practical paths forward [149]. Importantly, there may be useful approaches beyond dissociated precursors and retinal sheet paradigms described above. For example, given that there are a number of methodologies to produce sheets of PSC-derived RPE [101, 150], it may be possible to develop an RPE and retina co-transplant. The unrestricted abundance of human PSC-derived retinal organoids compared to human fetal retina will permit more exploration and optimization required to understand how best to restore vision.

5.2 Ganglion cell replacement

Glaucoma is characterized by a progressive loss of retinal ganglion cells due to genetic and environmental factors, the most important of which is intraocular pressure [3]. Replacement of retinal ganglion cells (RGCs) requires not only local wiring into the inner plexiform layer, but also the extension of axons into the optic nerve and to appropriate brain areas. This is clearly a more formidable challenge than the replacement of photoreceptors.

Following intravitreal transplantation of rat RGCs into adult rat eyes, about 1% of transplanted cells migrated into the nerve fiber layer and about 0.2% migrated to the retinal ganglion cell layer and survived for 7 days [96]. Survival was much higher, about 40%, when the ganglion cells were transplanted into a retinal explant and cultured in media optimized for retinal ganglion cell survival. In the latter conditions, about 65% of the surviving cells migrated into the retinal ganglion cell layer and 90% of those extended neurites, primarily into the nerve fiber layer but also into the IPL. Some axons extended over long distances and reached the optic nerve head. Cells isolated from several embryonic and postnatal ages displayed similar results, but adult retinal ganglion cells performed poorly. To direct transplanted retinal ganglion cell axons toward the optic nerve head, Kador et al. developed a scaffold engineered with a gradient of immobilized netrin, an axon guidance molecule present at the optic nerve head [151]. When retinal ganglion cells were grown on this device, the number of RGCs that exhibit polarized axon growth increased from 31 to 52%. However the field awaits further ingenuity to overcome all obstacles to functional ganglion cell replacement.

The challenge of transplanting retinal ganglion cells should benefit from the ability to grow human retinal organoid cultures, since they are a source of the large cell numbers that will be required to work out multifaceted conditions needed for success. While there are a number of reports describing direct differentiation of PSCs to retinal ganglion cells in vitro, there is a need to improve efficiency and cell purification [152] and, as argued previously, the three-dimensional culture methods are anticipated to yield more authentic cells. In day 40 retinal organoid cultures, approximately 14% of the cells are retinal ganglion cell precursors (Aparicio et al. unpublished data). These cells are of a suitable age, and, as noted above, can robustly extend axons capable of anterograde transport and generation of axon potentials [63, 66]. However, it will be critical to document their ability to respond appropriately to a variety of guidance and trophic cues and to devise methods to purify the cells. Whether the Thy1 surface marker used in rodent retinal ganglion cell purification is suitable is not clear [153]. Also uncharted territory is whether the retinal organoid ganglion cells are competent to form the panoply of ganglion cell subtypes [154].

5.3 Summary

In light of the extraordinary complexity of the human retina, cellular replacement therapy might appear to be an insurmountable challenge. Mature mammalian retinae comprise seven main cell types and ~80–90 subtypes [154]. These cells assemble into specific retinal layers with a precise distribution and form specific cell-cell interactions including but not limited to synaptic connections. In the end, all retinal cells must work together to absorb light, to integrate light signals received in diverse spatial-temporal patterns, and to transduce light position, color, movement, shapes, border qualities, and presumably other visual information to a retinotopic map in the lateral geniculate, which is then relayed to the visual cortex [155]. However, despite this complexity, the PSC-derived retina studies presented above demonstrate that PSC-derived differentiating tissues have a remarkable ability to organize and self-pattern. There is as yet no reason to expect that these cells or tissues lack the abilities to differentiate and make functional connections to neighboring neurons upon transplantation into an appropriate environment. Nevertheless, consideration of the diseased retina environment may be critical for successful transplantation since gliosis, cytoarchitectural remodeling of inner retinal neurons, and neural retinal thinning at late stages may impede the ability of transplanted cells or tissue to restore the precise neural circuitry that underlies the visual signals exiting the retina [110, 156]. Thus, maximal functional restoration may also require pre- or post-transplantation retinal remodeling.

6 Future Directions

Use of human PSC-derived retinal organoids to generate cells and tissue for therapeutic purposes is clearly on the horizon after surmounting obstacles like differentiation efficiency, reproducibility, validation, and GMP production. However, the future of retinal organoid culture is impossible to predict, and yet extensive in imaginable possibilities. Today's rudimentary retinal organoids may one day be viewed as just a humble beginning.

Retinal organoids are a new and rapidly evolving technology. Coupling organoid usage with other technologies amplifies their potential usefulness in clinical as well as in basic and translational research realms. For example, CRISPR-Cas9 genome editing technology can be used to engineer PSC lines almost at will to create retinal organoids with any desired genetic alteration [157–160]. As a system amenable to genome editing, retinal organoids are perfectly poised to take advantage of the increasingly sophisticated toolbox being developed by optogeneticists, synthetic biologists, or even sonogeneticists to engineer new circuits and functions with externally controllable switches in order to rationally control cellular behavior [113, 161–165].

For clinical purposes, the genetic manipulability of retinal organoids may help overcome obstacles to successful therapy. Achieving functional integration of photoreceptor precursors in sufficient quantities to restore vision in a plethora of disease settings is a challenge. Transplantation of cells engineered to secrete neurotrophic factors is already underway [166]. Might PSC-derived photoreceptors engineered to secrete such factors facilitate survival, or might lines designed to modify their environment facilitate integration?

One of the most critical challenges to broadening the use of retinal organoids is to improve their portrayal of later retinal development stages. New and powerful abilities to define cell type-specific transcriptomes and developmental signaling pathways may facilitate understanding retinal organoid biology and the underlying cell signaling differences from the developing fetal retina. Further, genome editing may enable these differences to be corrected, such as by timed induction of key regulatory genes. However, while such an approach might suffice to modulate selected pathways, a more holistic approach may be to reproduce the microenvironment of the in vivo maturing retina so that the cells respond in kind. For example, the function of in vitro-derived neurons is dramatically improved when media mimics the physiological concentration of ions, neuroactive amino acids, and energetic substrates of the brain [167].

The verisimilitude of in vivo retinal development may be further improved by developing cultures that include other eye structures. Mellough and colleagues have already made progress toward growing an eye from PSCs by producing rudimentary lens and cornea along with NR and RPE [39]. Perhaps a more feasible approach will be to coculture different tissues initially produced separately. Coculture of RPE and NR tissue will very likely improve photoreceptor maturation in vitro. Increasingly, different stem cell-derived tissues and cell types are being combined to build tissues and organs. "Organs on chips" already exist for such tissues as kidney, lung, and gut. Here different cell types, along with factors such as extracellular matrix, are combined in a microfluidics chamber that reproduces organ fluid flow and mechanical dynamics to yield a culture that recapitulates many organ functions [168]. Will it one day be possible to combine cells or tissues of appropriate developmental stages to coordinate their maturation and development into a functional eye?

Improved authenticity of retinal organoids and more complex in vitro organ cultures could provide new opportunities to study previously inaccessible retinal development mechanisms, such as those underlying neuronal connectivity or human fovea development. If more mature retinal development is achieved, retinal organoids will undoubtedly offer a way to better understand pathologic processes, ranging from retinitis pigmentosa to gliosis, thereby facilitating design of methods to manipulate connectivity, reverse gliotic changes, and restore functional architecture. Perhaps a future version of a retinal organoid will be useful to better understand the subcellular effects of retinal prostheses or to determine optimal stimulation patterns of these devices [5]. Engineering a route for retinal ganglion cell axons to exit the organoid and extend toward and innervate appropriate synaptic targets (maybe in other neural organoids) would permit analysis of neurite growth, guidance, synapse formation, and create an in vitro model to study axonal regeneration. hiPSC-derived retinal organoids may also be used to diagnose an early disease phenotype, to predict disease severity, or to optimize treatment for slowly progressing diseases.

Clearly much remains to be done before retinal organoid cultures can be used in diagnostic, prognostic, or therapeutic settings. Yet the ultimate benefit of such efforts—restoring vision to the blind—is so alluring that there will be no shortage of interest in this endeavor. Proceeding with an appropriate mix of creativity, ingenuity, ambition, caution, and rigor, the future will indeed be brighter for millions affected by blinding diseases.

Readers are encouraged to seek out the following very relevant manuscripts published after chapter completion but prior to book publication. They are listed after the chapter section to which they are most related. 3: PMID:27132890, PMID: 27471043, PMID:28220575; 5.1.1: PMID:27901042, PMID:27701381, PMID:27701378; 5.1.2: PMID:28076757; 5.2: PMID:26843334.

Acknowledgments The authors acknowledge the generous support of the A.B. Reins Foundation, the Neonatal Blindness Research Fund, the Larry and Celia Moh Foundation, Research to Prevent Blindness, NIH National Eye Institute grant R21EY025419 (JGA & DC), and NIH grant T32HD060549 to the USC Keck School of Medicine Development, Stem Cells, and Regenerative Medicine Program (DWHS) during the preparation of this chapter.

References

1. Zarbin M. Cell-based therapy for degenerative retinal disease. Trends Mol Med. 2016;22(2):115–34.
2. Huang Y, Enzmann V, Ildstad ST. Stem cell-based therapeutic applications in retinal degenerative diseases. Stem Cell Rev. 2011;7(2):434–45.
3. Weinreb RN, Aung T, Medeiros FA. The pathophysiology and treatment of glaucoma: a review. JAMA. 2014;311(18): 1901–11.
4. Jayakody SA, Gonzalez-Cordero A, Ali RR, Pearson RA. Cellular strategies for retinal repair by photoreceptor replacement. Prog Retin Eye Res. 2015;46:31–66.
5. Fine I, Cepko CL, Landy MS. Vision research special issue: sight restoration: prosthetics, optogenetics and gene therapy. Vis Res. 2015;111(Pt B):115–23.
6. Mellough CB, Collin J, Sernagor E, Wride NK, Steel DHW, Lako M. Lab generated retina: realizing the dream. Vis Neurosci. 2014;31(4–5):317–32.
7. Thomson JA, Itskovitz-Eldor J, Shapiro SS, Waknitz MA, Swiergiel JJ, Marshall VS, et al. Embryonic stem cell lines derived from human blastocysts. Science. 1998;282(5391):1145–7.
8. Takahashi K, Tanabe K, Ohnuki M, Narita M, Ichisaka T, Tomoda K, et al. Induction of pluripotent stem cells from adult human fibroblasts by defined factors. Cell. 2007;131(5):861–72.
9. Eiraku M, Takata N, Ishibashi H, Kawada M, Sakakura E, Okuda S, et al. Self-organizing optic-cup morphogenesis in three-dimensional culture. Nature. 2011;472(7341):51–6.
10. Haston KM, Finkbeiner S. Clinical trials in a dish: the potential of pluripotent stem cells to develop therapies for neurodegenerative diseases. Annu Rev Pharmacol Toxicol. 2016;56:489–510.
11. Chiba K, Hockemeyer D. Genome editing in human pluripotent stem cells using site-specific nucleases. Methods Mol Biol. 2015;1239:267–80.
12. Sterneckert JL, Reinhardt P, Schöler HR. Investigating human disease using stem cell models. Nat Rev Genet. 2014;15(9):625–39.
13. Fox IJ, Daley GQ, Goldman SA, Huard J, Kamp TJ, Trucco M. Stem cell therapy. Use of differentiated pluripotent stem cells as replacement therapy for treating disease. Science. 2014;345(6199):1247391.
14. Zhao T, Zhang Z-N, Westenskow PD, Todorova D, Hu Z, Lin T, et al. Humanized mice reveal differential immunogenicity of cells derived from autologous induced pluripotent stem cells. Cell Stem Cell. 2015;17(3):353–9.
15. Meyer JS, Shearer RL, Capowski EE, Wright LS, Wallace KA, McMillan EL, et al. Modeling early retinal development with human embryonic and induced pluripotent stem cells. Proc Natl Acad Sci. 2009;106(39):16698–703.
16. Phillips MJ, Wallace KA, Dickerson SJ, Miller MJ, Verhoeven AD, Martin JM, et al. Blood-derived human iPS cells generate optic vesicle-like structures with the capacity to form retinal laminae and develop synapses. Invest Ophthalmol Vis Sci. 2012;53(4): 2007–19.
17. Nakano T, Ando S, Takata N, Kawada M, Muguruma K, Sekiguchi K, et al. Self-formation of optic cups and storable stratified neural retina from human ESCs. Cell Stem Cell. 2012;10(6):771–85.
18. Zhong X, Gutierrez C, Xue T, Hampton C, Vergara MN, Cao L-H, et al. Generation of three-dimensional retinal tissue with functional photoreceptors from human iPSCs. Nat Commun. 2014; 5:4047.
19. Kuwahara A, Ozone C, Nakano T, Saito K, Eiraku M, Sasai Y. Generation of a ciliary margin-like stem cell niche from self-organizing human retinal tissue. Nat Commun. 2015;6:6286.
20. Vandamme TF. Use of rodents as models of human diseases. J Pharm Bioallied Sci. 2014;6(1):2–9.
21. Shanks N, Greek R, Greek J. Are animal models predictive for humans? Philos Ethics Humanit Med. 2009;4:2.
22. van der Worp HB, Howells DW, Sena ES, Porritt MJ, Rewell S, O'Collins V, et al. Can animal models of disease reliably inform human studies? PLoS Med. 2010;7(3):e1000245.
23. Xu XL, Fang Y, Lee TC, Forrest D, Gregory-Evans C, Almeida D, et al. Retinoblastoma has properties of a cone precursor tumor and depends upon cone-specific MDM2 signaling. Cell. 2009;137(6): 1018–31.
24. Avior Y, Sagi I, Benvenisty N. Pluripotent stem cells in disease modelling and drug discovery. Nat Rev Mol Cell Biol. 2016;17(3): 170–82.
25. Yu J, Vodyanik MA, Smuga-Otto K, Antosiewicz-Bourget J, Frane JL, Tian S, et al. Induced pluripotent stem cell lines derived from human somatic cells. Science. 2007;318(5858):1917–20.
26. Eiraku M, Sasai Y. Self-formation of layered neural structures in three-dimensional culture of ES cells. Curr Opin Neurobiol. 2012;22(5):768–77.
27. Fuhrmann S. Eye morphogenesis and patterning of the optic vesicle. Curr Top Dev Biol. 2010;93:61–84.
28. Wilson SW, Wilson SW, Houart C, Houart C. Early steps in the development of the forebrain. Dev Cell. 2004;6(2):167–81.
29. Fuhrmann S. Wnt signaling in eye organogenesis. Organogenesis. 2008;4(2):60–7.
30. Graw J. Eye development. Curr Top Dev Biol. 2010;90:343–86.
31. Zuber ME. Eye field specification in *Xenopus laevis*. Curr Top Dev Biol. 2010;93:29–60.
32. Lamb TD, Collin SP, Pugh EN. Evolution of the vertebrate eye: opsins, photoreceptors, retina and eye cup. Nat Rev Neurosci. 2007;8(12):960–76.
33. Nasonkin IO, Merbs SL, Lazo K, Oliver VF, Brooks M, Patel K, et al. Conditional knockdown of DNA methyltransferase 1 reveals a key role of retinal pigment epithelium integrity in photoreceptor outer segment morphogenesis. Development. 2013;140(6): 1330–41.
34. Strauss O. The retinal pigment epithelium in visual function. Physiol Rev. 2005;85(3):845–81.
35. Norden C, Young S, Link BA, Harris WA. Actomyosin is the main driver of interkinetic nuclear migration in the retina. Cell. 2009;138(6):1195–208.
36. Del Bene F, Wehman AM, Link BA, Baier H. Regulation of neurogenesis by interkinetic nuclear migration through an apical-basal notch gradient. Cell. 2008;134(6):1055–65.
37. Strzyz PJ, Lee HO, Sidhaye J, Weber IP, Leung LC, Norden C. Interkinetic nuclear migration is centrosome independent and ensures apical cell division to maintain tissue integrity. Dev Cell. 2015;32(2):203–19.

38. Meyer JS, Howden SE, Wallace KA, Verhoeven AD, Wright LS, Capowski EE, et al. Optic vesicle-like structures derived from human pluripotent stem cells facilitate a customized approach to retinal disease treatment. Stem Cells. 2011;29(8):1206–18.
39. Mellough CB, Collin J, Khazim M, White K, Sernagor E, Steel DHW, et al. IGF-1 signaling plays an important role in the formation of three-dimensional laminated neural retina and other ocular structures from human embryonic stem cells. Stem Cells. 2015; 33(8):2416–30.
40. Reichman S, Terray A, Slembrouck A, Nanteau C, Orieux G, Habeler W, et al. From confluent human iPS cells to self-forming neural retina and retinal pigmented epithelium. Proc Natl Acad Sci. 2014;111(23):8518–23.
41. Chambers SM, Fasano CA, Papapetrou EP, Tomishima M, Sadelain M, Studer L. Highly efficient neural conversion of human ES and iPS cells by dual inhibition of SMAD signaling. Nat Biotechnol. 2009;27(3):275–80.
42. Klimanskaya I, Rosenthal N, Lanza R. Derive and conquer: sourcing and differentiating stem cells for therapeutic applications. Nat Rev Drug Discov. 2008;7(2):131–42.
43. Lee KJ, Jessell TM. The specification of dorsal cell fates in the vertebrate central nervous system. Annu Rev Neurosci. 1999;22: 261–94.
44. Levine AJ, Brivanlou AH. Proposal of a model of mammalian neural induction. Dev Biol. 2007;308(2):247–56.
45. Fuhrmann S, Levine EM, Reh TA. Extraocular mesenchyme patterns the optic vesicle during early eye development in the embryonic chick. Development. 2000;127(21):4599–609.
46. Raymond SM, Jackson IJ. The retinal pigmented epithelium is required for development and maintenance of the mouse neural retina. Curr Biol. 1995;5(11):1286–95.
47. Pinzón-Duarte G, Kohler K, Arango-González B, Guenther E. Cell differentiation, synaptogenesis, and influence of the retinal pigment epithelium in a rat neonatal organotypic retina culture. Vis Res. 2000;40(25):3455–65.
48. Bumsted KM, Rizzolo LJ, Barnstable CJ. Defects in the MITF(mi/mi) apical surface are associated with a failure of outer segment elongation. Exp Eye Res. 2001;73(3):383–92.
49. Jablonski MM, Tombran-Tink J, Mrazek DA, Iannaccone A. Pigment epithelium-derived factor supports normal development of photoreceptor neurons and opsin expression after retinal pigment epithelium removal. J Neurosci. 2000;20(19):7149–57.
50. Sheedlo HJ, Nelson TH, Lin N, Rogers TA, Roque RS, Turner JE. RPE secreted proteins and antibody influence photoreceptor cell survival and maturation. Brain Res Dev Brain Res. 1998;107(1): 57–69.
51. Fujimura N, Taketo MM, Mori M, Korinek V, Kozmik Z. Spatial and temporal regulation of Wnt/beta-catenin signaling is essential for development of the retinal pigment epithelium. Dev Biol. 2009;334(1):31–45.
52. Liu H, Mohamed O, Dufort D, Wallace VA. Characterization of Wnt signaling components and activation of the Wnt canonical pathway in the murine retina. Dev Dyn. 2003;227(3):323–34.
53. Westenskow P, Piccolo S, Fuhrmann S. Beta-catenin controls differentiation of the retinal pigment epithelium in the mouse optic cup by regulating Mitf and Otx2 expression. Development. 2009;136(15):2505–10.
54. Eiraku M, Sasai Y. Mouse embryonic stem cell culture for generation of three-dimensional retinal and cortical tissues. Nat Protoc. 2011;7(1):69–79.
55. Hughes CS, Postovit LM, Lajoie GA. Matrigel: a complex protein mixture required for optimal growth of cell culture. Proteomics. 2010;10(9):1886–90.
56. Hoch RV, Rubenstein JLR, Pleasure S. Genes and signaling events that establish regional patterning of the mammalian forebrain. Semin Cell Dev Biol. 2009;20(4):378–86.
57. Furuta Y, Piston DW, Hogan BL. Bone morphogenetic proteins (BMPs) as regulators of dorsal forebrain development. Development. 1997;124(11):2203–12.
58. Pandit T, Jidigam VK, Patthey C, Gunhaga L. Neural retina identity is specified by lens-derived BMP signals. Development. 2015;142(10):1850–9.
59. Bielen H, Houart C. BMP signaling protects telencephalic fate by repressing eye identity and its Cxcr4-dependent morphogenesis. Dev Cell. 2012;23(4):812–22.
60. Horsford DJ, Nguyen M-TT, Sellar GC, Kothary R, Arnheiter H, McInnes RR. Chx10 repression of Mitf is required for the maintenance of mammalian neuroretinal identity. Development. 2005; 132(1):177–87.
61. Park CM, Hollenberg MJ. Basic fibroblast growth factor induces retinal regeneration in vivo. Dev Biol. 1989;134(1):201–5.
62. Martinez-Morales JR, Del Bene F, Nica G, Hammerschmidt M, Bovolenta P, Wittbrodt J. Differentiation of the vertebrate retina is coordinated by an FGF signaling center. Dev Cell. 2005;8(4): 565–74.
63. Tanaka T, Yokoi T, Tamalu F, Watanabe S-I, Nishina S, Azuma N. Generation of retinal ganglion cells with functional axons from human induced pluripotent stem cells. Sci Rep. 2015;5:8344.
64. Cohen-Cory S, Escandón E, Fraser SE. The cellular patterns of BDNF and trkB expression suggest multiple roles for BDNF during Xenopus visual system development. Dev Biol. 1996;179(1): 102–15.
65. Parrilla-Reverter G, Agudo M, Sobrado-Calvo P, Salinas-Navarro M, Villegas-PErez MP, Vidal-Sanz M. Effects of different neurotrophic factors on the survival of retinal ganglion cells after a complete intraorbital nerve crush injury: a quantitative in vivo study. Exp Eye Res. 2009;89(1):32–41.
66. Maekawa Y, Onishi A, Matsushita K, Koide N, Mandai M, Suzuma K, et al. Optimized culture system to induce neurite outgrowth from retinal ganglion cells in three-dimensional retinal aggregates differentiated from mouse and human embryonic stem cells. Curr Eye Res. 2016;41(4):558–68.
67. Xu C-R, Li L-C, Donahue G, Ying L, Zhang Y-W, Gadue P, et al. Dynamics of genomic H3K27me3 domains and role of EZH2 during pancreatic endocrine specification. EMBO J. 2014;33(19): 2157–70.
68. Xie R, Everett LJ, Lim H-W, Patel NA, Schug J, Kroon E, et al. Dynamic chromatin remodeling mediated by polycomb proteins orchestrates pancreatic differentiation of human embryonic stem cells. Cell Stem Cell. 2013;12(2):224–37.
69. Breslin S, O'Driscoll L. Three-dimensional cell culture: the missing link in drug discovery. Drug Discov Today. 2013;18(5–6):240–9.
70. Schmeichel KL, Bissell MJ. Modeling tissue-specific signaling and organ function in three dimensions. J Cell Sci. 2003;116(Pt 12):2377–88.
71. Mammoto A, Ingber DE. Cytoskeletal control of growth and cell fate switching. Curr Opin Cell Biol. 2009;21(6):864–70.
72. Baker BM, Chen CS. Deconstructing the third dimension: how 3D culture microenvironments alter cellular cues. J Cell Sci. 2012;125(Pt 13):3015–24.
73. Lancaster MA, Knoblich JA. Organogenesis in a dish: modeling development and disease using organoid technologies. Science. 2014;345(6194):1247125.
74. Hynds RE, Giangreco A. Concise review: the relevance of human stem cell-derived organoid models for epithelial translational medicine. Stem Cells. 2013;31(3):417–22.
75. Tucker BA, Park I-H, Qi SD, Klassen HJ, Jiang C, Yao J, et al. Transplantation of adult mouse iPS cell-derived photoreceptor precursors restores retinal structure and function in degenerative mice. PLoS One. 2011;6(4):e18992.
76. Paull D, Sevilla A, Zhou H, Hahn AK, Kim H, Napolitano C, et al. Automated, high-throughput derivation, characterization and dif-

ferentiation of induced pluripotent stem cells. Nat Methods. 2015;12(9):885–92.
77. Kaewkhaw R, Kaya KD, Brooks M, Homma K, Zou J, Chaitankar V, et al. Transcriptome dynamics of developing photoreceptors in three-dimensional retina cultures recapitulates temporal sequence of human cone and rod differentiation revealing cell surface markers and gene networks. Stem Cells. 2015;33(12):3504–18.
78. Kim K, Doi A, Wen B, Ng K, Zhao R, Cahan P, et al. Epigenetic memory in induced pluripotent stem cells. Nature. 2010;467(7313):285–90.
79. Ohi Y, Qin H, Hong C, Blouin L, Polo JM, Guo T, et al. Incomplete DNA methylation underlies a transcriptional memory of somatic cells in human iPS cells. Nat Cell Biol. 2011;13(5):541–9.
80. Hiler D, Chen X, Hazen J, Kupriyanov S, Carroll PA, Qu C, et al. Quantification of retinogenesis in 3D cultures reveals epigenetic memory and higher efficiency in iPSCs derived from rod photoreceptors. Cell Stem Cell. 2015;17(1):101–15.
81. Polo JM, Liu S, Figueroa ME, Kulalert W, Eminli S, Tan KY, et al. Cell type of origin influences the molecular and functional properties of mouse induced pluripotent stem cells. Nat Biotechnol. 2010;28(8):848–55.
82. Wutz A. Epigenetic alterations in human pluripotent stem cells: a tale of two cultures. Cell Stem Cell. 2012;11(1):9–15.
83. Kim K, Zhao R, Doi A, Ng K, Unternaehrer J, Cahan P, et al. Donor cell type can influence the epigenome and differentiation potential of human induced pluripotent stem cells. Nat Biotechnol. 2011;29(12):1117–9.
84. Bar-Nur O, Russ HA, Efrat S, Benvenisty N. Epigenetic memory and preferential lineage-specific differentiation in induced pluripotent stem cells derived from human pancreatic islet beta cells. Cell Stem Cell. 2011;9(1):17–23.
85. Buganim Y, Markoulaki S, van Wietmarschen N, Hoke H, Wu T, Ganz K, et al. The developmental potential of iPSCs is greatly influenced by reprogramming factor selection. Cell Stem Cell. 2014;15(3):295–309.
86. Wu T, Liu Y, Wen D, Tseng Z, Tahmasian M, Zhong M, et al. Histone variant H2A.X deposition pattern serves as a functional epigenetic mark for distinguishing the developmental potentials of iPSCs. Cell Stem Cell. 2014;15(3):281–94.
87. Gafni O, Weinberger L, Mansour AA, Manor YS, Chomsky E, Ben-Yosef D, et al. Derivation of novel human ground state naive pluripotent stem cells. Nature. 2013;504(7479):282–6.
88. Fonseca SAS, Costas RM, Pereira LV. Searching for naïve human pluripotent stem cells. World J Stem Cells. 2015;7(3):649–56.
89. Halfter W, Dong S, Schurer B, Ring C, Cole GJ, Eller A. Embryonic synthesis of the inner limiting membrane and vitreous body. Invest Ophthalmol Vis Sci. 2005;46(6):2202–9.
90. Halfter W, Dong S, Dong A, Eller AW, Nischt R. Origin and turnover of ECM proteins from the inner limiting membrane and vitreous body. Eye. 2008;22(10):1207–13.
91. Tao C, Zhang X. Development of astrocytes in the vertebrate eye. Dev Dyn. 2014;243(12):1501–10.
92. Hughes S, Yang H, Chan-Ling T. Vascularization of the human fetal retina: roles of vasculogenesis and angiogenesis. Invest Ophthalmol Vis Sci. 2000;41(5):1217–28.
93. Bähr M. Live or let die - retinal ganglion cell death and survival during development and in the lesioned adult CNS. Trends Neurosci. 2000;23(10):483–90.
94. MacLaren RE, Pearson RA, MacNeil A, Douglas RH, Salt TE, Akimoto M, et al. Retinal repair by transplantation of photoreceptor precursors. Nature. 2006;444(7116):203–7.
95. Seiler MJ, Aramant RB. Cell replacement and visual restoration by retinal sheet transplants. Prog Retin Eye Res. 2012;31(6):661–87.
96. Hertz J, Qu B, Hu Y, Patel RD, Valenzuela DA, Goldberg JL. Survival and integration of developing and progenitor-derived retinal ganglion cells following transplantation. Cell Transplant. 2014;23(7):855–72.
97. Andrabi M, Kuraku S, Takata N, Sasai Y, Love NR. Comparative, transcriptome analysis of self-organizing optic tissues. Sci Data. 2015;2:150030.
98. La Torre A, Hoshino A, Cavanaugh C, Ware CB, Reh TA. The GIPC1-Akt1 pathway is required for the specification of the eye field in mouse embryonic stem cells. Stem Cells. 2015;33(9):2674–85.
99. Abbott RD, Kaplan DL. Strategies for improving the physiological relevance of human engineered tissues. Trends Biotechnol. 2015;33(7):401–7.
100. Stringari C, Cinquin A, Cinquin O, Digman MA, Donovan PJ, Gratton E. Phasor approach to fluorescence lifetime microscopy distinguishes different metabolic states of germ cells in a live tissue. Proc Natl Acad Sci. 2011;108(33):13582–7.
101. Ramsden CM, Powner MB, Carr A-JF, Smart MJK, da Cruz L, Coffey PJ. Stem cells in retinal regeneration: past, present and future. Development. 2013;140(12):2576–85.
102. Stern JH, Temple S. Stem cells for retinal replacement therapy. Neurotherapeutics. 2011;8(4):736–43.
103. Tucker BA, Mullins RF, Stone EM. Stem cells for investigation and treatment of inherited retinal disease. Hum Mol Genet. 2014;23(R1):R9–R16.
104. Garcia JM, Mendonça L, Brant R, Abud M, Regatieri C, Diniz B. Stem cell therapy for retinal diseases. World J Stem Cells. 2015;7(1):160–4.
105. Al-Shamekh S, Goldberg JL. Retinal repair with induced pluripotent stem cells. Transl Res. 2014;163(4):377–86.
106. Sluch VM, Zack DJ. Stem cells, retinal ganglion cells and glaucoma. Dev Ophthalmol. 2014;53:111–21.
107. Cooke JA, Meyer JS. Human pluripotent stem cell-derived retinal ganglion cells: applications for the study and treatment of optic neuropathies. Curr Ophthalmol Rep. 2015;3(3):200–6.
108. Veleri S, Lazar CH, Chang B, Sieving PA, Banin E, Swaroop A. Biology and therapy of inherited retinal degenerative disease: insights from mouse models. Dis Model Mech. 2015;8(2):109–29.
109. Ambati J, Fowler BJ. Mechanisms of age-related macular degeneration. Neuron. 2012;75(1):26–39.
110. Pearson RA, Hippert C, Graca AB, Barber AC. Photoreceptor replacement therapy: challenges presented by the diseased recipient retinal environment. Vis Neurosci. 2014;31(4–5):333–44.
111. Lakowski J, Baron M, Bainbridge J, Barber AC, Pearson RA, Ali RR, et al. Cone and rod photoreceptor transplantation in models of the childhood retinopathy Leber congenital amaurosis using flow-sorted Crx-positive donor cells. Hum Mol Genet. 2010;19(23):4545–59.
112. Santos-Ferreira T, Postel K, Stutzki H, Kurth T, Zeck G, Ader M. Daylight vision repair by cell transplantation. Stem Cells. 2015;33(1):79–90.
113. Pearson RA, Barber AC, Rizzi M, Hippert C, Xue T, West EL, et al. Restoration of vision after transplantation of photoreceptors. Nature. 2012;485(7396):99–103.
114. Singh MS, Charbel Issa P, Butler R, Martin C, Lipinski DM, Sekaran S, et al. Reversal of end-stage retinal degeneration and restoration of visual function by photoreceptor transplantation. Proc Natl Acad Sci. 2013;110(3):1101–6.
115. Barber AC, Hippert C, Duran Y, West EL, Bainbridge JWB, Warre-Cornish K, et al. Repair of the degenerate retina by photoreceptor transplantation. Proc Natl Acad Sci. 2013;110(1):354–9.
116. Homma K, Okamoto S, Mandai M, Gotoh N, Rajasimha HK, Chang Y-S, et al. Developing rods transplanted into the degenerating retina of Crx-knockout mice exhibit neural activity similar to native photoreceptors. Stem Cells. 2013;31(6):1149–59.

117. West EL, Pearson RA, Tschernutter M, Sowden JC, MacLaren RE, Ali RR. Pharmacological disruption of the outer limiting membrane leads to increased retinal integration of transplanted photoreceptor precursors. Exp Eye Res. 2008;86(4):601–11.
118. West EL, Pearson RA, Barker SE, Luhmann UFO, MacLaren RE, Barber AC, et al. Long-term survival of photoreceptors transplanted into the adult murine neural retina requires immune modulation. Stem Cells. 2010;28(11):1997–2007.
119. West EL, Gonzalez-Cordero A, Hippert C, Osakada F, Martinez-Barbera JP, Pearson RA, et al. Defining the integration capacity of embryonic stem cell-derived photoreceptor precursors. Stem Cells. 2012;30(7):1424–35.
120. Pearson RA, Barber AC, West EL, MacLaren RE, Duran Y, Bainbridge JW, et al. Targeted disruption of outer limiting membrane junctional proteins (Crb1 and ZO-1) increases integration of transplanted photoreceptor precursors into the adult wild-type and degenerating retina. Cell Transplant. 2010;19(4):487–503.
121. Pearson RA. Advances in repairing the degenerate retina by rod photoreceptor transplantation. Biotechnol Adv. 2014;32(2):485–91.
122. Warre-Cornish K, Barber AC, Sowden JC, Ali RR, Pearson RA. Migration, integration and maturation of photoreceptor precursors following transplantation in the mouse retina. Stem Cells Dev. 2014;23(9):941–54.
123. Gonzalez-Cordero A, West EL, Pearson RA, Duran Y, Carvalho LS, Chu CJ, et al. Photoreceptor precursors derived from three-dimensional embryonic stem cell cultures integrate and mature within adult degenerate retina. Nat Biotechnol. 2013;31(8):741–7.
124. Lamba DA, Gust J, Reh TA. Transplantation of human embryonic stem cell-derived photoreceptors restores some visual function in Crx-deficient mice. Cell Stem Cell. 2009;4(1):73–9.
125. Lamba DA, McUsic A, Hirata RK, Wang P-R, Russell D, Reh TA. Generation, purification and transplantation of photoreceptors derived from human induced pluripotent stem cells. PLoS One. 2010;5(1):e8763.
126. Woch G, Aramant RB, Seiler MJ, Sagdullaev BT, McCall MA. Retinal transplants restore visually evoked responses in rats with photoreceptor degeneration. Invest Ophthalmol Vis Sci. 2001;42(7):1669–76.
127. Sagdullaev BT, Aramant RB, Seiler MJ, Seiler MJ, et al. Retinal transplantation-induced recovery of retinotectal visual function in a rodent model of retinitis pigmentosa. Invest Ophthalmol Vis Sci. 2003;44(4):1686–95.
128. Thomas BB, Seiler MJ, Sadda SR, Coffey PJ, Aramant RB. Optokinetic test to evaluate visual acuity of each eye independently. J Neurosci Methods. 2004;138(1–2):7–13.
129. Seiler MJ, Thomas BB, Chen Z, Arai S, Chadalavada S, Mahoney MJ, et al. BDNF-treated retinal progenitor sheets transplanted to degenerate rats: improved restoration of visual function. Exp Eye Res. 2008;86(1):92–104.
130. Seiler MJ, Sagdullaev BT, Woch G, Thomas BB, Aramant RB. Transsynaptic virus tracing from host brain to subretinal transplants. Eur J Neurosci. 2005;21(1):161–72.
131. Seiler MJ, Rao B, Aramant RB, Yu L, Wang Q, Kitayama E, et al. Three-dimensional optical coherence tomography imaging of retinal sheet implants in live rats. J Neurosci Methods. 2010;188(2):250–7.
132. Seiler MJ, Aramant RB. Intact sheets of fetal retina transplanted to restore damaged rat retinas. Invest Ophthalmol Vis Sci. 1998;39(11):2121–31.
133. Aramant RB, Seiler MJ, Ball SL. Successful cotransplantation of intact sheets of fetal retina with retinal pigment epithelium. Invest Ophthalmol Vis Sci. 1999;40(7):1557–64.
134. Aramant RB, Seiler MJ. Fiber and synaptic connections between embryonic retinal transplants and host retina. Exp Neurol. 1995;133:1–12.
135. Seiler MJ, Aramant RB. Transplantation of neuroblastic progenitor cells as a sheet preserves and restores retinal function. Semin Ophthalmol. 2005;20(1):31–42.
136. Seiler MJ, Aramant RB, Thomas BB, Peng Q, Sadda SR, Keirstead HS. Visual restoration and transplant connectivity in degenerate rats implanted with retinal progenitor sheets. Eur J Neurosci. 2010;31(3):508–20.
137. Seiler MJ, Jones BW, Aramant RB, Yang PB, Keirstead HS, Marc RE. Computational molecular phenotyping of retinal sheet transplants to rats with retinal degeneration. Eur J Neurosci. 2012;35(11):1692–704.
138. Aramant RB, Seiler MJ. Transplanted sheets of human retina and retinal pigment epithelium develop normally in nude rats. Exp Eye Res. 2002;75(2):115–25.
139. Peng Q, Thomas BB, Aramant RB, Chen Z, Sadda SR, Seiler MJ. Structure and function of embryonic rat retinal sheet transplants. Curr Eye Res. 2007;32(9):781–9.
140. Yang PB, Seiler MJ, Aramant RB, Yan F, Mahoney MJ, Kitzes LM, et al. Trophic factors GDNF and BDNF improve function of retinal sheet transplants. Exp Eye Res. 2010;91(5):727–38.
141. Seiler MJ, Aramant RB, Ball SL. Photoreceptor function of retinal transplants implicated by light-dark shift of S-antigen and rod transducin. Vis Res. 1999;39(15):2589–96.
142. Seiler MJ, Thomas BB, Chen Z, Wu R, Sadda SR, Aramant RB. Retinal transplants restore visual responses: trans-synaptic tracing from visually responsive sites labels transplant neurons. Eur J Neurosci. 2008;28(1):208–20.
143. Thomas BB, Seiler MJ, Sadda SR, Aramant RB. Superior colliculus responses to light - preserved by transplantation in a slow degeneration rat model. Exp Eye Res. 2004;79(1):29–39.
144. Radtke ND, Aramant RB, Petry HM, Green PT, Pidwell DJ, Seiler MJ. Vision improvement in retinal degeneration patients by implantation of retina together with retinal pigment epithelium. Am J Ophthalmol. 2008;146(2):172–82.
145. Hendrickson A, Bumsted-O'Brien K, Natoli R, Ramamurthy V, Possin D, Provis J. Rod photoreceptor differentiation in fetal and infant human retina. Exp Eye Res. 2008;87(5):415–26.
146. O'Brien KMB, Schulte D, Hendrickson AE. Expression of photoreceptor-associated molecules during human fetal eye development. Mol Vis. 2003;9:401–9.
147. Assawachananont J, Mandai M, Okamoto S, Yamada C, Eiraku M, Yonemura S, et al. Transplantation of embryonic and induced pluripotent stem cell-derived 3D retinal sheets into retinal degenerative mice. Stem Cell Reports. 2014;2(5):662–74.
148. Shirai H, Mandai M, Matsushita K, Kuwahara A, Yonemura S, Nakano T, et al. Transplantation of human embryonic stem cell-derived retinal tissue in two primate models of retinal degeneration. Proc Natl Acad Sci. 2016;113(1):E81–90.
149. Gamm DM, Wong R, and the AGI Workshop Panelists. Report on the National Eye Institute Audacious Goals Initiative: Photoreceptor Regeneration and Integration Workshop. Transl Vis Sci Technol. 2015;4(6):2.
150. Nazari H, Zhang L, Zhu D, Chader GJ, Falabella P, Stefanini F, et al. Stem cell based therapies for age-related macular degeneration: the promises and the challenges. Prog Retin Eye Res. 2015;48:1–39.
151. Kador KE, Alsehli HS, Zindell AN, Lau LW, Andreopoulos FM, Watson BD, et al. Retinal ganglion cell polarization using immobilized guidance cues on a tissue-engineered scaffold. Acta Biomater. 2014;10(12):4939–46.
152. Gill KP, Hewitt AW, Davidson KC, Pébay A, Wong RCB. Methods of retinal ganglion cell differentiation from pluripotent stem cells. Transl Vis Sci Technol. 2014;3(4):7.
153. Barres BA, Silverstein BE, Corey DP, Chun LL. Immunological, morphological, and electrophysiological variation among retinal ganglion cells purified by panning. Neuron. 1988;1(9):791–803.

154. Sanes JR, Masland RH. The types of retinal ganglion cells: current status and implications for neuronal classification. Annu Rev Neurosci. 2015;38:221–46.
155. Masland RH. The fundamental plan of the retina. Nat Neurosci. 2001;4(9):877–86.
156. Jones BW, Marc RE. Retinal remodeling during retinal degeneration. Exp Eye Res. 2005;81(2):123–37.
157. Smith C, Abalde-Atristain L, He C, Brodsky BR, Braunstein EM, Chaudhari P, et al. Efficient and allele-specific genome editing of disease loci in human iPSCs. Mol Ther. 2015;23(3):570–7.
158. Zhu Z, González F, Huangfu D. The iCRISPR platform for rapid genome editing in human pluripotent stem cells. Methods Enzymol. 2014;546:215–50.
159. Blasco RB, Karaca E, Ambrogio C, Cheong T-C, Karayol E, Minero VG, et al. Simple and rapid in vivo generation of chromosomal rearrangements using CRISPR/Cas9 technology. Cell Rep. 2014;9(4):1219–27.
160. González F, Zhu Z, Shi Z-D, Lelli K, Verma N, Li QV, et al. An iCRISPR platform for rapid, multiplexable, and inducible genome editing in human pluripotent stem cells. Cell Stem Cell. 2014;15(2):215–26.
161. Cameron DE, Bashor CJ, Collins JJ. A brief history of synthetic biology. Nat Rev Microbiol. 2014;12(5):381–90.
162. König H, König H, Frank D, Frank D, Heil R, Heil R, et al. Synthetic genomics and synthetic biology applications between hopes and concerns. Curr Genomics. 2013;14(1):11–24.
163. Wang Y-H, Wei KY, Smolke CD. Synthetic biology: advancing the design of diverse genetic systems. Annu Rev Chem Biomol Eng. 2013;4:69–102.
164. Ruder WC, Lu T, Collins JJ. Synthetic biology moving into the clinic. Science. 2011;333(6047):1248–52.
165. Ibsen S, Tong A, Schutt C, Esener S, Chalasani SH. Sonogenetics is a non-invasive approach to activating neurons in *Caenorhabditis elegans*. Nat Commun. 2015;6:8264.
166. Sun Y, Williams A, Waisbourd M, Iacovitti L, Katz LJ. Stem cell therapy for glaucoma: science or snake oil? Surv Ophthalmol. 2015;60(2):93–105.
167. Bardy C, van den Hurk M, Eames T, Marchand C, Hernandez RV, Kellogg M, et al. Neuronal medium that supports basic synaptic functions and activity of human neurons in vitro. Proc Natl Acad Sci. 2015;112(20):E2725–34.
168. Bhatia SN, Ingber DE. Microfluidic organs-on-chips. Nat Biotechnol. 2014;32(8):760–72.

Erratum to: Stem Cell-Derived RPE Transplantation: The Feasibility and Advantages of Delivery as Monolayers

Odysseas Georgiadis, Lyndon da Cruz, and Peter Coffey

Erratum to:
Chapter 2 in: S.D. Schwartz et al. (eds.), *Cellular Therapies for Retinal Disease*, DOI 10.1007/978-3-319-49479-1_2

The original version of this chapter contained a mistake. Co-author names were omitted in the originally published Chapter 2. Chapter 2 was written by Odysseas Georgiadis and Lyndon da Cruz with oversight and supervision of Peter Coffey. The author names and corresponding affiliations have been updated with this erratum.

The updated online version of this chapter can be found at
https://dx.doi.org/10.1007/978-3-319-49479-1_2

S.D. Schwartz et al. (eds.), *Cellular Therapies for Retinal Disease*, DOI 10.1007/978-3-319-49479-1_11

Index

S.D. Schwartz et al. (eds.), *Cellular Therapies for Retinal Disease*, DOI 10.1007/978-3-319-49479-1

MIX
Papier aus verantwortungsvollen Quellen
Paper from responsible sources
FSC® C105338

If you have any concerns about our products,
you can contact us on
ProductSafety@springernature.com

In case Publisher is established outside the EU,
the EU authorized representative is:
Springer Nature Customer Service Center GmbH
Europaplatz 3, 69115 Heidelberg, Germany

Printed by Libri Plureos GmbH
in Hamburg, Germany